W9-CNW-347

Geomicrobiology

Geomicrobiology

Henry Lutz Ehrlich

Professor of Biology
Rensselaer Polytechnic Institute
Troy, New York

MARCEL DEKKER, INC.　　　　　New York • Basel

Library of Congress Cataloging in Publication Data

Ehrlich, Henry Lutz,
    Geomicrobiology.

    Includes bibliographical references and indexes.
    1. Geomicrobiology. I. Title.
QR103.E37   551'.01'576   81-422
ISBN 0-8247-1183-1   AACR1

COPYRIGHT © 1981 by MARCEL DEKKER, INC. ALL RIGHTS RESERVED

Neither this book nor any part may be reproduced or transmitted in any form or
by any means, electronic or mechanical, including photocopying, microfilming,
and recording, or by any information storage and retrieval system, without
permission in writing from the publisher.

MARCEL DEKKER, INC.
270 Madison Avenue, New York, New York 10016

Current printing (last digit):
10 9 8 7 6 5 4 3 2 1

PRINTED IN THE UNITED STATES OF AMERICA

To my former and present students, from whom I have learned as much as I hope they have learned from me.

Preface

This book deals with geomicrobiology as distinct from microbial ecology and microbial biogeochemistry. Although these fields overlap to some degree, each emphasizes different topics (see Chapter 1). A reader of this book should not, therefore, expect to find extensive discussions of ecosystems, food chains, nutritional cycles, mass transfer, or man-made pollution problems as such, because these topics are not at the heart of geomicrobiology. Geomicrobiology is the study of the role that microbes play or have played in specific geological processes.

This book arose out of a strong need that I felt in teaching a course in geomicrobiology. As of this writing, no single text is available that deals with the group of topics presented in this book. Previously, students in my geomicrobiology course had to be referred to the many primary publications on the various topics. These publications are very numerous, and are scattered among a plethora of journals and books that are often not readily available. Some are written in languages other than English. This book is an attempt to glean the basic geomicrobiological principles from this literature and to illustrate these principles with many different examples.

Some readers of this book will have a stronger background in earth and marine science than in microbial physiology, while others will have a stronger background in microbial physiology than in earth and marine sciences. To enable all these readers to place the geomicrobial discussions in the later chapters in proper context, the introductory Chapters 2-5 were written. They are not meant to be definitive treatises on their subjects, and as a result any one of them will appear elementary to a person already knowledgeable in its field. However, I have found the material in these chapters to be essential in teaching my students.

As for the rest of the book, Chapter 6 summarizes the methods used in geomicrobiology, and Chapters 7-17 examine specific geomicrobial activities in relation to geologically important classes of substances or elements. A single basic theme pervades these last 11 chapters: biooxidation and bioreduction and/or bioprecipitation and biosolution. This may seem an unnecessary reiteration of a common set of principles, but closer examination will show that the manifestations of these principles in different geomicrobial phenomena differ so strikingly as to require separate examination. In discussing geomicrobial processes, I have tended to emphasize the physiological more than the geological aspects. This is in part because the former is my own area of greater expertise, but also, and more importantly, because I feel that the physiological and biochemical nature of geomicrobial processes has to be understood to fully appreciate why some microbes are capable of these activities.

In citing microorganisms in the text, the names employed by the investigators whose work is described are used. In the case of bacteria, these names may have changed subsequently. The currently used names of the bacteria may be found by referring to *Bergey's Manual of Determinative Bacteriology* (8th edition, edited by R. E. Buchanan and N. E. Gibbons, 1974, Williams and Wilkins, Baltimore) and to the *Index Bergeyana,* (R. E. Buchanan, J. G. Holt, and E. F. Lessel, 1966, Williams and Wilkins, Baltimore). In some instances, however, it may be impossible to find a bacterial organism listed in the *Manual* or the *Index* because the organism was never sufficiently described to achieve taxonomic status. The current names of renamed bacteria may also be found in the Index of Organisms at the end of this book.

It is hoped that this book will serve not only as a text but also as an introduction and guide to the geomicrobiological literature for microbiologists, ecologists, geologists, environmental engineers, and others interested in the subject.

The preparation of this book was greatly aided by discussions with, and by review of the manuscript in various stages of completion by, Galen E. Jones, R. Schweisfurth, William C. Ghiorse, Edward J. Arcuri, and Paul A. LaRock, and many students in my geomicrobiology course. Responsibility for the presentation and interpretation of the subject matter as found in this book rests, however, entirely with me. I am indebted to a number of persons and publishers for making available original photographs or allowing reproduction of previously published material for illustration. They are acknowledged in the legends of the individual illustrations furnished by them. I wish to thank Stephen Chiang for his preparation of the finished line drawings from the crude sketches I furnished him. I also wish to thank the editorial staff of Marcel Dekker, Inc. for their help in readying my manuscript for publication.

Henry Lutz Ehrlich

Contents

Contents

Geomicrobiology

1

Introduction

The subject of **geomicrobiology** examines the role that microbes have played and are playing in a number of fundamental geological processes: for example, in the weathering of rocks, in soil and sediment formation and transformation, in the genesis and degradation of minerals, and in the genesis and degradation of fossil fuels. Geomicrobiology should not be equated with microbial ecology or with microbial biogeochemistry. **Microbial ecology** is the study of interrelationships between different microorganisms; among microorganisms, plants, and animals; and between microorganisms and their environment. **Microbial biogeochemistry** is the study of microbially catalyzed reactions and their kinetics, often in the context of cycles, with emphasis on environmental mass transfer and energy flow. Obviously, these subjects overlap to some degree, as shown in Figure 1.1.

The origin of the word "geomicrobiology" is obscure. It obviously derived from the term "geological microbiology." Beerstecher (1954) defined geomicrobiology as "the study of the relationship between the history of the earth and the microbial life upon it." Kuznetsov et al. (1963) defined it as "the study of microbiological processes currently taking place in the modern sediments of various bodies of water, in ground waters circulating through sedimentary and igneous rocks, and in the weathered earth crust . . . [and also] the physiology of specific microorganisms taking part in presently occurring geochemical processes." Neither author traced the history of the word, but they pointed to the important roles that scientists such as Winogradsky, Waksman, and ZoBell have played in the development of the field.

The study of geomicrobiology is not new. Although certain early investigators in soil and aquatic microbiology may not have thought of themselves as geomicrobiologists, they have, nevertheless, had an influence on the subject. One of the first contributors to geomicrobiology was Ehrenberg (1838), who in the second quarter of the nineteenth century discovered an association of *Gallionella ferruginea* with ochreous deposits of bog iron. He believed that the organism, which he thought to be an infusorian but which we now recognize as a stalked bacterium, was important in the formation of such deposits. Another

1

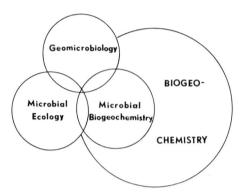

Figure 1.1 Interrelationships among geomicrobiology, microbial ecology, microbial biogeochemistry, and biogeochemistry in general.

important early contributor to geomicrobiology was Winogradsky (1887, 1888), who discovered that *Beggiatoa* could oxidize H_2S to elemental sulfur, and that *Leptothrix ochracea* could oxidize $FeCO_3$ to ferric oxide. He believed that both organisms gained energy from these processes. Still other important early contributors to geomicrobiology were Harder (1919), a researcher trained as a geologist and microbiologist, who studied the significance of microbial iron oxidation and precipitation to the formation of sedimentary iron deposits, and Stutzer (1911), Vernadsky (1908-1922), and others, whose studies led to recognition of the significance of microbial oxidation of H_2S and elemental sulfur in the formation of sedimentary sulfur deposits [see Ivanov (1967) for a discussion of early Russian geomicrobiology]. Our understanding of the role of bacteria in sulfur deposition in nature received a further boost from the discovery of bacterial sulfate reduction by Beijerinck (1895) and van Delden (1903). Starting with the Russian investigator Nadson (1903; see Nadson, 1928) at the end of the nineteenth century, and continuing with such investigators as Bavendamm (1932), the important role of microbes in $CaCO_3$ precipitation began to be noted. Microbial participation in manganese oxidation and precipitation in nature was first indicated by Beijerinck (1913), Soehngen (1914), Lieske (1919), and Thiel (1925). This was later related to sedimentary ore formation by Zappfe (1931). The microbial role in methane formation became apparent through the observation and studies of Béchamp (1868), Tappeiner (1882), Popoff (1875), Hoppe-Seyler (1886), Omeliansky (1906), and Soehngen (1906). The role of bacteria in rock weathering was first suggested by Muentz (1890) and Merrill (1895). Later, involvement of acid-producing microorganisms such as nitrifiers, and of crustose lichens and fungi was suggested (see Waksman, 1932). Thus, by the beginning of the twentieth century, many of the important areas of geomicrobiology had begun to receive serious attention from microbiologists. In general, it may be said that most of the geomicrobiologically important discoveries of the nineteenth century were made through physiological studies in the laboratory which revealed the capacity of specific organisms for geomicrobiologically important transformations, causing later workers to study the extent of the microbial activities in the field.

Geomicrobiology in the United States can be said to have begun with the work of E. C. Harder (1919) on iron-depositing bacteria. Other early American investigators of geomicrobial phenomena include J. Lipman, S. A. Waksman, R. L. Starkey, and H. O. Halvorson, all prominent in soil microbiology, and G. A. Thiel, C. Zappfe, and C. E. ZoBell, all prominent in aquatic microbiology.

Very fundamental discoveries in geomicrobiology continue to be made, some basic ones having been made comparatively recently. For instance, the concept of environmetnal limits of pH and E_h for microbes in their natural habitats was first introduced by Baas Becking et al. in 1960 (see Chapter 5). The discovery of acidophilic, iron-oxidizing thiobacilli was reported in 1950 by Colmer et al. as a result of their interest in the origin of acid coal-mine drainage. It led to the first demonstration by Bryner et al. in 1954 that these tiobacilli could also oxidize metal sulfides, such as those of iron and copper (Chapter 15). The first visual detection of Precambrian prokaryotic fossils in sedimentary rocks was reported by Tyler and Barghoorn in 1954, by Schopf et al. in 1965, and by Barghoorn and Schopf in 1965 (see Chapter 2). These paleontological discoveries were to have a profound influence on current theories about the evolution of life on earth.

As this book will show, many areas of geomicrobiology remain to be explored or developed further.

2

Major Features of the Earth:
Its Early History and the Origin of Life

2.1 THE BIOSPHERE

The surface of the earth can be divided into three distinct phases: the **lithosphere**, the crust of the earth; the **hydrosphere**, the aqueous portion of the earth's surface, including the oceans, seas, lakes, and rivers; and the **atmosphere**, the gaseous envelope around the globe. Portions or the whole of each of these phases are included in the **biosphere**, that portion of the earth's surface which is inhabited by living organisms.

The biosphere thus includes the upper portion of the lithosphere, or crust, to a depth of about 2,000 m. A claim for life to a depth of 4,000 m in the crust has been made by Pokrovskiy (see Kuznetsov et al., 1963, p. 26) but seems not yet to have been confirmed by others. An important factor limiting life to the upper portion of the crust is undoubtedly temperature, which increases with depth in the crust. At a depth of 1,000 m, the temperature may range from -20 to $+100°C$, depending on location (Kuznetsov et al., 1963, p. 25). The heat in the crust is the result of its diffusion from the interior of the earth, where it is generated primarily as a by-product of radioactivity. Lack of porosity in the rock, and lack of sufficient moisture and nutrients, are undoubtedly other limiting factors to life in the deep crust.

The biosphere also includes all of the hydrosphere, which harbors life even at its greatest depths (approximately 11,000 m). These great depths occur in special regions in the oceans where the bottom is rent by profound clefts called **trenches** (see Chapter 3).

Finally, the biosphere includes the lower portion of the atmosphere. Living microbes have been recovered from it at heights as great as 48-77 km above the earth's surface (Imshenetsky et al., 1978). Whether the atmosphere constitutes a true microbial habitat is very debatable. Although it harbors viable vegetative cells and spores, it is generally not capable of sustaining growth and multiplication of these organisms because of lack of sufficient moisture and nutrients and, especially at higher elevations, because of lethal radiation. At high humidity in the physiological temperature range, however, some bacteria may propagate to a limited extent (Dimmick et al., 1979; Straat et al., 1977). The residence time of microbes in air may also be limited, owing to eventual fallout. In the case of microbes associated with solid particles suspended in still air, the fallout rate may range from 10^{-3} cm sec^{-1} for 0.5-μm particle sizes to 2 cm sec^{-1} for 10-μm particle sizes (Brock, 1974, p. 541). Yet even if not a true habitat, the atmosphere is nevertheless important to microbes. It is a vehicle for spreading microbes from one site to another; and it is a source of oxygen for strict and facultative aerobes and a source of nitrogen for nitrogen-fixing microbes.

2.2   BRIEF CHARACTERIZATION OF THE PLANET EARTH

The earth as a whole can be divided into three major regions (Fig. 2.1) on the basis of seismic studies and inferred compositions. The innermost portion, the **core**, is estimated to have a radius of about 3,470 km. It is thought to consist of an Fe-Ni alloy with or without and admixture of Si and FeS (Mercy, 1972, p. 680). The inner portion of the core, having a radius of about 1,250 km, is solid, possessing a density of 13 g cm^{-3} and being subjected to a pressure of 3.7×10^{12} dyn cm^{-2}, but the outer portion of the core, having a thickness of 2,200 km, is molten, owing to the high temperature but lower pressure (1.3-3.2×10^{12} dyn cm^{-2}) to which it is subjected. The density of this portion is 9.7-12.5 g cm^{-3}. Overlying the core is the **mantle**, which is estimated to be 2,867 km thick. Its density ranges from 3.3 to 5.7 g cm^{-3}, and it is subject to pressures ranging from 0.01 to 1.3×10^{12} dyn cm^{-2}. The mantle is separated from the core by the Wickert-Gutenberg discontinuity. The composition of the mantle is thought to be dominated by the elements O, Mg, and Si, with lesser amounts of Fe, Al, Ca, and Na (Mercy, 1972, p. 680). Its consistency, although not molten, is thought to be plastic, especially its upper portion, the **asthenosphere**. Upper mantle rock of the ultramafic type, believed to be the principal constituent, seems to be cropping out as an uplifted block on the bottom of the western Indian Ocean (Bonatti and Hamlyn, 1978). Overlying the mantle is the **crust**, separated from the mantle by the Mohorovičić discontinuity. The thickness of the crust varies, being as great as 70 km under continental mountain ranges and as small as 5 km under ocean basins. Under an average continental shield area, it is 33 km (Mercy, 1972, p. 679). The composition of the crust is dominated by O, Si, Al, Fe, Mg, Na, and K. These elements make up 98.6% of the weight of the crust (Williams, 1962) and occur primarily in rocks and sediments. The bedrock under the ocean is generally basaltic, whereas that of the continent is granitic to an average crustal depth of about 25 km and basaltic below this to the Mohorovičić discontinuity (Ronov and Yaroshevsky, 1972, p. 243). Sediment covers most of the bedrock under the ocean. It ranges in

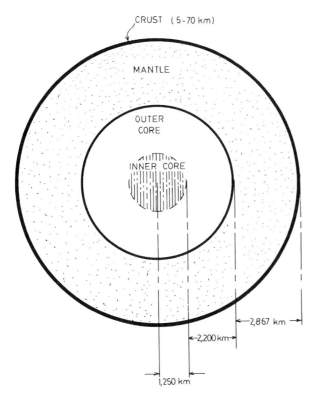

CRUST (5-70 km)

MANTLE

OUTER CORE

INNER CORE

2,867 km

2,200 km

1,250 km

Figure 2.1 Diagrammatic cross section of the earth. Radii of core and mantle drawn to scale.

thickness from 0 to 4 km. Sedimentary rock and sediment cover the bedrock of the continents; their thickness may exceed that of marine sediments (Kay, 1955, p. 655). The continents make up about 64% of the crustal volume, the oceanic crust makes up 21%, and the shelf and subcontinental crust make up the remaining 15% (Ronov and Yaroshevsky, 1972). Although the deep crust, the mantle, and the core do not harbor living organisms, the upper portion of the mantle is a source of juvenile minerals, water, and some gases, which may rise to the earth's surface during volcanic activity and thus become important to life.

Although until recently viewed as a coherent structure that rests on the mantle, the crust is now seen to consist of a series of moving and interacting plates of varying size and shape; some support the continents and parts of the ocean floor and others support only parts of the ocean floor. Figure 2.2 shows the outlines of the crustal plates of the ocean basins and adjacent continents.

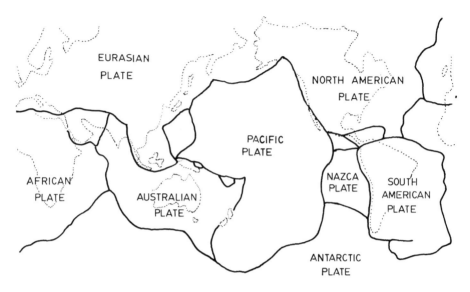

Figure 2.2   Major crustal plates.

Present estimates of the number of plates involved range from 6 to 20 (Strahler, 1977). The plates all float on the asthenosphere of the mantle. Oceanic plates are in a dynamic state. They seem to be growing along **oceanic ridges**, such as the Mid-Atlantic Ridge, the East Pacific Rise, and others (Fig. 2.3), and are destroyed in deep-sea trenches, such as the Kurile Trench and the Philippine Trench in the Pacific Ocean, and the Puerto Rico Trench in the Atlantic Ocean (Fig. 2.3). **Mid-ocean ridges** are examples of boundaries between two adjacent plates, where molten material rises from the region of the earth's mantle and adds to the preexistent crustal plates, thereby pushing them away from the ridges in opposite directions (Fig. 2.4). As a consequence of the resultant spreading of the plates, some portions of the older plate material disappears below the crustal surface. This happens at the site of oceanic trenches where portions of expanding crustal plates sink, or are **subducted**, below adjacent and overriding plates. In other places, adjacent plates move past each other along faults (e.g., the San Andreas fault in California). It is in regions of mid-ocean ridges, subduction zones (oceanic trenches), and faults that **tectonic activity** (crustal transformation) is most evident. Mountain building (**orogeny**) and volcanic activity on continents is associated primarily with sites of crustal overriding (i.e., zones of subduction) (Dewey and Bird, 1970). However, volcanic activity is not restricted to tectonically active continental margins. It is also observed at points on mid-ocean ridges (e.g., Iceland and Tristan da Cunha Island on the Mid-Atlantic Ridge) and other sites on the ocean floor, where it

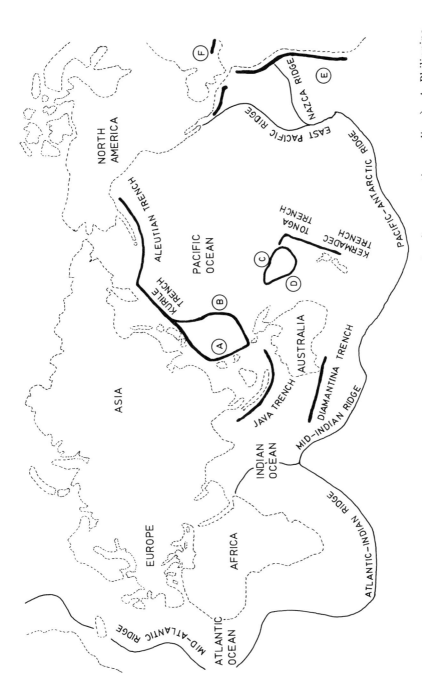

Figure 2.3 Major mid-ocean ridge systems (thin continuous lines) and trenches (heavy continuous lines): A, Philippine Trench; B, Mariana Trench; C, Vityaz Trench; D, New Hebrides Trench; E, Peru-Chile Trench; F, Puerto Rico Trench. The East Pacific Ridge is also known as the East Pacific Rise.

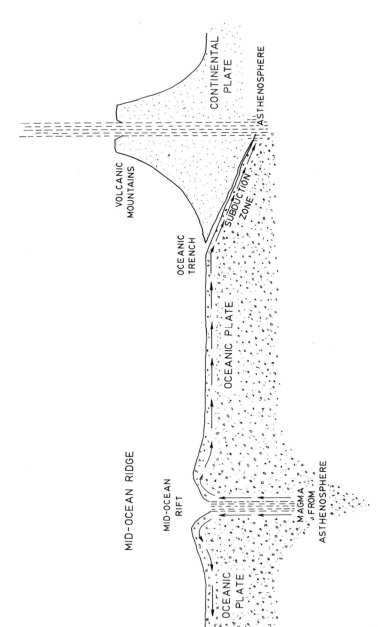

Figure 2.4 Schematic representation of sea floor spreading. New oceanic crust is formed at the mid-ocean rift. Old oceanic crust is consumed in the subduction zones near continental margins or island arcs.

manifests itself in submarine eruptions or as island volcanos such as those on the island of Hawaii.

2.3 CONTINENTAL DRIFT

The continents of the earth as they exist today are thought to have derived from a single continental mass, **Pangaea**, which broke apart less than 200 million years ago, first into **Laurasia** (which included present-day North America, Europe, and most of Asia) and **Gondwana** (which included present-day Africa, South America, Australia, Antarctica, and the Indian subcontinent), and these separated subsequently into the continents we know today, except that the Indian subcontinent did not join the Asian continent until some time after this breakup (Fig. 2.5) (Dietz and Holden, 1970; Fooden, 1972; Matthews, 1973; Palmer, 1974). The separation of Pangaea into the various continents is attributed to the movement of the crustal plates. Pangaea itself originated 250-260 million years ago from an aggregation of crustal plates bearing continental landmasses including Baltica (consisting of Russia west of the Ural Mountains, Scandinavia, Poland, and northern Germany), China, Gondwana, Kazakhstania (consisting of present-day Kazakhstan), Laurentia (consisting of most of North America, Greenland, Scotland, and the Chukotski Peninsula of the eastern U.S.S.R.) and Siberia (Bambach et al., 1980). The evidence for the evolution of the present-day continents rests on paleomagnetic and seismic studies of the earth's crust; on comparative sediment analyses of deep-sea cores obtained from drillings by the *Glomar Challenger*, an ocean-going research vessel; and on paleoclimatic considerations (Bambach et al., 1980; Nierenberg, 1978; Vine, 1970). Although the separation of the present-day continents had probably no significant effect on the evolution of prokaryotes (they had pretty much evolved to their present complexity by this time), it did have a profound effect on the evolution of metaphytes and metazoans (McKenna, 1972; Raven and Axelrod, 1972). Flowering plants, birds, and mammals, for example, had yet to establish themselves.

2.4 THE ORIGIN OF LIFE AND ITS EARLY EVOLUTION

The earth is thought to be about 4.6×10^9 years old (4.6 eons). One popular view holds that it condensed from a gaseous cloud of matter (a cosmic dust cloud), together with the rest of the solar system, over a period of some 60-200 million years previously (Miller and Orgel, 1974; p. 22). The original atmosphere surrounding the earth was reducing and probably included the gases CH_4, CO_2, CO, NH_3, and H_2O. It could have resulted from outgassing of the planet

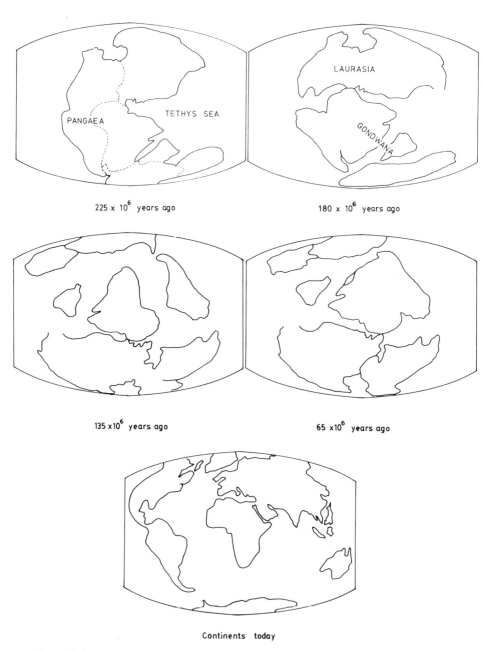

225 x 10^6 years ago

180 x 10^6 years ago

135 x10^6 years ago

65 x10^6 years ago

Continents today

Figure 2.5   Continental drift. Breakup of Pangaea up to the present. (From
Dietz and Holden, 1970; reproduced with authors' permission.)

in the early stages of its existence. Although abiogenic organic syntheses probably occurred since the earth's beginnings, life probably did not appear until about 0.5-1 billion years later (i.e., 3.6-4.0 $\times$ 10^9 years ago), according to present thinking.

Although our ideas about the sequence of events in the evolution of primitive (prokaryotic) life have changed in the last 40 years, it is now commonly agreed that the first cellular forms may have resembled present-day fermentative bacteria and lived at the expense of abiotically synthesized organic matter in the absence of oxygen (anaerobic heterotrophy). Abiotic synthesis is believed to have involved reactions of gas mixtures, including one or more of the following: methane, ammonia hydrogen, water vapor, hydrogen cyanide, carbon dioxide, and formaldehyde, with electrical discharge, helium spark, UV irradiation, or heat as a driving force, to yield a variety of organic compounds, including amino acids, purines, and sugars. The first living cells are postulated to have arisen by special aggregation of some of the abiotically synthesized compounds (e.g., Brock, 1974; Miller and Orgel, 1974; Oparin, 1938). These cells developed the ability to draw sustenance for synthesis of new protoplasm from other abiotically synthesized organic matter, and to self-reproduce.

As anaerobic life became more firmly established (i.e., as the number of living organisms increased), the organic nutrients must have begun to be depleted at a faster rate than they could be replenished by abiotic synthesis (Horowitz, 1945). Indeed, conditions on the earth's surface probably became progressively more unfavorable for abiotic synthesis, thereby contributing further to a potential shortage in organic nutrients. Hence, an alternative mechanism for producing organic matter was required to sustain life. This need may have been met through the emergence of photosynthesis as carried on by a special group of bacteria, such as present-day purple and green bacteria, which could produce organic matter by reducing CO_2 with inorganic electron donors such as H_2 or H_2S and with sunlight as the energy source. The emergence of these anaerobic photosynthesizing organisms required the appearance of **chlorophyll**, the light-harvesting and energy-transducing substance, in the process (see Chapter 5).*

---

*The obligate halophile *Halobacterium* can produce a light-harvesting and energy-transducing pigment called **bacteriorhodopsin** in its plasma membrane under oxygen limitation (Stoeckenius, 1976). The pigment is the carotenoid **retinal** combined with a protein. It assists in the conversion of light energy into chemical energy, much as chlorophyll acts in photosynthesis (Danon and Stoeckenius, 1974). Although this light-dependent energy-transducing system does not enable *Halobacterium* to assimilate CO_2 as a major source of carbon, it enables this aerobic organism to survive temporary anoxia, which occurs frequently in its hypersaline habitat, in which oxygen solubility is low. Its respiratory energy-generating system is inoperative in anoxia (see Stanier et al., 1976). At present it is not known when this light-harvesting, energy-transducing system evolved. It is an example of parallel evolution with respect to photosynthesis and apparently developed as the result of a special environmental need.

Once established, these organisms could maintain and even increase the supply of available organic carbon needed for the persistence of heterotrophic life. Another special microbial group capable of coping with an organically depleted, anaerobic environment may have been methane-producing bacteria, which derived energy by reducing CO_2 with H_2 to form methane (CH_4) (Fox et al., 1977). Indeed, some investigators have recently suggested that these organisms may have preceded the emergence of photosynthetic bacteria.

Further evolution of the photosynthetic machinery through the development of new types of chlorophyll must have led to the emergence of the prototypes of cyanobacteria,* which could substitute H_2O for H_2 or H_2S as a reducing agent of CO_2 and thereby introduced oxygen into the atmosphere:

$$H_2O + CO_2 \rightarrow (CH_2O) + O_2 \tag{2.1}$$

Initially, the oxygen that was evolved probably reacted rapidly with oxidizable inorganic matter such as iron [Fe(II)], forming iron oxides such as magnetite (Fe_3O_4) and hematite (Fe_2O_3) (see Chapter 12). But eventually, about 2.3 eons ago (Schopf, 1978), free oxygen began to accumulate in the atmosphere, changing it from a reducing environment to an oxidizing one. As the atmospheric oxygen concentration increased, organisms began to evolve which acquired mechanisms that included the **cytochrome system** for disposing of excess reducing power (electrons) to molecular oxygen instead of bound oxygen or partially reduced organic compounds (see Chapter 5). The cytochrome system provided a reaction sequence of discrete steps, the energy from some of which could be trapped in special chemical bonds, which, in turn, could be utilized in driving energy-requiring reactions (syntheses, polymerizations). Since toxic superoxide (O_2^-) may be readily formed from molecular oxygen, especially when O_2 is reduced in one-electron steps involving free intermediates (Fridovich, 1977), the organisms also had to evolve a special protective system against it. Such a system was the **superoxide dismutase**, which catalyzes the reduction of O_2^- to hydrogen peroxide (H_2O_2),

$$O_2^- + 2H^+ + e^- \rightarrow H_2O_2 \tag{2.2}$$

in conjunction with **catalase**, which catalyzes the destruction of H_2O_2 to water and oxygen,

$$2H_2O_2 \rightarrow 2H_2O + O_2 \tag{2.3}$$

*Cyanobacteria were formerly called blue-green algae. Because it is now clearly established that these organisms are prokaryotic, like bacteria in general, lacking a true nucleus, chloroplasts, and mitochondria, the term "blue-green algae" has been abandoned. Nevertheless, in the ecological literature, cyanobacteria are usually considered as part of algal populations.

or **peroxidase**, which catalyzes the reduction of H_2O_2 by an oxidizable organic molecule (RH_2):

$$RH_2 + H_2O_2 \rightarrow 2H_2O + R \tag{2.4}$$

These evolutionary developments resulted in the emergence of aerobic heterotrophic organisms which could use their organic substrates as energy sources much more efficiently than the anaerobic heterotrophs (see Chapter 5).

The evolution of cytochromes could have occurred before, after, or simultaneously with the evolution of chlorophyll while reducing conditions still prevailed, since both types of chemical structures include a porphyrin ring. However they evolved, the early cytochromes could have functioned in a more primitive electron transport system in which bound oxygen rather than free oxygen was the terminal electron acceptor (Brock, 1974).

The accumulation of oxygen in the atmosphere also must have led to the buildup of an ozone (O_3) shield which screened out the UV component of sunlight. This woud have stopped any abiotic synthesis dependent on this form of radiation, and at the same time would have allowed the emergence of life onto the land surfaces of the planet, which are directly exposed to sunlight. This emergence would have been impossible before because of the lethality of UV radiation (Brock 1974).

With the emergence of oxygen-producing cyanobacteria and aerobic heterotrophs, the stage was set for cellular compartmentalization of such vital processes as photosynthesis and respiration. New types of photosynthetic cells emerged in which photosynthesis was carried on in special organelles, the **chloroplasts**, and new types of respiring cells emerged in which respiration was carried on in special organelles, the **mitochondria**. It is very likely that these organelles arose by **endosymbiosis**, a process in which a primitive cell was invaded by cyanobacteria and/or aerobically respiring bacteria which established a permanent symbiosis with their host (see the discussions by Margulis, 1970; Dodson, 1979). Eventually, the relationship of the host cell with its endosymbionts became one of absolute interdependence, the endosymbionts having lost their capacity for an independent existence. The result has been the emergence of the **eukaryotic** cell, whose organization contrasts with the less-compartmentalized **prokaryotic** cells (bacteria, including cyanobacteria). Supporting evidence for a probable endosymbiotic origin of chloroplasts and mitochondria is the discovery in them of genetic substance—deoxyribonucleic acid—and of cell particles called ribosomes, in forms otherwise present only in prokaryotes and distinct from the deoxyribonucleic acid in the chromosomes of the nucleus and the ribosomes of the cytoplasm of the eukaryotic cell. The emergence of eukaryotic cells was evidently needed for the evolution of more complex forms of life, such as the fungi, plants, and animals.

2.5   MICROPALEONTOLOGY AND ITS CLUES
FOR THE EVOLUTIONARY TIME SCALE

Clues to when and how different forms of life evolved can be obtained from the fossil record. At present, the earliest acceptable evidence* of life on earth is represented by fossilized mats or filamentous microorganisms, either bacterial (e.g., *Chloroflexus*) or cyanobacterial. These mats, called **stromatolites**, are preserved in ferruginous dolomitic chert of the Pilbara Block of Western Australia (Lowe, 1980; Walter et al., 1980). They are estimated to be 3,400-3,500 million years old. Other microfossils, which are almost as old, have been reported in the Swartkoppie Formation of the Swaziland System of sedimentary rocks in the Transvaal of South Africa. These fossil remnants are estimated to be about 3,400 million years old and resemble cyanobacterial fossils in younger rock. They contain organic matter, and some of their remains show evidence of binary fission (Knoll and Barghoorn, 1977) (Fig. 2.6). Since cyanobacteria are considered to have evolved from bacteria, the latter must have existed even earlier. The fossil remains of the oldest bacteria are not likely to be found because the sedimentary rock in which they would have become trapped and preserved has long since disappeared as a result of weathering. Somewhat younger microfossils, some of which resemble bacteria (Fig. 2.7), have been

*Yeastlike microfossils have been reported from the Isua Quartzite in southwestern Greenland (about 3,800 million years old) (Pflug, 1978; Pflug and Jaeschke-Boyer, 1979). However, questions about this find remain. These organisms do not fall into the presently conceived pattern of cellular origin and evolution on earth.

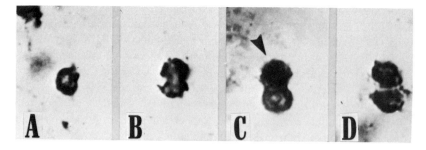

Figure 2.6   Microfossils in carbonaceous laminae of the Swartkoppie Formation, South Africa (1,600X). (A-C) Stages in cell division preserved in the Swartkoppie population. The arrow in (C) points to dark organic contents within the upper half of the dividing cell. (D) Dyads in the Swartkoppie population. (From Knoll and Barghoorn, 1977. Copyright 1977 by the American Association for the Advancement of Science.)

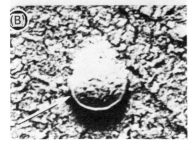

Figure 2.7 Negative prints of platinum-carbon surface replicas of chert from the Fig Tree Series, showing *Eobacterium isolatum,* n.gen., n.sp. The line in the figure represents 1 μm. (A) Organically preserved cell of *E. isolatum,* showing the short, broad, rod-shaped morphology of the fossil organism. Note the similarity of shape to common bacillary bacteria. (B) Circular structure interpreted as a transverse section through *E. isolatum.* Although the cellular contents appear to have been replaced by silica, the cell wall is organically preserved. The wall, about 0.015 μm thick, has a two-layered organization (shown at point of arrow) and is comparable in thickness and structure to the cell walls of many modern bacteria. (From Barghoorn and Schopf, 1966. Copyright 1966 by the American Institute for the Advancement of Science.)

reported in the Onverwacht Group (3,200 million years old) and the Fig Tree Series (3,100 million years old) in the Transvaal of South Africa (Barghoorn and Schopf, 1966; see also Margulis, 1970).

The appearance of cyanobacteria in the fossil record has been thought to mark the time of onset of introduction of oxygen into the atmosphere, because ancient cyanobacteria, on the model of their predominant present-day behavior, could be expected to have carried on oxygen-producing photosynthesis, thereby slowly converting the premordial reducing atmosphere to an oxidizing one. It is likely, however, that the earliest cyanobacteria photosynthesized like present-day photosynthetic bacteria—that is, anaerobically using H_2S rather than H_2O as their source of reducing power for CO_2 assimilation, and producing elemental sulfur rather than oxygen (O_2)—because modern cyanobacteria have been found that can carry out this type of photosynthesis in reducing environments in addition to oxygen-producing photosynthesis in oxidizing environments (Garlick et al., 1977). Only later would these microorganisms acquire the ability to evolve oxygen. The complete change from a reducing to an oxidizing atmosphere ultimately must have made possible the emergence of the eukaryotic cell. The earliest eukaryotic algal microfossils have been reported from the Precambrian sediments of the Bitter Springs Formation in central Australia (Schopf and Oehler, 1976). These fossils have been estimated to be 1,400 million years old.

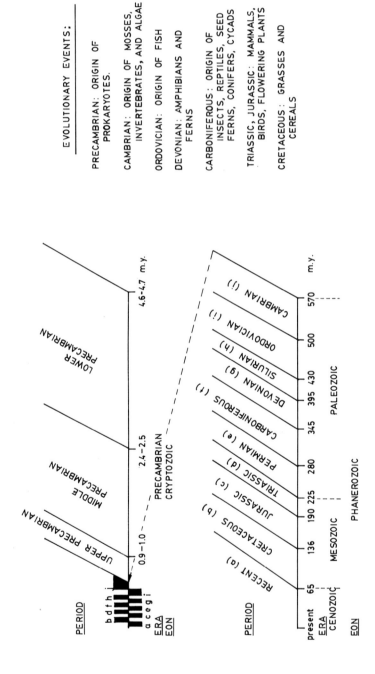

Figure 2.8 Geologic time scale and some corresponding evolutionary events. Note that the period called "Recent" in this figure is conventionally divided into Tertiary (65 million to 1 million years before present) and Quaternary (1 million years ago to present).

Knoll and Barghoorn (1975) have, however, questioned whether any of the microfossils from the Bitter Springs are indeed eukaryotic. They stated in 1975 that no evidence existed for eukaryotic microfossils older than 900 million years.

From the foregoing it may be inferred that the microbial world was well established on earth more than 1 billion years ago (Cloud, 1974). Oxygen-producing photosynthesis may have made its appearance about 3 billion years ago, if we accept the Swartkoppie microfossils as representing oxygen-producing cyanobacteria. If not, a later date may have been 2,700 million years ago, because cyanobacterial microfossils have been found in the Bulawayan Limestone of Rhodesia (Zimbabwe), which is of approximately that age (Schopf and Barghoorn, 1967). The stromatolites of the Pilbara Block of Western Australia could have been constructed by anaerobic photosynthetic organisms (e.g., *Chloroflexus*) (Walter et al., 1980). Figure 2.8 summarizes some of the events in the evolution of life on earth on a time scale measuring backward from the present.

2.6 SUMMARY

The surface of the earth includes the lithosphere, hydrosphere, and atmosphere, all of which are habitable by microbes to a greater or lesser extent and constitute the biosphere of the earth.

The structure of the earth can be separated into the core, the mantle, and the crust. Of these, only the uppermost portion of the crust is habitable by living organisms. The crust is not a continuous solid layer over the mantle, but consists of a number of crustal plates afloat on the mantle, or more specifically on the asthenosphere of the mantle. Some of these plates lie entirely under the oceans. Others carry parts of continents or of continents and oceans. Oceanic plates are growing along mid-ocean ridges, while old portions are being destroyed by subduction or collision with continental plates. The crustal plates are in constant, albeit slow, motion, accounting for continental drift.

The earth is about 4.6 eons old. Primitive life probably first appeared 0.5 to 1 eon after its condensation. A reducing atmosphere surrounded the earth until oxygen-generating photosynthetic organisms in the form of cyanobacteria evolved, probably more than 2.7 eons ago. Eukaryotic algae probably emerged between 0.9 and 1.4 eons ago. The dates of first appearance of various microbial forms have been deduced from the microfossil record in geologically dated sedimentary rock.

Life on earth probably arose de novo. Cells are believed to have arisen from the aggregation of critical abiogenically synthesized molecules. These cells had the ability to assimilate abiogenic molecules from the external environment and thus grew and multiplied. Extensive metabolism evolved when needed abiotically formed molecules became scarce and conditions became progressively less favorable for new abiotic synthesis. The first life forms were anaerobes. With the development of oxygen-producing photosynthesis and the emergence of an oxidizing atmosphere, a more efficient type of metabolism and more complex organisms could evolve, but only after a protective mechanism against toxic superoxide radicals was developed and an ozone shield blocking ultraviolet radiation arose in the upper atmosphere.

3

Rock and Soil

3.1 ROCK AND MINERALS

To understand the influence that microbes exert on the formation or transformation of rocks and minerals, we must familiarize ourselves with some general chemical and physical features of rocks and minerals.

Geologically speaking, the term **rock** refers to massive, solid, inorganic matter consisting usually of two or more intergrown minerals (Fig. 3.1). Rock may be **igneous** in origin; that is, it may arise by cooling of **magma** (molten rock material) from the interior of the earth. The cooling may be a slow or a fast process. In slow cooling, different minerals crystallize at different times, owing to their different melting points. In the process of crystallization they intergrow and thereby evolve into rock with relatively large crystals such as granite (Fig. 3.1). In fast cooling, rapid crystallization occurs, and rock forms which contains only tiny crystals. Basalt is an example of rock formed in this way.

Rock may also be **sedimentary** in origin; that is, it may arise through the accumulation and compaction of sediment, which consists mainly of mineral matter derived from other rock. It may also arise as a result of cementation of accumulated sediment by carbonate, silicate, aluminum oxide, or ferric oxide. The cementing substance may result from microbial activity. Sedimentary rock often exhibits a layered structure in vertical section. The transformation of loose sediment into sedimentary rock is called **lithification**. Examples of sedimentary rock are limestone, sandstone, and shale.

Finally, rock may be of **metamorphic** origin; that is, it may be produced through alteration of igneous or sedimentary rock by the action of heat and pressure. Examples of metamorphic rock are marble, derived from limestone; slate, derived from shale; quartzite, derived from sandstone; and gneiss, derived from granitic rock.

Geochemically speaking, **minerals** are usually defined as inorganic compounds, usually crystalline but sometimes amorphous, of specific chemical composition and structure. Sometimes the term "mineral" is also applied to

Figure 3.1 Pieces of granite, showing phenocrysts—visible crystals of mineral in a fine-crystalline ground mass. An igneous rock. The fragment in the back is 5 cm long.

certain organic compounds in nature, such as asphalt and coal. Inorganic minerals may be very simple in composition, as in the case of sulfur (S^0) or quartz (SiO_2), or very complex, as in the case of the igneous mineral biotite [$K(Mg, Fe,-Mn)_3 AlSi_3 O_{10}(OH)_2$].   Minerals that result from crystallization during cooling of magma are **primary** or **igneous minerals**. Minerals that result from chemical alteration (weathering or diagenesis) of primary minerals are known as **secondary minerals**. Microbes play a role in this transformation of primary to secondary minerals (Chapter 8). Examples of primary and secondary mineral groups are listed in Table 3.1. Minerals may also result from precipitation from solution, in which case they are called **authigenic minerals**.

Table 3.1 Minerals Classified as to Mode of Formation

Primary minerals
    Feldspars
    Pyroxenes and amphiboles
    Olivines
    Micas
    Silica
Secondary minerals
    Clay minerals
        Kaolinites
        Montmorillonites
        Illites
    Hydrated iron and aluminum oxides
    Carbonates

Source: Lawton (1955), pp. 54ff.

3.2 SOIL

Origin of Mineral Soil   Rock may be transformed into mineral soil as a result of weathering. **Weathering** is a process whereby rock is broken down into smaller and smaller particles and ultimately into constituent minerals. Some minerals may subsequently become chemically altered. Weathering processes may be physical. For example, freezing and thawing of water in cracks and crevices may cause expansion and hence exert pressure in the rock fissures. Sand carried by wind may cause sand blasting of rock. Alternate heating by the sun's rays and cooling at night may also cause detrimental expansion and contraction of rock. Waterborne abrasives or rock collisions may cause rock to break. Seismic activity may cause crumbling of rock. Evaporation of hard water in cracks and fissures of rock and resultant formation of crystals from the solutes may cause rock to break because the crystals occupy a larger volume than the original water solution from which they formed, thereby widening the cracks and fissures through the pressure they exert. Mere alternate wetting and drying may also cause such breakup.

Weathering processes may also be chemical. They include solvent action of water, chemical action by reagents such as CO_2 or mineral acids in water which may react with rock components, and oxidation-reduction reactions.

Biological agents may play an important role in weathering. Roots of plants penetrating cracks and fissures in rock may force it apart. Rock surfaces and the interior of porous rock are frequently inhabited by a flora of algae, fungi, lichens, and bacteria. Some of these microbes may dissolve rock through production of reactive metabolic products such as NH_3, HNO_3, H_2SO_4, CO_2 (forming H_2CO_3 in water), and oxalic, citric, and gluconic acids. Waksman and Starkey (1931) cited the following reactions as examples:

$$2KAlSi_3O_8 + 2H_2O + CO_2 \rightarrow H_4AlSi_2O_9 + K_2CO_3 + 4SiO_2 \qquad (3.1)$$
$$\text{orthoclase} \qquad\qquad\qquad \text{kaolinite}$$

$$12MgFeSiO_4 + 26H_2O + 3O_2 \rightarrow 4H_4Mg_3Si_2O_9 + 4SiO_2 + 6Fe_2O_3 \cdot 3H_2O$$
$$\text{olivine} \qquad\qquad\qquad\qquad \text{serpentine} \qquad\qquad\qquad \text{limonite}$$
$$(3.2)$$

The first reaction is promoted by CO_2 production in the metabolism of heterotrophic microbes, and the second reaction is promoted by O_2 production in photosynthesis, as by algae and lichens.

Some Structural Features of Mineral Soil   Mineral soil is an ultimate product of rock weathering and microbial development. It accumulates from transported or sedentarily formed mineral residue. In the former case it accumulates at a site other than where it was formed. In the latter case it accumulates at the site of its formation. Mineral soil will vary in composition, depending on the source of rock, the extent of weathering, the amount of organic matter, and the amount of moisture. Its texture is affected by the particle sizes of its inorganic constituents (stones, >2 mm; sand grains, 0.05-2 mm; silt, 0.002-0.05 mm; clay particles, <0.002 mm), which determine its porosity and thus its permeability to water and gases.

Many but not all soils tend to be more or less obviously stratified. In general, three strata or **horizons** are recognized in a **soil profile**, a vertical section through soil (Fig. 3.2). They are called the A, B, and C horizons. The A and B horizons constitute the true soil, and the C horizon is the parent material from which the soil was formed. It may be bedrock or an earlier soil. Each of the horizons may be further subdivided, although these divisions are somewhat arbitrary. The A horizon is the biologically most active zone, containing most of the root systems

(A)

Figure 3.2   Soil profiles. (A) Spodosol (Podzol). (B) Mollisol (Chernozem). (Courtesy of U.S.D.A., Soil Conservation Service.)

Figure 3.2 *(Continued)*

of plants growing on it and the microbes and other life forms that inhabit soil. As is to be expected, the carbon content is also highest in this horizon. The A horizon may be overlain by a litter zone or O horizon, consisting of much undecomposed and partially decomposed organic matter. The biological activity in the A horizon has the effect of solubilizing organic and inorganic components, some or all of which, especially the inorganic matter, is carried by soil water into the B horizon. The B horizon may be significantly enriched in Fe and Al in this way. The A horizon is, therefore, known at times as the **leached layer**, and the B horizon is at times known as the **enriched layer**. Both biological and abiological factors play a role in soil profile formation.

Effects of Plants and Animals on Soil Evolution   Biological agents such as plants assist soil evolution by contributing organic matter through excretions from their root systems and as dead organic matter. During their lifetime, plants remove some minerals and contribute to water movement by water absorption through their roots. Their root system may also help to prevent destruction of the soil by wind and water erosion. Burrowing animals, from small mites to large earthworms, also help break up soil, keep it porous, and redistribute organic matter.

Effects of Microbes on Soil Evolution   Microbes contribute to soil evolution by mineralizing some or all of any added organic matter during the decay process. Some of the metabolic products of this decay, such as organic and inorganic acids, CO_2, or NH_3, interact slowly with soil minerals and cause their alteration or solution, an important step in soil profile formation (Berthelin, 1977). Thus, the mineral chlorite has been reported to be bacterially altered through loss of Fe and Mg and an increase in Si, and the mineral vermiculite has been reported to be bacterially altered through solubilization of Si, Al, Fe, and Mg, thereby forming montmorillonite (Berthelin and Boymond, 1978). Certain microbes may interact directly (i.e., enzymatically) with certain inorganic soil constituents by oxidizing or reducing them (see Chapters 10 and 12-16), resulting in their solution or precipitation (Berthelin, 1977). Microbes also play an important role in **humus** formation. Humus is an important constituent of soil, consisting of humic and fulvic acids as well as amino acids, lignin, amino sugars, and other compounds of biological origin (Campbell and Lees, 1967, p. 209). These constituents represent, in part, refractory components of soil organic matter and, in part, products of microbial attack of the metabolizable constituents of soil organic matter. Humus gives proper texture to soil and plays a significant role in regulating the availability of minerals important in plant nutrition by complexing them. Humus also contributes to the water-holding capacity of soil.

Effects of Water on Soil Evolution   Water, in the form of rain or melting snow, causes diffusion of some soluble soil components and precipitation of others, and contributes to horizon development as it permeates the soil. Precipitates, especially inorganic ones, may promote soil clumping. Water may also affect the distribution of soil gases by displacing the rather insoluble ones, such as nitrogen and oxygen, and by absorbing the more soluble ones, such as CO_2, NH_3, and H_2S.

Some Major Soil Types   Distinctive soil types may be recognized and correlated with climatic conditions and with the vegetation they support (Bunting, 1967). The climatic conditions determine, of course, the kind of vegetation that may develop. Thus, in the high northern latitudes, **Tundra soil** prevails, which in that cold climate is often frozen and therefore supports only limited plant and microbial development. It has a poorly developed profile. Its reaction may range from slightly acidic to slightly alkaline. Examples are Arctic Brown soil and Bog soil. In the cool (i.e., temperate) humid zones, **Spodosols** (Fig. 3.2) prevail, which support extensive forests, particularly of the coniferous type. Spodosols tend to be acidic, having a strongly leached, grayish A horizon depleted in colloids, iron, and aluminum, and a brown B horizon enriched in the iron, aluminum, and colloids leached from the A horizon. In regions of moderate rainfall in temperate climates, **Mollisols** (Fig. 3.2) prevail. These are soils that support grasslands (i.e., they are prairie soils). They exhibit a rich black top soil and show lime accumulation in the B horizon because of neutral to alkaline pH conditions. **Oxisols** are found in tropical, humid climates. They are poorly zonated jungle soils. Owing to the climatic conditions, these soils are intensely active microbiologically and require constant replenishment of organic matter by the vegetation and by animal excretions to remain fertile. The neutral-to-alkaline pH conditions of jungle soils promote leaching of silica and precipitation of iron and aluminum. When jungle soils are denuded of their arboreal vegetation, as in slash-and-burn agriculture, they quickly lose their fertility as a result of the intense microbial activity which rapidly destroys soil organic matter. Since little organic matter is returned to the soil in its agricultural exploitation, conditions favor **laterization**, a process in which iron and aluminum oxides, silicates, and carbonates are precipitated which cement the soil particles together and greatly reduce its water-holding capacity and make it generally unfavorable for plant growth.

   Aridisols and **Entisols** are desert soils that occur mostly in hot, arid climates. Aridisols have an ochreous surface soil and may show one or more subsurface horizons as follows: argillic horizon (a layer with silica clay minerals dominating), cambic horizon (an altered, light-colored layer, low in organic matter, with carbonates usually present), natric horizon (dominant presence of sodium in exchangeable cation fraction), salic horizon (enriched with water-soluble salts), calcic horizon (secondarily enriched with $CaCO_3$), gypsic horizon (secondarily enriched with $CaSO_4 \cdot 2H_2O$), and duripan horizon (primarily cemented by silica and secondarily by iron oxides and carbonates) (see Fuller, 1974). Entisols are poorly developed, immature desert soils without subsurface development. They may arise from recent alluvial deposits or from rock erosion (Fuller, 1974). These desert soils are not infertile. It is primarily the lack of sufficient moisture that prevents the development of a lush vegetation. However, nitrogen as a major nutrient, and zinc, iron, and sometimes copper, molybdenum, or manganese as minor nutrients may also be limiting to plant growth. Desert soils support a specially adapted macroflora and fauna that can cope with the stressful conditions existing there. They also harbor a characteristic microflora of bacteria, fungi, algae, and lichens. Actinomycetes, algae, and lichens may sometimes be dominant. Cyanobacteria seem to be more important in nitrogen fixation in desert soils than are bacteria. Desert soils can sometimes be converted to productive agricultural soils by irrigation. Such watering often results in extensive leaching of salts which have accumulated during soil-forming episodes.

Water Distribution in Mineral Soil   Only about 50% of the volume of mineral soil is solid matter. The other 50% is occupied by water and gases such as CO_2, N_2, and O_2. As might be expected, owing to the biological activity in the soil, the CO_2 concentration in the soil atmosphere usually exceeds that of air, whereas the O_2 concentration is less than that in air. According to Lebedev (see Kuznetsov et al., 1963, pp. 39-41), soil water arranges itself around soil particles in distinct zones. Surrounding a soil particle is **hygroscopic water**, a thin film 3×10^{-2} μm in thickness around a 25-mm-diameter particle. This water never freezes and never moves as a liquid. It is adsorbed by soil particles from water vapor in the soil atmosphere. In a water-saturated atmosphere, **pellicular water** surrounds the hygroscopic water. It moves by intermolecular attraction but not by gravity. It may have dissolved salts which may depress its freezing point to $-1.5°C$. **Gravitational water** surrounds pellicular water when moisture in excess of the soil atmospheric capacity is present. It moves by gravity and responds to hydrostatic pressure, unlike hygroscopic and pellicular water. So far, no clear evidence seems to have been obtained as to which of these forms of water is available to microorganisms. The water need of microorganisms has usually been studied in terms of moisture content, water activity, or water potential (Dommergues and Mangenot, 1970).

   Water potential is the difference in free energy between the system under consideration and pure water at the same temperature. The more negative the value of water potential, the lower it is. A zero potential is equivalent to pure water. Either adsorption to surfaces (matric effect) or the presence of solutes (osmotic effect) can lower the water potential. Osmotic water potential can be calculated from the formula of Lang (1967):

Water potential (J/kg^{-1}) = 1.332 $\times$ freezing-point depression

where 100 J/kg^{-1} is equal to 1 bar. Matric water potential requirements can be determined by the method of Harris et al. (1970), in which NaCl or glycerol solutions of desired water potentials, solidified with agar, are used to equilibrate matric material on which microbial growth is to be measured.

The water potential requirement has recently been determined for two strains of the acidophilic iron oxidizer *Thiobacillus ferrooxidans* (Brock, 1975). Using NaCl as osmotic agent, strain 57-5 exhibited minimum water potential at -18 to -32 bars, whereas strain 59-1 exhibited it at -18 to -20 bars. Using glycerol as osmotic agent, strain 57-5 exhibited minimum water potential at -8.8 bars, whereas strain 59-1 exhibited it at -6 bars (Table 3.2). In the same study, it was shown that coal refuse material with water potentials between -8 and -29 bars allowed for significant CO_2 assimilation by *T. ferrooxidans,* whereas coal refuse material with a water potential of >-90 bars did not.

Organic Soils   In addition to mineral soils, there are **organic soils** or **Histosols**. They form from accumulation and slow decomposition of organic matter, especially plant matter, engendered by water displacement of air, which prevents

Table 3.2   Effect of Osmotic Water Potential (Glycerol) on Growth of *T. ferrooxidans* [a]

Glycerol (g liter^{-1})	Total water potential (bars)	Strain 57-5	Strain 59-1
184	-61	-	-
147	-49	-	-
92	-32	-	-
74	-26	-	-
55	-20	-	-
37	-15	-	-
18.4	-8.8	+	-
9.2	-6	+	+
3.7	-4.2	+	+
0	-3	+	+

[a] All experiments were done with replicate tubes which showed the same results. The incubation period was 2 weeks. Iron concentration in the medium, 10 g of $FeSO_4 \cdot 7H_2O$ per liter. +, Visible iron oxidation and microscopically visible growth; -, no iron oxidation and microscopically visible growth.
Source: Brock (1975); reproduced by permission.

rapid and extensive microbial decomposition of the organic matter. They are thus frequently of sedimentary origin and are never due to rock weathering. These soils consist of 20% or more organic matter (Lawton, 1955). Their formation is associated with the evolution of swamps, tidal marshes, bogs, and even shallow lakes. An organic soil such as peat may have an ash content of 2-50% and contain cellulose, hemicellulose, lignin and derivatives, heterogeneous complexes, fats, waxes, resins, and water-soluble substances such as polysaccharides, sugars, amino acids, and humins (Lawton, 1955). The pH of organic soils may range from 3 to 8.5. Examples of such soils are peats and "mucks." They accumulate to depts of from less than a meter to more than 8 m (Lawton, 1955) and are not stratified like mineral soils. They are rare.

Types of Microbes and Their Distribution in Soil   Microorganisms found in any soil include bacteria, fungi, protozoans, and algae, and viruses associated with these groups. A great variety of different types of bacteria may be encountered. Morphological types include gram-positive rods and cocci, gram-negative rods and spirals, sheathed bacteria, stalked bacteria, mycelial bacterial (actinomycetes), budding bacteria, and others. Physiological types include cellulytic, pectinolytic, saccharolytic, proteolytic, ammonifying, nitrifying, denitrifying, nitrogen-fixing, sulfate-reducing, iron-oxidizing and -reducing, manganese-oxidizing and -reducing, and other types. Morphologically, the dominant forms seem to be gram-positive cocci, probably representing the coccoid phase of *Arthrobacter* or possibly microaerophilic cocci related to *Mycococcus* (Casida, 1965). At one time, nonsporeforming rods were held to be the dominant form. Sporeforming rods are not a very prevalent form. Numerical dominance of a given type does not speak for its biochemical importance in soil. Thus, nitrifying and nitrogen-fixing bacteria are less numerous than some others but of vital importance to the nitrogen cycle in soil. A given soil under a given set of conditions will harbor an optimum number of individuals of each resident bacterial group. Total counts generally range from 10^5 per gram in poor soil to 10^8 per gram in garden soil.

The bacteria in soil are primarily responsible for mineralization of organic matter, for fixation of nitrogen, for nitrification, for denitrification, for mineral mobilization and immobilization, and for other processes. They often reside in microcolonies on soil particles because the soil particles, through adsorption, concentrate the microbial nutrients on their surface. Conditions of nutrient supply, oxygen supply, moisture supply, pH, and E_h may vary widely from particle to particle, owing in part to the activity of different bacteria or other micro- and macroorganisms. Thus, soil contains many different microenvironments. The colonization of soil particles by bacteria may cause some particles to adhere to each other (Martin and Waksman, 1940, 1941), and the bacteria thus affect soil texture.

Fungi reside mainly in the upper A horizon of soil, because they are principally strict aerobes and find their richest food supply there. In numbers they represent a much smaller fraction of the total microbial population than do the bacteria. Their mycelial growth habit causes them to grow over soil particles and penetrate the pore space of soil. They may also cause clumping of soil particles. The soil fungi include members of all the major groups: Phycomycetes, Ascomycetes, Basidiomycets, and Deuteromycetes, and also the slime molds. The last named are usually classified separately from fungi and protozoans, although they have attributes of both. Total numbers of fungi in soil, expressed as propagules (spores, hyphae, hyphal fragments), may range from 10^4 to 10^6 per gram. Fungi are important in soil because their ability to attack cellulose is greater than that of the bacterial flora. Some fungi are also predaceous and help to control the protozoan (Alexander, 1977, p. 67) and nematode population (Pramer, 1964) of soil.

Protozoans are also found in soil. Like the fungi, they are less numerous than the bacteria, ranging typically from 7×10^3 to 4×10^5 per gram of soil (Alexander, 1977). They are represented by flagellates (Mastigophora), amoebae (Sarcodina), and ciliates (Ciliata). Although both saprozoic and holozoic types occur, it is the latter that are of ecological importance in soil. Being predators, the holozoic forms help to keep the bacteria and, to a much lesser extent, protozoans, fungi, and algae in check. The protozoans inhabit mainly the upper portion of soil where their food source (prey) is most abundant. The types and numbers of protozoans in a given soil depend on soil type and soil condition.

Although algae are associated mostly with aquatic environments, they do occur in significant numbers in the uppermost portion of A horizons in soils (Alexander, 1977). They are usually the least numerous of the various microbial groups, ranging from 10^2 to 10^4 per gram just below the soil surface. The three most important groups in soil are the green algae (chlorophytes), diatoms (chrysophytes), and cyanobacteria. Xanthophytes may also be present. Since algae are photosynthetic organisms, they grow mostly on or just below the soil surface, where the light can penetrate. However, algae are also found below the light zone, where they may grow heterotrophically. Some green algae, diatoms, and cyanobacteria have been shown to be capable of heterotrophic growth in the dark. Some of the cyanobacteria in soil are capable of nitrogen fixation and may in some cases be more important in enriching the soil in fixed nitrogen than bacteria. The growth of algae in soil is dependent on adequate moisture and CO_2 supply. The latter is rarely limiting. The pH will influence which types of algae will predominate. While cyanobacteria prefer neutral to alkaline soil, green algae will also grow in acid soil. When growing photosynthetically, algae are primary producers, generating at least some of the reduced carbon on which heterotrophs depend. Because of their role as primary producers, algae are the pioneers in the formation of new soils, for instance in volcanic areas.

Table 3.3   Bacterial Densities at Different Depths of Different Soil Types
on July 7, 1915

Soil depth (in.)	Bacterial densities (bacteria per gram of air-dried soil $\times$ 10^6)			
	Garden soil	Orchard soil	Meadow soil	Forest soil
1	7.76	6.23	6.34	1.00
4	6.22	3.70	5.20	0.34
8	2.81	1.01	3.80	0.27
12	0.80	0.82	1.11	0.060
20	0.31	0.075	0.10	0.040
30	0.30	0.052	0.70	0.023

Source: Waksman (1916).

Bacterial distribution in mineral soil is illustrated in Table 3.3. Characteristically, the largest number of organisms occur in the upper A horizon and the smallest in the B horizon. Aerobic bacteria are generally more numerous than are anaerobic bacteria, actinomycetes, fungi, or algae. Anaerobes decrease with depth to about the same degree as aerobes. This seems contradictory but may reflect the fact that most of the anaerobes enumerated are facultative.

3.3 SUMMARY

The lithosphere of the earth consists of rock which may be igneous, metamorphic, or sedimentary. Rock is composed of intergrown minerals. Its surface, and in the case of porous rock its interior, may be a habitat for microbes. It may be roken down by weathering, which may ultimately lead to formation of mineral soil. Some of the rock minerals become chemically altered in the process. Weathering may be biological, especially microbiological, as well as chemical and physical.

Progress of mineral soil development is recognizable in the soil profile. A vertical section through mineral soil may reveal more or less well-developed horizons. Typical horizons of Spodosols and Mollisols include a litter zone (O horizon), a leached layer (A horizon), an enriched layer (B horizon), and the parent material (C horizon). The appearance of these horizons varies with soil type. Climate is one of several important determinants of soil type. The horizons are the result of intense biological activity in the litter zone and A horizon. Much of the organic matter in the litter zone is microbially solubilized and at least partially degraded. Soluble components are washed into the A horizon, where they may be further metabolized and where they contribute directly or indirectly to transformation of some of the mineral matter. Soluble products, especially inorganic ones, formed in the A horizon may be washed into the B horizon. The more refractory organic matter in the soil accumulates as humus, which contributes to soil texture, water-holding capacity, and general fertility. Soil may be 50% solid matter and 50% pore space. The pore space is occupied by gases, such as N_2, CO_2, and O_2, and by water. Water is also fixed to varying degrees by the soil particles. Microbes, including bacteria, fungi, protozoans, and algae, may inhabit the soil pores or live on the soil particles. They are most numerous in the upper layers of soil.

Not all soils can be classified as mineral soils. A few are organic and have a different origin. They arise from the slow decomposition of organic matter, mainly plant residues, which accumulates by sedimentation, as in swamps, marshes, or shallow lakes. They are not stratified and usually have a low mineral content.

Soil represents a most important microbial habitat in the earth's lithosphere.

4

The Hydrosphere

4.1 THE OCEANS

The oceans are a habitat for various forms of life, ranging from the largest anywhere on earth to the smallest. The **macrofauna** includes various vertebrates (mammals, birds, reptiles, fish) as well as a wide range of invertebrates. The **flora** includes the algae—from the macroscopic kelps to the small unicellular forms. The **plankton** of the sea includes the floating biota. The **phytoplankton** includes free-floating algae such as the diatoms and dinoflagellates, and the **zooplankton** includes the free-floating microscopic invertebrates and protozoans. The **bacterioplankton** includes the free, unattached bacterial forms.

The oceans cover about 70% of the earth's surface, occupying a surface area of 3.6×10^8 km^2 and a volume of 1.37×10^9 km^3, which amounts to 1.41×10^{21} kg of water. (The total mass of the earth is estimated to be 5.98×10^{24} kg.) Because of the unequal distribution of the continents between the northern and southern hemispheres of the present earth, only 60.7% of the northern hemisphere is covered by the oceans, whereas 80.9% of the southern hemisphere is covered by them. The world's major oceans include the Atlantic Ocean (16.2% of the earth's surface), the Pacific Ocean (32.4% of the earth's surface), the Indian Ocean (14.4% of the earth's surface), and the Arctic Ocean (2.8% of the earth's surface). The average depth of all oceans is 3,795 m. The average depth of the Atlantic Ocean is 3,926 m, that of the Pacific is 4,282 m, that of the Indian Ocean is 3,693 m, and that of the Arctic Ocean is 1,205 m. The greatest depth in the oceans occurs in **ocean trenches**. For instance, in the Pacific Ocean, the Mariana Trench achieves a depth of 10,912 m, and the Philippine Trench achieves a depth of 10,497 m. In the Atlantic Ocean, the Puerto Rico Trench achieves a depth of 8,648 m, and in the Indian Ocean, the Java Trench achieves a depth of 7,450 m. Shallow depths occur in marginal seas along the coasts of the continents. They are usually less than 2,000 m and frequently less than 1,000 m deep. Figure 4.1 shows the oceans of the world.

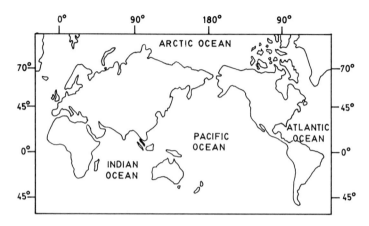

Figure 4.1   Oceans of the world.

The ocean basin is surrounded by the **continental margin**, which has several structural features. Projecting from each continental shore is the **continental shelf**, encompassing about 7.5% of the ocean area. It slopes gently downward at an average angle of 7 min to a water depth of about 130 m. Its average width is about 65 km, but may range from 0 to 1,290 km, the greatest distance being represented by the shelf projecting from the coast of Siberia into the North Polar Sea. The waters over the continental shelf are a biologically important part of the ocean, being the site of high productivity. This is easily explained by the contribution of nutrients in general runoff from land and from rivers empty-ing into the waters of this area.

At the edge of the continental shelf, the ocean floor drops sharply at an average angle of 4° (range 1-10°) to abyssal depths of about 3,000 m. This is the region of the **continental slope** and constitutes about 12% of the ocean area. In some places the slope is cut by deep canyons, which sometimes occur at the mouths of large rivers (e.g., Hudson River, Amazon River). Most but probably not all canyons were cut in past ages by **turbidity currents**. Such currents con-sist of strong water movements which carry a high sediment load picked up as a result of slumping of an unstable sediment deposit on a portion of a continental slope. On occasion, the continental slope may be interrupted by a terraced region, as in the case of the Blake Plateau off the southern Atlantic coast of the United States. This particular shelf is about 302 km wide and drops from a depth of 732 m to one of 1,100 m over this distance.

At the foot of the continental slope lies the **continental rise**, consisting of accumulations of sediment carried downslope by turbidity currents. Such deposits may be a few kilometers thick, forming fanlike structures in some places and wedges in others. An idealized section of a continental margin is shown in Figure 4.2.

The **ocean basin** takes up 80% of the ocean area. Its floor, far from being a flat expanse, as some once believed, often exhibits a rugged topography. Mountain ranges cut by fracture zones and rift valleys stretch over thousands of kilometers as the mid-ocean ridge systems. Elsewhere, submarine mountains, some active and some dormant or extinct volcanoes, dot the ocean floor. Some of the seamounts exhibit flattened tops and have been given the special name of **guyots**. Some of the tops of seamounts, especially in the Pacific, penetrate into surface waters which are about 50-100 m deep and have a temperature of about 21°C. There they serve as the substratum for colonization by corals (coelenterates) and coralline algae, forming atolls and reefs.

Covering the ocean floor almost everywhere are **sediments**. They range from 0 to 4 km in thickness, with an average thickness of 300 m. Their rate of accumulation varies, being slowest in mid-ocean (less than 1 cm per 10^3 years) and fastest on continental shelves (10 cm per 10^3 years). These rates may be even greater in some inland seas and gulfs (e.g., 1 cm per 10-15 years in the Gulf of California and 1 cm per 50 years in the Black Sea). Sediments in the deep

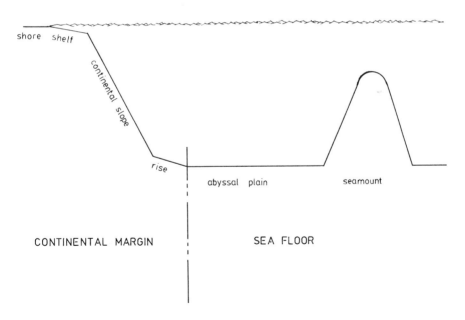

Figure 4.2   Cross section of part of an idealized ocean basin.

ocean consist in some regions mainly of deposits of siliceous and calcareous remains of marine organisms. The siliceous remains are derived from the tests of diatoms (algae) and radiolarians (protozoans), and the calcereous remains are derived from the tests of foraminifera (protozoans), pteropods (mollusks), and coccolithophores (algae). **Diatomaceous oozes** predominate in colder waters, whereas **radiolarian oozes** predominate in warmer waters. **Calcareous oozes** are found mainly in warmer waters and on ocean bottoms no deeper than 4,550-5,000 m. At greater depths the CO_2 concentration of water is large enough to cause dissolution of carbonate. Other vast areas of the ocean floor are covered by clays (**red clays** or **brown mud**) which are probably of terrigenous origin, and washed into the sea by rivers and general runoff from continents and islands and carried into the ocean basins by ocean currents, mudflows, and turbidity currents. At high latitudes in both hemispheres, particularly on and near continental shelves, ice-rafted sediments are found. They were dropped into the ocean by melting icebergs which had previously separated from glacier fronts that had picked up terrigenous debris during glacial progression. Except for ice-rafted detritus, only the fine portion of terrigenous debris (clays and fine silts) is carried out to sea. The clay particles are defined as less than 0.004 mm in diameter, and the silt particles are defined in a size range from 0.004 to 0.1 mm in diameter. Figure 4.3 shows the appearance of some Pacific Ocean sediments under the microscope.

The Ocean in Motion   A significant portion of the ocean is in motion at all times. The causes of this motion are (1) wind stress on surface waters, (2) Coriolis force arising from the rotation of the earth, (3) density variations of seawater resulting from temperature and salinity variations, and (4) tidal movement due to gravitational influences on the water by the sun and moon. Surface currents (Fig. 4.4) are prominent in regions of prevailing winds, such as the trade winds, which blow from east to west about 20° north and south latitude; the westerlies, which blow from west to east between 40 and 60° latitude in the nothern and southern hemispheres; and the easterly polar winds, which blow in a westerly direction south of the Arctic Ocean. The effect of these winds, together with the deflecting influence of the continents and the Coriolis force, is to set up surface circulations in the form of gyrals between the north and south poles in each major ocean. They are the North Subpolar Gyral (small), North Subtropical Gyral (large), North Tropical Gyral (small), South Tropical Gyral (small), South Subtropical Gyral (large), and the Antarctic Current, which circulates around Antarctica from west to east (Fig. 4.4A). Thus, the Gulf Stream, together with the Canary Current and the North Equatorial Current, is part of the North Subtropical Gyral of the Atlantic Ocean (Fig. 4.4B). The flow rates vary. The flow rate of the water in the Gulf Stream is the fastest for any surface current, 250 cm sec^{-1}. Other currents have flow rates that fall mostly in the range 25-65 cm sec^{-1}.

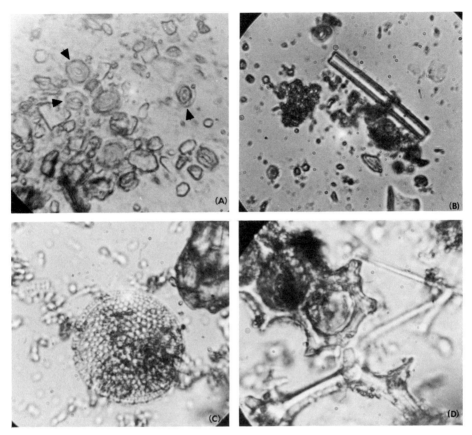

Figure 4.3 Microscopic appearance of marine sediments (1,750X). (A) Atlantic sediment showing cocoliths (CaCO$_3$) (arrows) and clay particles. (B) Atlantic sediment showing diatom frustules (SiO$_2$), cocoliths (CaCO$_3$), and other debris. (C) Pacific sediment showing a centric diatom frustule (SiO$_2$) and other debris. (D) Pacific sediment showing fragments of radiolarian tests (SiO$_2$).

Deep water is also in motion. Its movement has been thought to be caused by slow diffusion of water masses through broad zones in the oceans, but a full understanding of the causes of deep currents is still to come. Some deep currents that have been measured in the Atlantic Ocean have a velocity between 1 and 2 cm sec^{-1} (Dietrich and Kalle, 1965; pp. 399, 407).

In regions where the movement of a water mass diverges into different directions, deep water rises into the zone of **divergence**, resulting in **upwelling**. The same upwelling of deep water may also arise when winds blow large surface

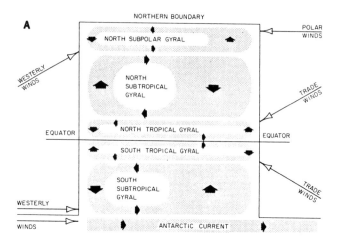

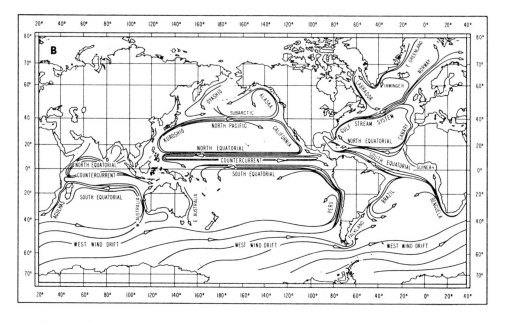

Figure 4.4 Oceanic surface currents. (A) Schematized representation of the prevailing winds and their effects on the surface currents of an imaginary rectangular ocean. (B) Average surface currents of the world's oceans. (From Williams, 1962; by permission.)

water masses away from coastal regions (Smith, 1968). Deep water is relatively rich in mineral nutrients, including nitrate and phosphate, and thus upwelling is of great ecological consequence because it replenishes biologically depleted nutrients in the surface waters. Regions of upwelling are, therefore, very fertile. In regions where two water masses of different density meet in a **convergence**, the heavier water will sink to a level where it meets water of its own density. This phenomenon is important in the oxygenation of deep waters (discussed further later in this section).

Chemical and Physical Properties of Seawater   Seawater is saline. Some important chemical components of seawater, listed in decreasing order of concentration, are presented in Table 4.1. Of these components, chloride (55.2%),

Table 4.1   Some Constituents of Seawater (μg liter^{-1})

Major constituents		Minor constituents	
Cl	1.9×10^7	Si	3×10^3
Na	1.1×10^7	N	6.7×10^2
Mg	1.3×10^6	Li	1.7×10^2
S (SO_4)	9.0×10^5	Rb	1.2×10^2
Ca	4.1×10^5	P	90
K	3.9×10^5	I	60
Br	6.7×10^4	Ba	20
C (CO_3, HCO_3)	2.8×10^4	Mo	10
B	4.5×10^3	Zn	10
		Ni	7
		Cu	3
		Fe	3
		U	3
		As	2.6
		Mn	2
		Al	1
		Co	0.4
		Se	9×10^{-2}
		Pb	3×10^{-2}
		Ra	1×10^{-7}

Source: Values taken from *Marine Chemistry* (1971).

sodium (30.4%), sulfate (7.7%), magnesium (3.7%), calcium (1.16%), potassium (1.1%), bromide (0.1%), strontium (0.04%), and borate (0.07%) account for 99.5% of the total salts in solution. Because these components generally occur in constant proportions relative to each other in true ocean waters, it has been possible to estimate salt concentration in seawater samples by merely measuring their chloride concentration. The chloride concentration in grams per kilogram (chlorinity) is related to the total salt concentration in grams per kilogram (salinity, S) by the following empirical relationship*:

$$S \permil = 0.300 + 1.8050 \, Cl \permil \tag{4.1}$$

The salinity so determined is an estimate of the total amount of solid material in a unit mass of seawater in which all carbonate has been converted to oxide and all bromide and iodide has been replaced by chloride, and in which all organic matter has been completely oxidized. For reference purposes, the salinity of standard seawater has been taken as 34.3 $\permil$. The actual salinity of different parts of the world oceans can vary from less than 34 $\permil$ to almost 36 $\permil$ (Dietrich and Kalle, 1965; p. 156). Table 4.2 lists the salinities of some different marine waters as well as of some saline lakes. It must be pointed out that some inland hypersaline water bodies such as the Dead Sea at the mouth of the Jordan River have a different salt composition than waters that are part of the marine system.

Bicarbonate and carbonate ions constitute 0.35% of the solute components in seawater. Together with borates and silicates, they form the buffering system of seawater, keeping the pH in the range 7.5-8.5. Surface seawater pH tends to fall into the narrow range 8.0-8.5. The variation in pH of seawater with depth may

*The symbol $\permil$ represents parts per thousand or grams per kilogram.

Table 4.2   Salinities ($\permil$) of Some Marine Waters and Salt Lakes

Gulf of Bothnia	0
Baltic Sea	3.5-12
Black Sea	16-23
Red Sea	40
Dead Sea	320
Great Salt Lake	320
Ocean bottoms	34.6-35

Source: Data from ZoBell (1946).

be related to oxygen utilization by marine organisms to a major extent and to carbonate dissolution to a lesser extent (Park, 1968). Figure 4.5 illustrates pH variation with depth at one particular station in the Pacific Ocean.

The salts dissolved in seawater impart a special osmotic property to it. The **osmotic pressure** of seawater is of the order of magnitude of the internal pressure of bacterial cells or of the cell sap of eukaryotic cells. Thus, at a salinity of 35 ‰ and a temperature of 0°C, seawater has an osmotic pressure of 23.07 atm, whereas at the same salinity but at 20°C it has an osmotic pressure of 24.69 atm. Clearly, then, the osmotic pressure of seawater is not deleterious to living cells.

With increases in depth in the water column, **hydrostatic pressure** becomes increasingly significant. Related to the weight of overlying water at a given depth, it ranges from 0 to more than 1,000 atm. It increases by about 1 atm for every 10 m of depth. Thus, the highest pressures are experienced in the deep ocean trenches. Living organisms in the sea are strongly affected by high hydrostatic pressure. Among the fauna, some are adapted to live only in surface waters, others at intermediate depths, and still others at abyssal depths. Generally, none are known that can live over the whole depth range of the open ocean. Microorganisms such as bacteria, on the other hand, appear to be more adaptable.

Salinity and temperature affect the **density** of seawater. At 0°C, seawater with a salinity of 30-37 ‰ has a corresponding density range of 1.024-1.030. A variation in seawater density due to variation in salinity may be a cause of water movement in the ocean, because denser water will sink below lighter water, or, conversely, lighter water will rise above denser water. The following processes may cause changes in salinity and, therefore, density: (1) dilution of seawater by runoff or by less saline water; (2) dilution by rain or snow; (3) concentration through surface evaporation by the sun's heat; (4) freezing, which excludes salts from ice and thus leaves any residual water more saline; or (5) thawing of ice, which dilutes already existent saline water.

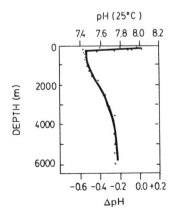

Figure 4.5 Vertical profile of pH at station 54°46′N, 138°36′W in the northeastern Pacific Ocean. (Adapted from Park, 1968. Copyright 1968 by the American Association for the Advancement of Science.)

As already stated, the variation of salinity of seawater is not the sole cause of its variation in density. The other important cause of density variation of seawater is temperature. Unlike fresh water, whose density is greatest at about 4°C (Fig. 4.6B), seawater with a salinity of 24.7 ‰ or greater has its maximum density at its freezing point (Fig. 4.6A). A body of fresh water thus freezes from its surface downward because fresh water at its freezing point is lighter than water at temperatures up to 4°C. Ocean water in the Arctic and Antarctic seas also freezes from the surface downward; in this case, because ice, which

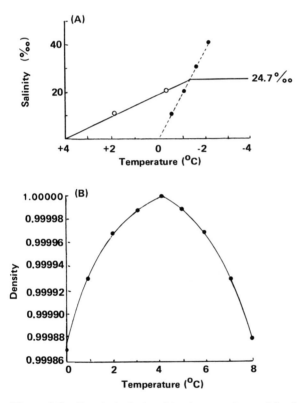

Figure 4.6 Density relationships in seawater and fresh water. (A) Relationship of seawater salinity to freezing point. Symbols: ○, temperature at maximum density at a given seawater salinity; ●, freezing-point temperature at a given salinity. Note that above a salinity of 24.7 ‰, seawater freezes at its maximum density since its temperature at maximum density cannot be lower than its freezing point. (B) Relationship of freshwater density to temperature. Data points for chemically pure water are shown. Note that in the case of fresh water, its density at its freezing point is lower than its density at 4°C.

excludes salt as it forms from seawater, will be lighter than the seawater, and will thus float on it.

The **temperature range** of seawater is from about $-2°C$ (the freezing point at 36 ‰ salinity) to $+30°C$, in contrast to the temperature range of air over the ocean, which is from about -65 to $+65°C$. The narrower temperature range for seawater can be related to (1) its high heat capacity, (2) its latent heat of evaporation, and (3) the heat transfer from lower to higher latitudes by surface currents on both hemispheres. The major source of heat in the ocean is solar radiation. More than half of the surface waters of the ocean are at $15\text{-}30°C$. Only 27% of the surface waters are below $10°C$. From about 50°N latitude to 50°S latitude, the ocean is thermally stratified. In this range of latitudes, the water temperature below about 1,000 m is below $4°C$. At depths from about 300 to 1,000 m, the temperature drops rapidly with depth. The zone of this rapid temperature change is called the **thermocline**. The thickness and position of the thermocline varies with geographic location and season of the year. Above the thermocline lies the warm surface water, the **mixed layer**, which is continually agitated by wind and water currents.

North of about 50°N latitude and south of about 50°S latitude, seawater is not thermally stratified. The waters around Antarctica, being cold ($-1.9°C$) and hypersaline (34.62 ‰) due to ice formation, are hyperdense and thus sink below warmer, less dense water and flow northward along the bottom of the ocean basin. This is an example of convergence. Similarly, Atlantic waters from the subarctic region having a temperature in the range $2.8\text{-}3.3°C$ and a salinity in the range 34.9-34.96 ‰ sink and flow southward at deep to bottom levels of the ocean. Because the Arctic Ocean bottom is separated from the other oceans by barriers such as the shallow Bering Strait in the case of the Pacific Ocean, and a shallow ridge in the case of the Atlantic Ocean, it does not influence the water masses of the Atlantic and Pacific oceans directly. Other convergences occur in both hemispheres in the world's oceans because of interaction of waters of different densities. Here, however, the heavier waters sink to lesser depths because they have lower densities than the heavier waters at high latitudes.

The water convergences alluded to above help to explain why ocean water is oxygenated at all depths (Fig. 4.7). Of all ocean waters, only some coastal waters (e.g., estuarine waters) may, as a result of intensive biological activity, be devoid of oxygen at depth. This intensive biological activity is often caused by human pollution. Surface waters of the open ocean tend to be saturated with oxygen because of oxygenation by the atmosphere and, equally important, by the photosynthetic activity of the phytoplankton. Oxygenation by phytoplankton can occur to depths of about 100 m (200 m in exceptional cases), where light penetration is 1% of the surface illumination. Seawater at a salinity of 34.352 ‰ is saturated at 5.86 ml or 8.40 mg of oxygen per liter at 760 mm Hg and 15°C. The higher the salinity and the higher the temperature, the lower is the oxygen solubility in seawater.

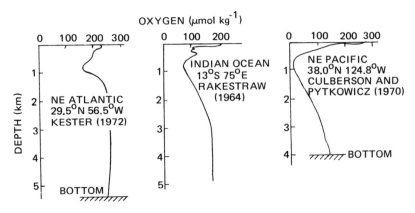

Figure 4.7 Vertical distribution of oxygen in the ocean. Profiles from three ocean basins. [From Kester, 1975 with permission from *Chemical Oceanography* (J. P. Riley and G. Skirrow, Eds.), Vol. 1, 2nd ed., 1975. Copyright by Academic Press Inc. (London) Ltd.]

The oxygen concentration in seawater will decrease with depth, owing mainly to oxygen consumption by the respiration of living organisms (Fig. 4.7). Since many life forms in the oceans tend to be concentrated in the upper waters, oxygen concentration will fall to a minimum at about 600-900 m of depth, where respiration (oxygen consumption) by zooplankton and other animal forms as well as bacterioplankton, but not photosynthesis (oxygen production) by phytoplankton, will occur. Below this depth, because of rapidly decreasing biological activity, the oxygen will at first increase again and then slowly decrease to a minimum toward the bottom. The water at the ocean bottom may still be half-saturated with oxygen relative to surface water. This oxygen does not come from photosynthesis, which cannot occur in the absence of light at these depths, or from significant oxygen diffusion from the atmosphere to such great depths. As previously indicated, these oxygenated waters derived from the Antarctic and Subarctic convergences. The oxygen-carrying waters from the Antarctic convergence flow northward along the bottom and at intermediate depths of the ocean basin, whereas the waters from the subarctic convergence in the Atlantic tend to flow southward at more intermediate depths. The oxygen content of these waters is only slowly depleted because of the low numbers of oxygen-consuming organisms in these deep regions of the oceans and the low rates of oxygen consumption in the upper sediments.

Photosynthetic activity of phytoplankton is dependent on sunlight penetration since it derives its energy almost exclusively from this source. It has been shown that light absorption by pure water in the visible range between 400 and

700 μm increases greatly toward the red end of the spectrum. It has also been shown that light penetrating transparent water has been 60% absorbed at a depth of 1 m. The same light has been 80% absorbed at 10 m and 99% absorbed at 140 m. In less-transparent coastal water, 95% of the light may have been absorbed at 10 m. Although the photosynthetic process of phytoplankters can use light over the whole visible spectrum, action spectra show peaks in the red and blue end of the spectrum, where chlorophylls absorb optimally. Accessory pigments, such as carotenoids, absorb light at intermediate wavelengths. Clearly, light penetration limits the depth at which phytoplankton can grow to about 80-100 m on the average (200 m maximally), and often to much lesser depths in less-transparent waters. The water layer to this depth constitutes the **euphotic zone**. Zooplankton and bacteria may abound to somewhat lower depths than phytoplankton (to about 750 m), being scavengers and able to feed on dying and dead phytoplankters and their remains in the process of settling.

Microbial Distribution in the Water Column and Sediments   Microbial distribution in the open oceans is not uniform geographically throughout the water column (Fig. 4.8). Factors affecting this distribution are energy- and carbon-source availability, nitrogen and phosphorus availability, temperature, hydrostatic pressure, and salinity. Accessory growth factors, such as vitamins, may also be limiting to those microbes that cannot synthesize them themselves. Phytoplankton distribution is primarily limited to the euphotic zone of the water column by available sunlight, the energy source. However, phytoplankton distribution in the euphotic zone may also be limited by temperature and to some extent by salinity, and by the availability of dissolved nitrogen (nitrate) and phosphorus (phosphate). Nonphotosynthetic microorganisms are limited to certain zones in the water column of the oceans by nutrient availability, and in addition by temperature, salinity, and hydrostatic pressure. The nonphotosynthetic organisms include predators (zooplankton), scavengers (zooplankton, fungi), and decomposers (bacteria and fungi and, to a possibly small extent, zooplankton). Zooplankters therefore dominate the euphotic zone, where they can feed optimally on phytoplankton, zooplankton, and bacteria. Bacteria and fungi are also very prevalent here, because they find sufficient sources of nutrients produced by the phytoplankton and zooplankton. Very high bacterial populations also occur at the air-water interface of the ocean as a result of a concentration of organic carbon in the surface film, where it is as much as 1,000-fold greater than in the water column below. The bacterial population in this film is known as **bacterioneuston** (Sieburth, 1976; Sieburth et al., 1976).

Sediments contain significant numbers of bacteria, fungi, and other benthic microorganisms, as well as higher forms of life. Viable bacteria have been recovered from 350 cm below the sediment surface (Rittenberg, 1940). Their numbers usually decrease with increasing depth in the sediment column. Fungi

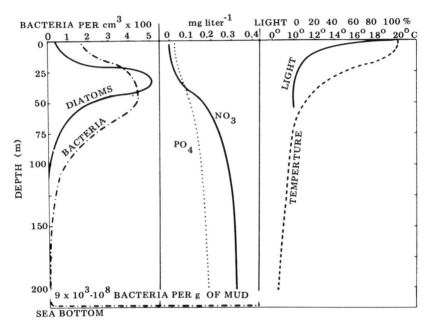

Figure 4.8 Vertical distribution of bacteria (number per cubic centimeter of water), diatoms (number per liter of water), PO_4, NO_3 (milligrams per liter), light, and temperature in the sea based upon the average results at several different stations off the coast of southern California. (From ZoBell, 1942; by permission.)

seem commonly to be restricted to sediments at shallow water depths, whereas bacteria, protozoans, and metazoans are found associated with sediments of shallow as well as abyssal depths. The chief function of the microbes is to aid in scavenging or decomposing of the organic matter that has settled undecomposed or partially decomposed from the overlying regions of biological productivity. Most of the organic matter from the euphotic zone settles to the bottom in the form of fecal pellets from metazoans. It should be pointed out that not all settled organic matter is utilizable by microbes, for reasons that are not yet clearly understood. This unutilizable organic matter constitutes a significant part of sedimentary humus. The metabolic activity of free-living bacteria of deep-sea sediments has been shown to be at least 50 times lower than that of microorganisms in shallow waters or on sediments at shallow depths (Jannasch et al., 1971; Jannasch and Wirsen, 1973; Wirsen and Jannasch, 1974). Environmental factors contributing to this slow rate of bacterial metabolism seem to be the low temperature ($<5°C$) and, especially, hydrostatic pressure. In a recent experiment, samples of wood were left on sediment in the Atlantic Ocean at a depth of 1,830 m for 104 days and found to be rapidly attacked by two species of wood borers (mollusks) (Turner, 1973). This observation has led to the suggestion that the primary attackers of organic matter in the deep sea, including sediment, are metazoans, and that bacteria and other microbes harbored in the digestive tract of these metazoans decompose this organic matter only after ingestion by the metazoans. It was recently reported that the bacterial flora of the intestines of amphipods (crustaceans) collected in the Pacific Ocean at a depth of 7,050 m was able to grow and metabolize nearly as rapidly at 780 atm and 3°C as at 1 atm and 3°C in laboratory experiments (Schwarz and Colwell, 1976). It became apparent in this study that these gut bacteria behave very differently from free-living bacteria from the same depths.

The generally low rate of metabolism of the biological community (benthos) on deep-sea sediments is also reflected by respiratory measurements carried out at 1,850 m. The measurements revealed a rate of oxygen consumption that was orders of magnitude less than in sediments of shallow shelf depths (Smith and Teal, 1973).

Phytoplankton, zooplankton, bacteria, and fungi are not found in very significant numbers at intermediate depths, chiefly because of a lack of an adequate nutrient supply, including a suitable source of energy, or because of the low temperature. Kriss (1970), having examined a north-south transect of the Atlantic Ocean, found an uneven distribution of bacteria at intermediate depths, which he associated with the different origins and characteristics of particular water masses. He interpreted his findings on the basis of available metabolizable nutrients, claiming that higher concentrations occur in water masses of equatorial-tropical origin, owing to **autochthonous** (of native, e.g., planktonic, origin) and **allochthonous** (from runoff from continents and islands) contributions, than in water masses of Arctic and Antarctic origin.

Bacterial and fungal growth and reproduction in ocean water also occurs on surfaces of some living organisms and on the surface of suspended organic and inorganic detrital particles (**epiphytes**) because at these sites essential nutrients may be very concentrated (Sieburth, 1975, 1976). Detritus, even if by itself not a nutrient, usually has adsorption capacity, which helps to concentrate nutrients on its surface and thus make for a preferred microbial habitat. The beneficial effect that the buildup of nutrients by adsorption to particulare surfaces has on microbial growth is great because the concentration of these nutrients in solution in seawater is very low (0.35-0.7 mg liter^{-1}; Menzel and Ryther, 1970). ZoBell (1946) long ago showed a significant increase in the bacterial population in natural seawater during 24 hr of storage in an Erlenmeyer flask. He attributed this to the adsorption of essential nutrients in the seawater to the walls of the flask where the bacteria actually grew.

Effects of Temperature and Pressure on Microbial Distribution   Temperature and pressure may have profound influence on where a given nonphotosynthetic microbe may live in the ocean. Some will grow only in the temperature range 15-45°C (**mesophiles**); others will grow only in the range 0°C or slightly below to 20°C, with an optimum at 15°C or below (**psychrophiles**). and still others will grow in the range 0-30°C, with an optimum near 25°C (**psychrotrophs**) (Morita, 1975). Thus, mesophiles would be expected to grow only in waters of the mixed zone, whereas psychrophiles would grow only below the thermocline and in the polar seas, and psychrotrophs would grow above and below the thermocline, although they would do better above it. Mesophiles may be recovered from cold waters and deep sediments, where they may be able to survive but cannot grow (**psychrotolerant**). Many bacteria that normally grow at atmospheric pressure are not inhibited by hydrostatic pressures up to about 200 atm, but are retarded at 300 atm and will not grow above 400 atm. Many bacteria isolated from waters at 500 and 600 atm were found to grow better at these pressures under laboratory conditions than at atmospheric pressure. Such organisms are called **barophiles**. Some organisms recovered from extreme depths (10,000 m) were reported which may be obligate barophiles (ZoBell and Morita, 1957). Yayanos et al. (1979) have isolated a barophilic spirillum from 5,700 m depth which grows fastest at about 500 atm and 2-4°C, with a generation time of 4-13 hr. The most pressure-sensitive biochemical process in prokaryotes, which determines the degree of barotolerance and limits growth under pressure, is protein synthesis—more specifically the translation phase in protein synthesis (Smith et al., 1975). (For a more complete discussion of ecological implications of temperature and pressure in the marine environment, see Morita, 1967.)

Microorganisms, especially bacteria recovered from marine samples, vary in their salinity requirements. Those which can grow only in a narrow range of salinities are called **stenohaline**, and those which can grow in a wide range of salinities are called **euryhaline**. Both types can be found in the open ocean. The salt requirement is not explained on the basis of osmotic pressure but by a specific requirement for one or more of the ions Na^+, K^+, Mg^{2+}, and Ca^{2+}. These ions may affect cell permeability or specific enzyme activities or both (MacLeod, 1965). They may also affect cell integrity.

Dominant Phytoplankters and Zooplankters   Diatoms, dinoflagellates, coccolithophores, and other flagellates are the dominant phytoplankters of the sea. The first two are the chief source of food for the herbivorous organisms in the seas. Diatoms are also important agents in the control of Si and Al concentrations in seawater (Mackenzie et al., 1978). Kelps and other sessile algae are mostly restricted to the shelf areas of the seas since they cannot grow at depths below about 30 m. A few kelps can float in the open ocean (e.g., sargasso weed). The dominant members of zooplankton include not only protozoans but

also invertebrates such as coelenterates, pteropods, and crustaceans, some of
which are not found free-floating as adults but have planktonic larval stages.
Among protozoans of the zooplankton, dominant forms include foraminifers
and radiolarians. Some of these forms are the food for other zooplankters and
for higher predatory animal life. All ultimately depend on the phytoplankters,
the primary producers (i.e., the original synthesizers of organic carbon by
photosynthesis). (For a further discussion, see Gross, 1972.)

Plankters of Geomicrobial Interest   The plankters of special geomicrobial
interest include the diatoms coccolithophores, and silicoflagellates among the
phytoplankters, and the foraminifera and radiolarians among the zooplankters.
It is these organisms that precipitate much of the $CaCO_3$ and SiO_2 in the open
sea that upon their death becomes incorporated into the sediments (see Chapters
7 and 8).

4.2   FRESHWATER LAKES

Structural Features of Lake Basins   Fresh waters are also important habitats for
certain life forms. Among these habitats are lakes. They are part of the **lentic**
environments, the standing water series, which consists of lakes, ponds, and
swamps. These water bodies arise in various ways. Some result from glacial
action—an advancing glacier gouges out a basin which when the glacier retreats
fills with water from the melting ice, and is later kept filled by runoff from the
surrounding watershed. Other basins result from landslides which obstruct
valleys and block the outflow from their watershed, or from crustal up-and-
down movement (dip-slip faulting) which forms dammed basins for the collection of runoff water. Still others result from solution of underlying rock,
especially limestone, which leads to the formation of basins in which water can
collect. Lakes are also formed by the collection of water from glacier melts
in extinct volcanic craters and by the obstruction of river flow or changes in
river channels.
    Lakes vary greatly in size. The combined Great Lakes in the United States
cover an area of 328,000 km^2, an unusually great expanse. More commonly,
lakes cover areas of 26-520 km^2, but many are smaller. Most lakes are less than
30 m deep. However, the deepest lake in the world, Lake Baikal in southern
Siberia, has a depth of 1,700 m. The average depth of the Great Lakes is 700 m
and that of Lake Tahoe on the California-Nevada border is 487 m. The elevation
of lakes ranges from below sea level (e.g., the Dead Sea at the mouth of the
Jordan River) to elevations as high as 3,600 m (Lake Titicaca in the Andes on
the border between Bolivia and Peru).

Some Physical and Chemical Features of Lakes   Some of the water of lakes may be in motion, at least intermittently. Most prevalent are horizontal currents, which result from wind action and the deflecting action of shorelines. Vertical currents are rare in lakes of average or small size. They may result from thermal, morphological, or hydrostatic influences. Thermal influence can result in changes in water density such that heavier (denser) water sinks below lighter water. Morphological influence can result from rugged topography of the bottom, which may deflect horizontal water flow downward or upward. Hydrostatic influence can result from springs at the lake bottom which force water upward into the lake. Besides horizontal and vertical currents, return currents may occur as a result of water being blown against a shore and piling up. Depending on the type of lake and the season of the year, only a portion of the total water mass of a lake, or all of it, may be circulated by the wind. (For a further discussion, see Welch, 1952.)

The waters of lakes vary in composition from a very low salt content (e.g., Lake Baikal) to a very high salt content (e.g.,Dead Sea; Lake Natron in Africa), and from low organic content to high organic content. Salt accumulation in lakes is the result of input of runoff from the watershed, including stream flow, slow solution of sediment components and rock minerals in the lake bed, and evaporation.

The waters of lakes may or may not be thermally stratified, depending upon various factors: geographic location, the season of the year and lake depth and size. Stratification, when it occurs, may or may not exist permanently, depending on whether the lake waters turn over (i.e., whether they are fully mixed by complete circulation) at certain times of the year. Lakes may be classified according to whether and when they turn over (Reid, 1961). The categories can be defined as follows. **Amictic** lakes are bodies of water that never turn over, being permanently covered by ice. Such lakes are found in Antarctica and at high altitudes. **Cold monomictic** lakes are bodies of water that contain waters never exceeding $4°C$, which turn over once during the summer, being thermally stratified the rest of the year. **Dimictic** lakes turn over twice each year, in spring and fall. They are thermally stratified at other times. These are typically found in temperate climates and at higher altitudes in subtropical regions. **Warm monomictic** lakes have water that is never colder than $4°C$. They turn over once a year in winter and are thermally stratified the rest of the year. **Oligomictic** lakes contain water that is significantly warmer than $4°C$ and turns over irregularly. Such lakes occur mostly in tropical zones. **Polymictic** lakes have water just over $4°C$ that turns over continually. Such lakes are found on high mountains in equatorial regions. **Meromictic** lakes are deep, narrow lakes whose bottom waters never mix with the waters above. The bottom waters usually have a relatively high concentration of dissolved salts, which makes them dense and separates them from the overlying waters by a chemocline. The upper waters in temperate climates may be thermally stratified in summer and winter and may undergo turnover in spring and fall.

A dimictic lake in a temperate zone during spring thaw accumulates water near $0°C$, which, because of its lower density, floats on the remaining denser water which is near $4°C$. As the season progresses, the colder surface water is slowly warmed by the sun to near $4°C$. At this point, all water has a more-or-less uniform temperature and thus uniform density. This allows the water to be completely mixed or turned over by wind agitation. As the surface water undergoes further warming by the sun, segregation of water masses recurs as warmer, lighter water comes to lie over colder, denser water. A thermocline is established between the two water masses, separating them into epilimnion and hypolimnion. The temperature of the water in the epilimnion may be greater than $10°C$ and vary little with depth (perhaps $1°C \ m^{-1}$). The water in the thermocline, on the other hand, will show a rapid drop in temperature with depth. This drop may be as drastic at $5.5°C$ per 0.3 m but is more usually $2.4°C$ per 0.3 m. The thickness of the thermocline varies with the position in the lake and between different lakes, an average value being around 1 m. The water in the hypolimnion will have a temperature well below the epilimnion and show a small drop in temperature with depth, usually less than $1°C \ m^{-1}$. The water in the epilimnion but not in the hypolimnion is subject to wind agitation and is thus fairly well mixed at all times. It is the greater density of the waters of the hypolimnion that prevents their agitation by the wind. Continual warming by the sun and mixing by wind produces horizontal currents and, in larger lakes, return currents over the thermocline, causing some exchange with water of the thermocline. This water exchange progressively increases the volume of the epilimnion and causes a progressive drop in the position of the thermocline. At fall turnover, the thermocline will have touched bottom of the lake and disappeared, the water now having a uniform warm temperature. With the approach of winter, the lake water will cool. Once the surface water has cooled below $4°C$, a thermocline will be reestablished. Ice may form on the epilimnion if the water temperature reaches the freezing point. The winter thermocline will usually remain near the lower surface of any ice cover on the lake. Figure 4.9 shows in idealized form the seasonal cycle of thermal stratification of a dimictic lake.

The thermocline of a lake acts as a barrier between epilimnion and hypolimnion. It prevents easy exchange of salts, dissolved organic matter, and gases because the two water masses that it separates, owing to their different density, do not readily mix. The oxygen content in the epilimnion is usually around the saturation level. At times of intense photosynthetic activity of phytoplankton, oxygen supersaturation may be achieved. The source of oxygen in the epilimnion is photosynthesis and air introduced into the water, especially during wind agitation. The oxygen concentration in fresh water at saturation at $0°C$ is 14.62 ppm or 10.23 ml liter^{-1}; at $15°C$ it is 10.5 ppm or 7.10 ml liter^{-1}; and

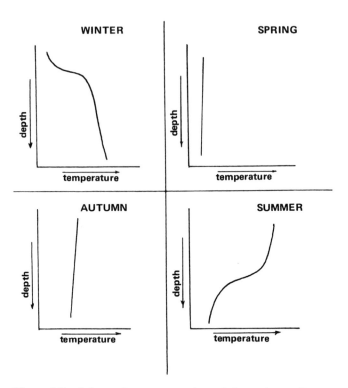

Figure 4.9   Schematic representation of thermal stratification in a dimictic lake.

at 20°C it is 9.2 ppm or 6.5 ml liter^{-1}. It is the optimal conditions of light and oxygen concentration together with adequate nutrient supply that cause phytoplankton, zooplankton, bacteria, fungi, and fish life to occur in greatest numbers in the waters of the epilimnion. The phytoplankters are the primary producers on which the remaining life forms depend directly or indirectly for food.

The oxygen that had been introduced into waters of the hypolimnion of a fertile (eutrophic) lake during turnover will be depleted as a result of biological activity, especially on and in the sediment. Thus, the hypolimnion may be anoxic during a shorter or longer period of time before fall turnover. Only anaeorobic or facultative organisms will carry on active life processes under these anoxic conditions. Such organisms include bacteria and protozoans as well as certain nematodes, annelids, immature stages of some insects, mollusks, and some fishes (Welch, 1952; pp. 183-186). Figure 4.10 illustrates the oxygen distribution measured in a dimictic lake during summer stratification.

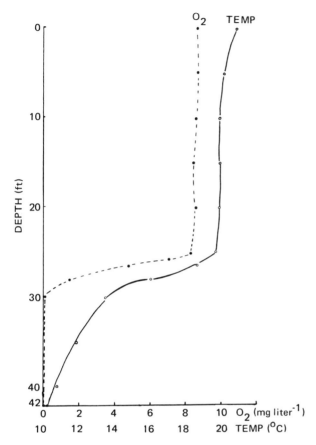

Figure 4.10 Oxygen profile in Tomhannock Reservoir (near Troy, N.Y.) on September 5, 1967. (From LaRock, 1969; by permission.)

Lake Bottoms The nature of lake bottoms is highly variable, depending on the location and history of the lake. The basins of many smaller lakes are flat expanses of sediment overlying bedrock. On the other hand, the basins of larger lakes (e.g., the Great Lakes) have a more rugged topography in many places. The bottom of lakes may be dominated by sand and grit, by clay, by a brown mud rich in humus, by diatom oozes, by ochreous mud rich in limonitic iron oxide, or by calcareous deposits. The organic components of any sediment may derive from dead and dying plankters that have sunk to the bottom, or from plant or animal remains. Some inorganic and some organic components may have been introduced into the lake and its sediment by the wind. Much silt and clay in the sediment is washed into lakes by runoff. Some is also contributed by erosion from the shoreline. The sediments are a major habitat of microbes.

Lake Fertility   Lakes may be classified in terms of their fertility or their nutritional status (i.e., their ability to support a flora and fauna). **Oligotrophic** lakes have an impoverished nutrient supply in which phosphorus or nitrogen are limiting and in which the oxygen concentration is high at all depths. **Eutrophic** lakes, on the other hand, are fertile lakes with nonlimiting supplies of nitrogen and phosphorus. **Mesotrophic** lakes are intermediate between oligotrophic and eutrophic lakes. **Dystrophic** lakes are lakes with an oversupply of organic matter which cannot be completely decomposed because of an insufficiency of oxygen. Nitrogen or phosphorus concentrations may be limiting. The water of such lakes is turbid and often acid. The origin of the dystrophic condition may be the encroachment of the shoreline by plants, including reeds, shrubs, and trees.

Lake Evolution   Lakes have an evolutionary history. Once fully matured, they age progressively. Their basins slowly fill with sediment, in part contributed by the surrounding land through runoff and erosion and in part by the biological activity in the lake. The size of the contribution that each process makes depends on the fertility of the lake. Changes in climate may also contribute to lake evolution (e.g., through lessened rainfall, which can cause a drop in water level, or through increase in temperature, which can cause more rapid evaporation). These effects make themselves felt slowly. Ultimately, a lake may change into a swamp.

Microbial Populations in Lakes   The microbial population in eutrophic lakes tends to be orders of magnitude greater than in the seas. Bacterial numbers may range from 10^2 to 10^5 per milliliter of lake water and be in the order of 10^6 per gram of lake sediment. The size of the bacterial population may be affected by runoff, which contributes soil bacteria. The bacterial population of lakes consists predominantly of gram-negative rods (Wood, 1965, p.36), although gram-positive, sporeforming rods can be readily isolated, especially from sediments. None of the bacterial types are represented by exclusively limnetic organisms. Such organisms appear not to exist. Actinomycetes can also be found along with fungi and protozoans. Algal types include blue-green (cyanobacterial) and green forms, as well as diatoms and pyrrhophytes. **Algal blooms** may occur at certain times when one species suddenly multiplies explosively and becomes the dominant algal form temporarily, often forming a carpetlike layer on the water surface. After having reached a population peak, most of the algal cells die off.

4.3   RIVERS

Rivers are a part of the **loctic** environment in which water moves in channels
on the land surface. Such flowing water may start as a brook, then widen into
a stream and ultimately into a river. The source of the water is surface runoff
and groundwater reaching the surface through springs or, more important,
through general seepage. A riverbed is shaped and reshaped by the flowing water
which scours the bottom and walls, especially with the help of suspended
particles from clays to small stones. Young rivers may feature rapids and steep
valley slopes. Mature rivers lack rapids and feature more uniform stream flow,
owing to a smoothly graded river bottom and an ever-widening riverbed. Old
rivers meander in their wide, flat floodplains. The flow of the water is caused by
gravity because the head of a river always lies above its mouth. Average flow
rates of rivers range from 0 to 9 m sec^{-1}. However, the flow of water in a river
cross section is not uniform. Some portions in such a section flow much faster
than others. This can be attributed to frictional effects related to the riverbed
topography as well as to density differences of different parts of the water mass.
Density variations may arise from temperature differences or from solute-
concentration differences between parts of the river. Portions of river water
may exhibit strong turbulence. Water velocity, turbulence, and terrain determine the size of particles the river may sweep along.

Where rivers empty into the sea, the less-saline water will flow over the denser
saline water from the sea with incomplete mixing, forming an **estuary**. Here
tidal effects of the sea will affect the water level of the river. Tidal effects may
be noted a considerable distance upstream in some rivers.

Because of constant water movement, the water temperature of rivers tends
to be rather uniform (i.e., rivers generally are not thermally stratified when
examined in cross section). Only where a tributary at a different water temperature enters a stream may there be local temperature stratification. Different
segments of a river may, however, differ in temperature. The pH of river water
can range from very acid (pH 3), for instance in streams receiving acid mine
drainage, to alkaline (pH 8.6) (Welch, 1952; p. 413). Unless heavily polluted by
human activities, rivers are generally well aerated. It has been thought that in
unpolluted rivers most organic and inorganic nutrients supporting microbial as
well as higher forms of life are largely introduced by runoff (allochthonous).
Recently, it was suggested that a significant portion of fixed carbon in such
streams and rivers may be contributed by autotrophs, mainly algae, growing in
quiet waters (autochthonous) (Minshall, 1978). Pollution may cause organic
overloading, which because of excessive oxygen demand will result in anoxic
conditions with the consequent elimination of many micro- as well as macroorganisms.

Planktonic organisms tend to be found in greater numbers in the more stagnant or slower-flowing waters of a river than in fast-flowing portions. The plankters include algae such as diatoms, cyanobacteria, green algae, protozoans, and rotifers. The proportions depend on the condition of a particular river and its sections. Sessile plants or algae tend to develop to significant extents only in sluggish streams or in the backwaters of otherwise rapidly flowing streams. Bacteria are also represented in significant numbers, but no unique flora occurs in unpolluted rivers.

4.4 GROUNDWATERS

Below the land surface, in soil, sediment, and permeable rock strata, **groundwater** is found. It derives mainly from **surface water** whose origin is meteoric precipitation such as rain and melted snow. It also derives from the water of rivers, lakes, and the like (Fig. 4.11). A minor amount of groundwater derives from **connate water, water of dehydration,** or **juvenile** water. **Connate water,** often marine in origin, and therefore saline, is water trapped in rock strata in the geologic past and isolated as a stagnant reservoir by up- or downwarping or

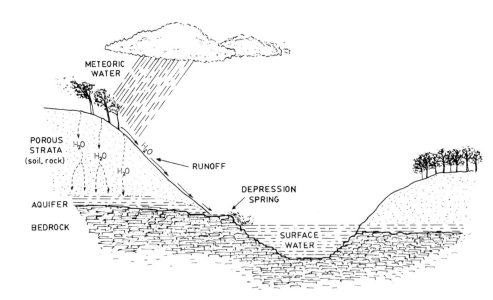

Figure 4.11   Interrelation of meteoric, surface, and ground waters.

faulting of a rock formation. Its salt composition is often greatly different from modern marine waters, presumably as of result of interaction with the enclosing rock. Connate waters are frequently associated with oil formations. **Waters of dehydration** are derived from waters of crystallization, which are part of the structure of certain crystalline minerals. They are released as a result of the action of heat and pressure in the lithosphere. **Juvenile waters** are waters associated with magmatism, which causes them to escape from the interior of the earth. They had never before reached the earth's surface.

Surface water slowly infiltrates the permeable ground as long as the gound is not already saturated. It will generally seep downward until it meets a water-saturated stratum or an impermeable stratum. Where groundwater gathers over an impermeable stratum, it forms an **aquifer.** The rate of infiltration of permeable strata depends not only on the surface water supply but also on the porosity of the permeable strata. Similarly, the water-holding capacity of the aquifer depends on the porosity of the rock material. The cause of water movement below ground level is not only gravity but also intermolecular attraction between water molecules, capillary action, and hydrostatic head (see Chapter 3).

Groundwater may escape to the surface or into the atmosphere through **springs** or by evaporation with or without the mediation of plants **(transpiration).** Some water will be accumulated by vegetation itself. Depending on relative rates of water infiltration and water loss, the level of the water table in the ground may rise, fall, or remain constant. Groundwater that reaches the surface through springs may do so under the influence of gravity, which may create sufficient head to force the water to the surface through a channel, as in an artesian spring. Groundwater may also reach the surface as a result of an intersection of the water table with the land surface, as in a depression spring. Finally, groundwater may reach the surface in springs under the influence of thermal energy applied to reservoirs deep underground. Such **hot springs** in their most spectacular form are geysers from which hot water spurts forth intermittently. Some hot springs emit not only water derived from infiltration of surface water but, in addition, juvenile water.

Surface water, as it infiltrates permeable soil and rock strata, will undergo changes in the composition of dissolved and suspended organic and inorganic matter. These changes are the result of adsorption and ion exchange by surfaces of soil and rock particles and the biochemical actions of microbes such as bacteria, fungi, and protozoans, which metabolize the adsorbed organic and (to a limited extent) inorganic matter. Polluted water infiltrating the ground may thus become thoroughly purified, provided that it moves through a sufficient depth and does not encounter major cracks and fissures which, because of reduced surface area, would exert only limited "filtering" action. Under some circumstances, groundwater may also become highly mineralized during infiltration or after reaching the water table. Such mineralized water when reaching the land surface may leave extensive deposits of calcium carbonate, iron oxide, and other material.

4.5 SUMMARY

The hydrosphere is mainly marine. It occupies more than 70% of the earth's surface. The world's oceans reside in basins surrounded by the margins of continental land masses which project into the sea by way of the continental shelf, the continental slope, and the continental rise, and bottoming out at the ocean floor. The ocean floor is traversed by mountain ranges cut by fracture zones and rift valleys—the mid-ocean ridges, where new ocean floor is being formed. The ocean floor is also cut by deep trenches, the zones of subduction, where the margin of oceanic crustal plate slips beneath continental crustal plate. Parts of the ocean floor also feature isolated mountains, which are live or extinct volcanoes. They may project above sea level as islands. The average world ocean depth is 3,975 m, the greatest depth about 11,000 m in the Mariana Trench.

Most of the ocean floor is covered by sediment of 300 m average thickness, accumulating at rates of less than 1 cm to greater than 10 cm per 10^3 years. Ocean sediments may consist of sand, silt, and/or clays of terrigenous origin, or of oozes of biogenic origin, such as diatomaceous, radiolarian, or calcareous oozes.

Different parts of an ocean are in motion at all times, driven by wind stress, the earth's rotation, density variations, and gravitational effects exerted by the sun and moon. Surface, subsurface, and bottom currents have been found in various geographic locations.

Where water masses diverge, upwelling of deeper waters, which replenish nutrients for plankton in the surface waters, occurs. Where water masses converge, surface water sinks and carries oxygen to deeper levels of the ocean, ensuring some degree of oxygenation at all depths.

Seawater is saline (average salinity about 35‰) owing to the presence of chloride, sodium, sulfate, magnesium, calcium, potassium, and some other ions. Variations in total salt concentration affects the density of seawater as does variation in water temperature. The ocean is thermally stratified between 50°N and 50°S latitude into a mixed layer (about 300 m deep), with water at more or less uniform temperature between 15 and 30°C depending on latitude; a thermocline (about 700 m deep) in which the temperature drops to about 4°C with depth; and the deep water (from the thermocline to the bottom), where the temperature is uniform between less than 0 and 4°C. Hydrostatic pressure in the water column increases by about 1 atm for every 10 m of increase in depth. Light penetrates to an average depth of about 100 m, which restricts phytoplankton to shallower depths. Zooplankton and bacterioplankton can exist at all depths but are found in greatest numbers at the seawater-air interface, near where phytoplankton abounds, and on the ocean sediment. Intermediate depths are at most sparsely inhabited because of limited nutrient supply.

Phytoplankton is constituted of algal forms, mainly diatoms, dinoflagellates, and coccolithophores, while zooplankton is constituted mainly of flagellated and amoeboid protozoans as well as some small invertebrates. Bacterioplankton is composed of bacteria, chiefly heterotrophs. The phytoplankton are the primary producers, the zooplankton and fungi are the predators and scavengers, and the bacteria are the decomposers. The metabolic rate of microorganisms decreases markedly with depth, probably as a result of the effects of high hydrostatic pressure and low temperature. Different life forms in the ocean show different tolerances to salinity, temperature, and pressure.

On land, fresh water is found in lakes and streams above ground and in aquifers below ground. Lakes are standing bodies of water, usually of low salinity which may be thermally stratified into epilimnion, thermocline, and hypolimnion. The degrees of stratification may vary with the season of the year. Water below the thermocline (i.e. the hypolimnion) may develop anoxia because the thermocline is an effective barrier to the diffusion of oxygen into it. Only after the disappearance of the thermocline do these waters become reoxygenated again due to total mixing by wind agitation. Lakes vary in their nutrient quality. Phosphorus is usually the most limiting element to lake life. Phytoplankton, zooplankton, and bacterioplankton are important life forms in lakes. Phytoplankton is restricted to the epilimnion, whereas zooplankton and bacterioplankton together with fungi are found in the entire water column and in the sediment.

Rivers constitute moving fresh waters, which are generally not thermally stratified. Abundant life forms, such as phytoplankton, zooplankton, and bacterioplankton, are concentrated mainly in the quieter portions of the steams, especially when unpolluted.

Groundwaters are derived from surface waters that seep into the ground and accumulate above impervious rock strata as aquifers. Water from the aquifer may come to the surface again by way of springs. In passing through the ground, water is purified. Microorganisms as well as organic and inorganic chemicals are removed by adsorption to rock and soil particles. Organic matter may be mineralized by the microbial decomposers. Water from deep aquifers is likely to exhibit a very low microbial population.

5

Geomicrobial Processes: A Physiological Treatment

Various microorganisms, both **prokaryotes** (bacteria, including cyanobacteria) and **eukaryotes** (higher algae, protozoans, and fungi), play an active role in certain geological processes, a fact that seems not always to be sufficiently appreciated by some microbiologists and geologists. A geological process that may be influenced by microbes and occurs in some types of sedimentary rock formation is **lithification**, a process in which microbes may cause production of substances such as $CaCO_3$, iron oxide, aluminum oxide, or silicate that bind sedimentary particles together. Another geological process that may be influenced by microbes is **mineral formation** in which microbes bring about precipitation of inorganic substances, such as iron sulfides, iron oxides, manganese oxides, calcium carbonates, and silica, as authigenic minerals. Still another geological process influenced by microbes is **mineral diagenesis**, in which microbes may contribute to the alteration of rock and to the transformation of a primary mineral to a secondary mineral, as in the conversion of orthoclase to kaolinite (see Chapter 3). **Sedimentation**, for example the acculumation of $CaCO_3$ tests of coccolithophores or forminifera or the accumulation of SiO_2 frustules of diatoms or SiO_2 tests of radiolarians or actinopods in pelagic or limnetic oozes, is a geolocial process to which microbes contribute substantially. The evolution of aging of lakes is also influenced by microbes (see Chapter 4).

Geological processes that are not influenced by microbes include **magmatic activity** or **volcanism**; **rock metamorphism**, resulting from heat and pressure; **tectonic activity**, related to crustal formation and transformation; and the allied process of **orogeny** or **mountain building**. **Wind** and **water erosion** should also be included, although these activities may be facilitated by prior microbial weathering activity. Although these geological processes are not influenced by microbes, microbes may be influenced by them as a result of the creation of new environments which may be more or less favorable for microbial growth and activities than before these happenings.

5.1 SOME IMPORTANT PHYSIOLOGICAL PRINCIPLES CONCERNING MICROBES

The influence that microbes may exert on geological processes is physiological. Microbes may act as catalysts of geomicrobial processes, or they may act as consumers or producers of geochemically active substances. In either case, they act through their **metabolism**. This metabolism has two aspects. One, **catabolism**, is that portion which provides the cell with needed energy and with some compounds that can serve as building blocks for polymers. It generally involves the oxidation of a suitable nutrient or metabolite (a compound metabolically derived from a nutrient). The other, **anabolism**, is that portion of metabolism that deals with **assimilation** and leads to the formation of organic polymers such as nucleic acids, proteins, lipids, and others, and of "inorganic polymers" such as the silicates of diatom frustules and radiolarian tests. Anabolism, by contributing to an increase in cellular mass and duplication of vital molecules, makes growth and reproduction possible. Catabolism and anabolism are linked to each other in that catabolism provides the energy and some of the building blocks that make anabolism, which is an energy-requiring process, possible. Both catabolism and anabolism may play a geomicrobiological role. Catabolism is involved, for instance, in large-scale iron, manaanese, and sulfur oxidation. Anabolism is involved in the formation of the organic compounds from which fossil fuels (peat, coal, petroleum) are generated and in the formation of silica frustules of diatoms and silica tests of radiolarians.

Catabolism may take the form of aerobic respiration, anaerobic respiration, or fermentation. Catabolism may thus be carried on in the presence or absence of oxygen in air. Indeed microorganisms can be grouped as **aerobes** (oxygen-requiring organisms), **anaerobes** (oxygen-shunning organisms), and **facultative** organisms. The facultative microbes can adapt their catabolism to operate in the presence or absence of air.

Metabolic Redox Reactions of Geomicrobial Significance: Respiration and Fermentation   In the process of **aerobic respiration**, reducing power in the form of hydrogen atoms or electrons, removed during oxidation of organic or inorganic compounds, is conveyed stepwise by hydrogen and electron carriers and enzymes—collectively known as the **electron transport system**—to oxygen to form water. In this process some of the energy liberated during electron transport to oxygen is trapped in special chemical bonds called high-energy bonds, for example the anhydride phosphate bonds of adenosine $5'$-triphosphate (ATP). These bonds (Fig. 5.1), upon hydrolysis, yield 7.3 kcal of free energy per mole at pH 7.0 and $25°C$ (Lehninger, 1975), as opposed to ordinary phosphate ester bonds, which release only about 2 kcal mol^{-1} of free energy under these conditions. The energy in high-energy bonds is used by cells for driving forward energy-consuming reactions, such as syntheses and polymerizations.

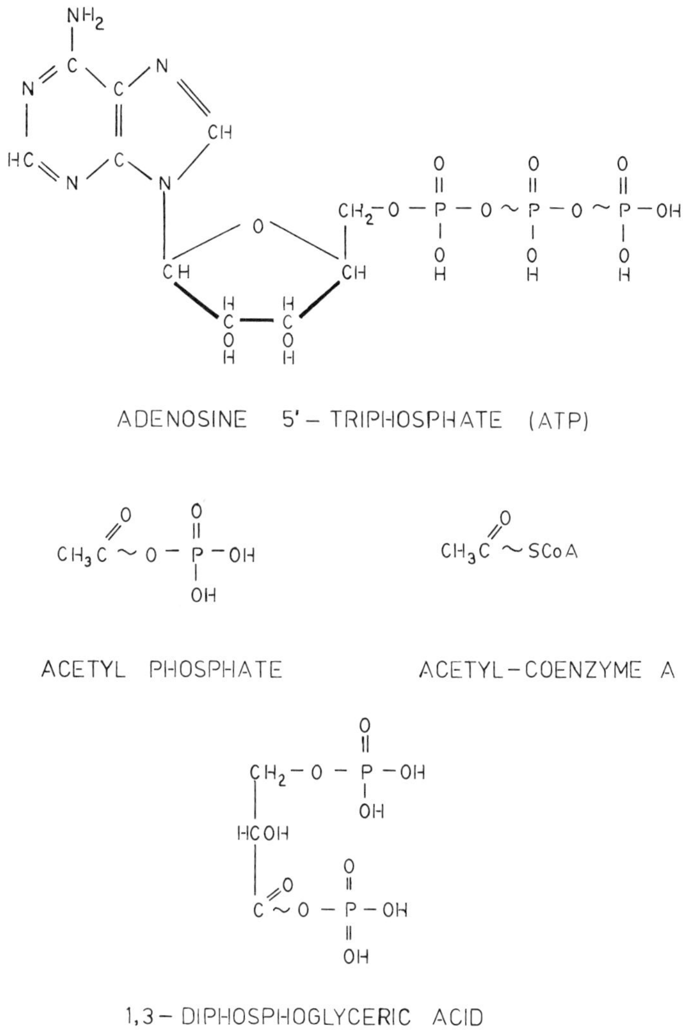

ADENOSINE   5' – TRIPHOSPHATE   (ATP)

ACETYL   PHOSPHATE   ACETYL – COENZYME   A

1,3 – DIPHOSPHOGLYCERIC   ACID

Figure 5.1   Examples of compounds containing a high-energy phosphate bond ($\sim$).

Typical components of the electron transport system include nicotinamide adenine dinucleotide (NAD), flavoprotein (FP), quinone (CoQ), cytochromes (cyt Fe) and cytochrome oxidase (cyt oxid). They interact as shown in Figure 5.2. Hydrogen or electrons are fed into this electron transport system at a level corresponding to the E_h value for the oxidative half-reaction of the

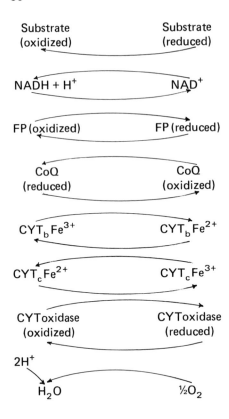

Figure 5.2   Schematic representation of a typical electron transport system to O_2. NAD^+ and $NADH + H^+$, oxidized and reduced forms of nicotinamide adenine dinucleotide, respectively; FP, flavoprotein; CoQ, coenzyme Q, a quinone; Cyt_b and Cyt_c, cytochromes b and c, respectively.

organic or inorganic substrate being oxidized. For example, the electrons from the oxidation of H_2 or pyruvate enter the transport chain at the level of NAD^+, whereas those from the oxidation of ferrous iron enter the transport chain at the level of cytochrome c. Table 5.1 lists the E_h for some geomicrobially important enzyme-catalyzed oxidations, the level at which their hydrogens or electrons are fed into the electron transport system upon their oxidation, and also the maximum number of high-energy phosphate bonds (ATP) that may be generated in the transfer of hydrogen or electron pairs to oxygen. It is important to recognize that in prokaryotic cells the electron transport system is located in the **plasma membrane** (Fig. 5.3A), whereas in the eukaryotic cell it is located internally in special organelles called **mitochondria** (Fig. 5.3B). As a result, appropriate bacteria are able to attack insoluble substrates which cannot be taken into the cell, such as elemental sulfur, iron sulfide, iron oxide, and manganese oxide, because the necessary enzymes, located in their plasma membrane, can make direct contact with the substrate. By the same token, eukaryotic organisms are

Table 5.1   Microbially catalyzed oxidations of geological significance; some
characteristics of their interaction with the electron transport system

Reaction	E_h' at pH 7.0 (V)	Entrance level into ETS	ATP/2e$^-$ or 2H
$Fe^{2+} \rightarrow Fe^{3+} + e^-$	0.77	Cytochrome c	1
$S^0 + 4H_2O \rightarrow SO_4^{2-} + 8H^+ + 6e^-$	0.357	Cytochrome c or b	1½ or 2½^a
$H_2S \rightarrow S^0 + 2H^+ + 2e^-$	-0.28	NAD$^+$ (?)	3 (?)
$H_2 \rightarrow 2H^+ + 2e^-$	-0.42	NAD$^+$ or FP	3 or 2
$Mn^{2+} + 2H_2O \rightarrow MnO_2 + 4H^+ + 2e^-$	0.86	FP (?)	1 or 2 (?)

^{a}It is assumed that ½ mol of ATP can be generated per mole of SO_3^{2-} oxidized to SO_4^{2-} via
substrate-level phosphorylation.

unable to carry out such reactions because their redox enzymes and carriers are
located on mitochondrial membranes and in the cytoplasm below the cell
surface (Ehrlich, 1978a).

Anaerobic respiration is a process in which certain reducible inorganic compounds, such as nitrate, sulfate, sulfur, carbon dioxide, ferric oxide and other
ferric compounds, and manganese dioxide, substitute for oxygen as terminal
electron acceptors. The process is normally associated only with some bacteria
and some cyanobacteria. In most cases of anaerobic respiration, oxygen com-
petes with the other possible electron acceptors, and so must be absent or
present at lower concentration than in normal air. Anaerobic respiration usually
utilizes some of the hydrogen and electron carriers of aerobic respiration but
frequently substitutes a suitable terminal reductase for cytochrome oxidase to
convey electrons to the terminal electron acceptor which replaces oxygen. The
best characterized of these systems are those in which sulfate and nitrate are
reduced. Like aerobic respiration, anaerobic respiration takes place in the
bacterial plasma membrane.

Fermentation is a process in which hydrogen atoms or electrons that are
removed in the oxidation of an organic compound *inside* a cell are transferred
to another partially oxidized organic compound, reducing it thereby. The
organic hydrogen or electron acceptor in this case is usually formed from the
substrate in steps subsequent to the oxidation step in a metabolic sequence.
This can be illustrated by the conversion of the sugar glucose to lactic acid
(Fig. 5.4). Fermentation occurs in the cytoplasm of the cell. It is common
among a number of bacteria but relatively rare among eukaryotic microbes.
Certain fungi, such as the yeast *Saccharomyces cerevisiae,* are an exception.

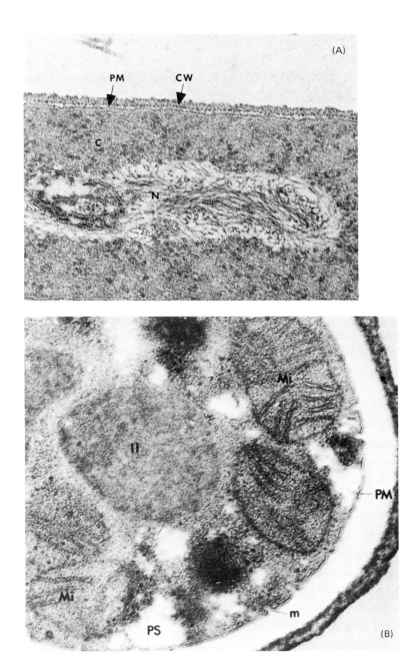

Figure 5.3

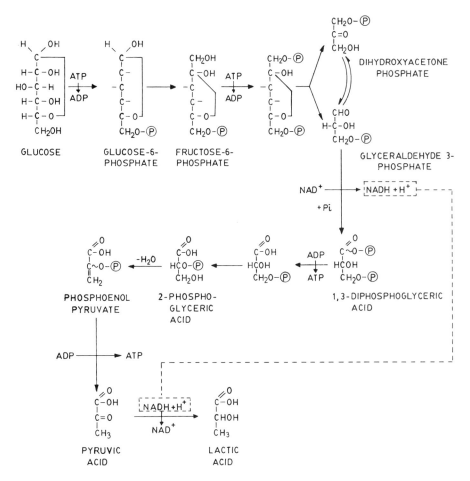

Figure 5.4 Conversion of glucose to lactic acid by glycolysis. Note that in this reaction sequence, biochemically useful energy in the form of ATP is generated exclusively by substrate-level phosphorylation.

Figure 5.3 Location of the electron transport system (ETS) in prokaryotic and eukaryotic cells. (A) Longitudinal section of *Bacillus* 29, a MnO_2-reducing prokaryote (128,000X). ETS is located in the plasma membrane. (CW, cell wall; PM, plasma membrane; C, cytoplasm; N, bacterial nucleus (nucleoid). (B) Cross section of a dormant conidium (spore) of *Aspergillus fumigatus*, a fungus and eukaryote (64,000X). ETS is located in the mitochondria. (Mi, mitochondrion; PM, plasma membrane; m, thin layer of electron dense material; PS, polysaccharide storage material; II, membrane-bound storage body.) (Courtesy of W. C. Ghiorse.)

In aerobic and anaerobic respiration, most of the useful energy is trapped in high-energy phosphate bonds during passage of the electrons along the electron transport pathway in a process called **oxidative phosphorylation**. The reaction of high-energy bond formation may be summarized by*

$$ADP + P_i \rightarrow ATP \qquad (5.1)$$

A maximum of 3 molecules of ATP per electron pair may be formed in aerobic respiration, and a probable maximum of 2 molecules of ATP may be formed in anaerobic respiration. In fermentation, useful energy is formed mostly by substrate-level phosphorylation, a process in which a high-energy bond, which traps some of the total free energy released during oxidation, is formed on the substrate. An example is the oxidation of 3-phosphoglyceraldehyde to 1,3-diphosphoglycerate in glucose fermentation, illustrated in Figure 5.4. Substrate-level phosphorylation may also occur during aerobic and anaerobic respiration, but it contributes only a small portion of the total energy trapped in high-energy bonds by cells. Clearly, aerobic and anaerobic respiration are much more efficient energy-yielding processes than fermentation for a cell. It takes much less substrate to satisfy a fixed energy requirement of a cell if the substrate is oxidized by aerobic or anaerobic respiration than if oxidized by fermentation.

Whereas the majority of microbes that oxidize inorganic substances to obtain energy are aerobes, a few are not. All methane-forming bacteria oxidize hydrogen gas (H_2) anaerobically by transferring electrons from H_2 to CO_2 to form methane (CH_4), generating ATP in the process by oxidative phosphorylation. Some oxidizers of sulfur compounds can transfer electrons from a reduced sulfur substrate such as thiosulfate or elemental sulfur to nitrate in the absence of oxygen. In the presence of oxygen, these sulfur-oxidizing organisms transfer the electrons from the reduced sulfur compound to oxygen. The maximum ATP yield in methane formation from H_2 reduction of CO_2 and in the oxidation of reduced sulfur compounds by nitrate, two examples of anaerobic respiration, has not yet been established.

Classification of Some Geomicrobially Active Microbes by Type of Nutrition

Metabolism is carried out at the expense of certain nutrients taken into the cell. For an organism to grow and multiply, all nutrients have to be supplied in an appropriate mix, satisfying carbon, energy, nitrogen, and mineral requirements. It is possible to group geomicrobiologically important bacteria which act as

---

*Explanation of abbreviations: ADP, adenosine 5'-diphosphate; P_i, orthophosphate; ATP, adenosine 5'-triphosphate.

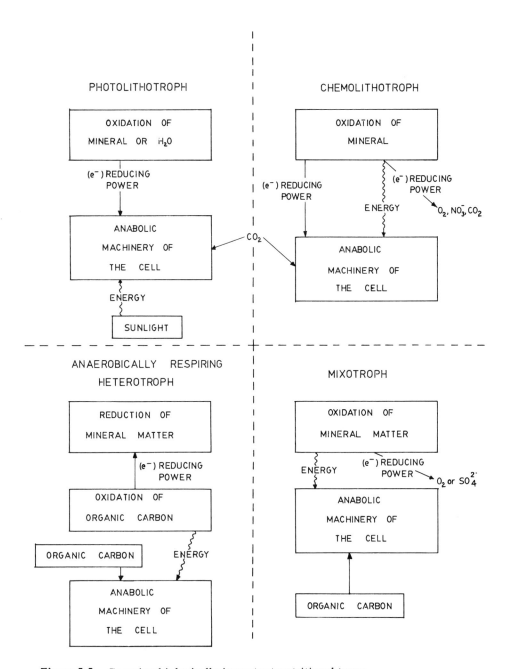

Figure 5.5   Geomicrobiologically important nutritional types.

catalysts of geomicrobial processes into four categories according to their energy source, carbon source, and electron acceptor requirements (Fig. 5.5). The four categories are chemolithotrophs, photolithotrophs, mixotrophs, and anaerobically respiring heterotrophs. **Chemolithotrophs**, also known as **chemosynthetic autotrophs**, use an oxidizable inorganic compound as a source of energy and reducing power. Substances that can serve in this capacity include H_2, H_2S, or other reduced forms of inorganic sulfur compounds, NH_3, HNO_2, Fe^{2+}, or Mn^{2+}, and others (Ehrlich, 1978a). As a carbon source, chemolithotrophs use inorganic carbon in the form of CO_2, HCO_3^-, or CO_3^{2-}, which they convert to organic carbon compounds with the energy and reducing power from the oxidation of the inorganic energy source (Fig. 5.5). These physiological processes can be illustrated by considering the case of *Thiobacillus ferrooxidans*. This organism oxidizes ferrous iron to ferric iron at acid pH:

$$2Fe^{2+} \rightarrow 2Fe^{3+} + 2e^- \tag{5.2}$$

Some of the reducing power generated in this way is transferred to oxygen,

$$\tfrac{1}{2}O_2 + 2e^- + 2H^+ \rightarrow H_2O \tag{5.3}$$

with the simultaneous production of a molecule of ATP by oxidative phosphorylation for every electron pair $(2e^-)$ transferred to oxygen:

$$ADP + P_i \rightarrow ATP \tag{5.4}$$

The remaining reducing power from the oxidation of the ferrous iron is used to reduce pyridine nucleotide.* Since, however, the electrons in this case have to travel against a redox potential gradient from +0.38 V (cytochrome c of *T. ferrooxidans* at pH 2.9; Vernon et al., 1960) to −0.34 V (NAD^-/NADH + H^+ redox couple at pH 2.9), energy as stored in high-energy phosphate bonds has to be consumed.

$$NAD^+ + 2e^- + 2H^+ + 2ATP \rightarrow NADH + H^+ + 2ADP + 2P_i \tag{5.5}$$

$$NADH + H^+ + NADP^+ \rightarrow NAD^+ + NADPH + H^+ \tag{5.6}$$

---

*Explanation of abbreviations: NAD^+, nicotinamide adenine dinucleotide (oxidized); NADH + H^+, nicotinamide adenine dinucleotide (reduced); $NADP^+$, nicotinamide adenine dinucleotide phosphate (oxidized); NADPH + H^+, nicotinamide adenine dinucleotide phosphate (reduced). These are all oxidized or reduced forms of pyridine nucleotide.

The NADPH + H⁺, together with some of the remaining ATP, are used in the assimilation of CO_2:

$$\text{Ribulose 5-phosphate} + \text{ATP} \rightarrow \text{ribulose 1,5-diphosphate} + \text{ADP} \qquad (5.7)$$

$$\text{Ribulose 1,5-diphosphate} + CO_2 \rightarrow 2(\text{3-phosphoglycerate}) \qquad (5.8)^*$$

$$2(\text{3-Phosphoglycerate}) + 2NADPH + 2H^+ + 2ATP \rightarrow$$
$$2(\text{glyceraldehyde 3-phosphate}) + 2NADP^+ + 2ADP + 2P_i \qquad (5.9)$$

From glyceraldehyde 3-phosphate the various organic constituents of the cell are then manufactured, including the building blocks for proteins, nucleic acids, lipids, polysaccharides, and so on, and subsequently combined into the corresponding polymers. Also, ribulose 5-phosphate is regenerated to permit continued CO_2 fixation (reaction 5.7-5.9). Although generally, chemolithotrophs can grow in the complete absence of organic matter under laboratory conditions, many, if not all of these organisms can assimilate some types of organic compounds, such as amino acids and vitamins. Although some chemolithotrophs are also able to use **organic carbon** as a sole energy source (**facultative chemolithotrophs**), others cannot.

Photolithotrophs, also known as **photosynthetic autotrophs** (Fig. 5.5), such as purple sulfur bacteria, green sulfur bacteria, and several cyanobacteria, also oxidize appropriate inorganic compounds, such as H_2, H_2S, and other reduced forms of sulfur, but solely as a source of reducing power to generate reduced pyridine nucleotide. Unlike the case with most chemolithotrophs, this process is always anaerobic. With H_2S, for example,

$$H_2S + NAD^+ \rightarrow NADH + H^+ + S^0 \qquad (5.10)$$

$$NADH + H^+ + NADP^+ \rightarrow NAD^+ + NADPH + H^+ \qquad (5.11)$$

No energy useful to the cell is released in these reactions. Instead, photolithotrophs obtain useful energy by converting radiant energy from the sun into chemical energy with the help of bacteriochlorophyll (purple sulfur bacteria) or chlorobium chlorophyll (green sulfur bacteria) or chlorophyll a (cyanobacteria) and accessory pigments in a process known as **photophosphorylation**.

---

*Reaction 5.8 is catalyzed by ribulose biphosphate carboxylase. Under conditions of CO_2 limitation, this enzyme can act as an oxidase, catalyzing an oxygenation (addition of oxygen) of ribulose diphosphate, forming phosphoglycolate and 3-phosphoglycerate. The phosphoglycolate is hydrolyzed to glycolate and phosphate. Microbial autotrophs probably excrete much of this glycolate. These reactions can occur in photosynthetic as well as chemosynthetic autotrophs. Energetically, they are a wasteful process.

$$ADP + P_i \xrightarrow[\text{light}]{\text{chlorophyll,}} ATP \tag{5.12}$$

Carbon dioxide is assimilated by photolithotrophs through the same mechanism as in chemolithotrophs (reaction 5.7-5.9), although exceptions are known (reverse tricarboxylic acid cycle; see Kelly, 1971). It is in reaction 5.9 that the NADPH + H$^+$ and some of the ATP is consumed.

Whereas purple sulfur bacteria and green sulfur bacteria photosynthesize anaerobically, the photolithotrophic cyanobacteria, which are also prokaryotic, usually photosynthesize aerobically. Recently it was discovered, however, that *Oscillatoria limnetica* is an exception: It can photosynthesize anaerobically in the presence of H$_2$S (Cohen et al., 1975). Some other cyanobacteria have also been found which are capable of anaerobic photosynthesis (Castenholz, 1976, 1977; Garlick et al., 1977; Padan, 1979). Cyanobacteria, like other algae and higher plants, produce O$_2$ when reducing NADP$^+$ with water. Since the NADP$^+$/NADPH + H$^+$ redox couple is much more reduced (−0.324 V) than the O$_2$/H$_2$O redox couple (+0.815 V), energy is required for the reduction of NADP$^+$. It is derived from sunlight with the help of chlorophyll and accessory pigments:

$$NADP^+ + H_2O \xrightarrow[\text{light}]{\text{chlorophyll,}} NADPH + H^+ + \tfrac{1}{2}O_2 \tag{5.13}$$

The same enzymatic mechanism involved in NADP$^+$ reduction is also used in generating ATP, another example of photophosphorylation (see reaction 5.12). The CO$_2$ assimilation mechanism is the same as in chemolithotropic and photolithotrophic bacteria (reactions 5.7-5.9).

Another nutritional group of bacteria of geomicrobial importance is represented by **mixotrophs** (Fig. 5.5). These organisms may obtain all or part of their energy from the oxidation of inorganic compounds and their carbon from organic compounds (Rittenberg, 1969). Most of the mixotrophs studied so far have been aerobes, but at least one, *Desulfovibrio desulfuricans*, is an anaerobe (Mechalas and Rittenberg, 1960).

From a geomicrobial standpoint, **anaerobically respiring heterotrophs** (Fig. 5.5) are also an important nutritional group. Most of these organisms use organic compounds as an energy and carbon source, but use inorganic compounds as electron acceptors in place of oxygen. Such acceptor compounds may be CO$_2$ (converted to methane), SO$_4^{2-}$ (converted to H$_2$S), NO$_3^-$ (converted to N$_2$O, NO, N$_2$), Fe$_2$O$_3$ (converted to Fe^{2+}), or MnO$_2$ (converted to Mn^{2+}). Only the organisms reducing CO$_2$ or SO$_4^{2-}$ are strict anaerobes. The others are facultative organisms, or even aerobes. It is now beginning to be recognized that this group may play an important role in the breakdown of organic matter (mineralization) under anaerobic conditions.

5.2   PHYSICAL INFLUENCE ON METABOLISM AND GROWTH

Effect of Temperature   Metabolism represents the biochemical activity of
organisms and, like all chemical reactions, is subject to physical influences, such
as temperature, pH and E_h (redox potential). Growth and reproduction are the
outcome of metabolic activity and thus are subject to these same influences. In
the microbial world we thus recognize distinct groups according to the temperature range they favor for growth and reproduction. These groups include
psychrophiles, psychrotrophs, mesophiles, and thermophiles. **Psychrophiles**
grow in a range from slightly below 0°C to about 20°C with an optimum at
15°C or lower (Morita, 1975). **Psychrotrophs** grow over a wider temperature
range than do psychrophiles (e.g., 0-30°C), with an optimum near 25°C. **Mesophiles** are microbes that grow in the range 10-45°C, with an optimum range for
some about 25-30°C and for others about 37-40°C. **Thermophiles** are microbes
that live in the temperature range 42-99°C or higher. Their optima depend
on the kind of microbe and its normal habitat. In general, thermophilic
photosynthetic prokaryotes cannot grow at temperatures higher than 73°C,
whereas thermophilic eukaryotic algae cannot grow at temperatures higher than
56°C (Brock, 1967, 1974, 1978). Thermophilic fungi generally exhibit temperature maxima around 60°C and thermophilic protozoans around 50°C. Only
nonphotosynthetic, thermophilic bacteria have temperature maxima as high as
99°C or higher.

The heat sensitivity of microbes is explained by changes in the lipid phase of
cell membranes and by heat lability of some key cellular components, such as
enzyme proteins, which become denatured (structurally randomized) at exces-
sively high temperatures. Key molecules in the different types of microorganisms evidently have different heat labilities (Brock, 1967; Tansey and Brock,
1972). Increasing hydrostatic pressure may reverse some of the denaturing
effects of heat (Haight and Morita, 1962). The low-temperature limit of growth
may be explained by maximal sensitivity of initiation of protein synthesis
(Broeze, 1978).

Effect of pH and E_h   Growth and reproduction of microbes, and thus their biochemical activity, is also dependent on favorable pH and E_h. Although the
enzymes inside a cell are known to be affected by pH and E_h, the pH and E_h
of the external environment of the cell should not have a direct effect on them
unless the pH or E_h values are extreme, since the cell is able to control its
internal environment. External pH and E_h effects are, therefore, more likely to
affect membrane-associated processes.

Environmental pH and E_h control the range of distribution of microorganisms. Figure 5.6 summarizes the pH and E_h ranges in which certain microbial
groups are able to grow. An important feature shown in the diagram is the

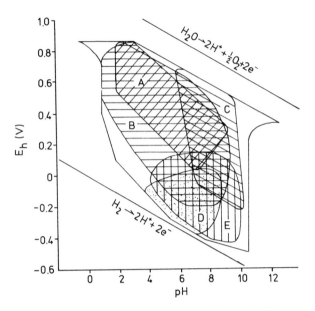

Figure 5.6 Environmental limits of E_h and pH for some bacteria; A, "iron bacteria"; B, thiobacteria; C, denitrifying bacteria; D, heterotrophic bacteria; E, sulfate-reducing bacteria. (Adapted from Baas Becking et al., 1960, from *J. Geol.* 68:243-284. Copyright 1960 by The University of Chicago Press.)

prevalence of iron-oxidizing bacteria and, to some extent, of thiobacteria in environments of relatively reduced potential and elevated pH.

5.3 ORGANIC CONVERSIONS IN THE LITHOSPHERE AND HYDROSPHERE

Microbes are involved in major ways in the transformation of organic and inorganic matter in the upper lithosphere and in the hydrosphere. Some of their activities manifest themselves in geologically recognizable processes.

The degradation of organic matter may occur in two ways. One is the complete breakdown of organic molecules by oxidation to CO_2, H_2O, NO_3^-, SO_4^{2-}, and so on, under aerobic conditions, and to CH_4, CO_2, NH_3, H_2S, and so on, under anaerobic conditions. This complete degradation is called **mineralization**. The other way is the incomplete degradation of organic matter by partial oxidation to simpler organic molecules. Aerobic mineralization requires a nonlimiting supply of atmospheric oxygen, whereas anaerobic mineralization requires an ample supply of inorganic electron acceptors such as CO_2, NO_3^-, or SO_4^{2-} in place of oxygen. Mineralization is an essential process in the recycling of matter needed for the continuation of life in the biosphere. Incomplete microbial oxidation of organic matter may be the result of limiting oxygen supply or a complete absence of oxygen or a limiting supply of NO_3^- or SO_4^{2-}, which favor fermentation, or it may be due to an inability of certain microbial populations to oxidize certain organic compounds completely because of enzymatic deficiencies. In nature, incomplete oxidation of organic compounds contributes to the formation of humus in both soil and sediments. **Soil humus** is a mixture of substances specifically derived from the partial decomposition of plant, animal, and microbial remains and from microbial syntheses. It is usually a brownish-black organic complex, only portions of which are soluble in water. A larger fraction is soluble in alkali. As mentioned in Chapter 3, soil humus includes aromatic molecules, often in polymerized form, whose origin is mostly lignin, bound and free amino acids, uronic acid polymers, free and polymerized purines and pyrimidines, and other forms of bound phosphorus.

Organic matter accumulating in marine sediments has been called **marine humus** because of a similarity to soil humus in its C/N ratio (Waksman, 1933) and because of its relative resistance to aerobic microbial decomposition (Waksman and Hotchkiss, 1937; Anderson, 1940). However, more detailed chemical analysis of marine humus indicates differences from soil humus, which are not surprising in view of differences in origin (Jackson, 1975; Moore, 1969, p. 271). Whereas soil humus is formed from plant remains, typical marine humus in sediments far from land derives mainly from phytoplankton remains. Marine humus from three northern Pacific sediments contains from 0.14 to 0.34% organic matter, including 20-1,145 ppm alkaline-soluble humic acids, 40-55% benzene-soluble bitumen, and 50-180 ppm amino acids. The remainder is kerogen, a material insoluble in aqueous organic solvents (Palacas et al., 1966). The organic matter of deep-sea sediments contains a fraction which, although refractory to microbial attack in situ, is readily attached by microbes when brought to the surface (see, e.g., Ehrlich et al., 1972). Presumably high hydrostatic pressure (>300 atm) and low temperature (<4°C) prevent rapid microbial in situ decomposition (Jannasch and Wirsen, 1973; Wirsen and Jannasch, 1975). Metabolizable organic matter in shallow-water sediments will undergo more complete decomposition provided that it does not accumulate too rapidly.

When organic matter is completely degraded under anaerobic conditions, a succession of different microorganisms are involved. One group of microorganisms provides hydrolytic enzymes to depolymerize proteins, carbohydrates, and so on. The same group or a different one then converts the products of hydrolysis to organic acids and alcohols. In the presence of sufficient SO_4^{2-} or NO_3^-, sulfate-reducing or denitrifying bacteria, respectively, may then oxidize these products to CO_2. In the absence of sufficient SO_4^{2-} or NO_3^-, another group of microorganisms may convert the organic acids and alcohols to CO_2 and H_2. Methane-forming bacteria may then convert the CO_2 and H_2 as well as residual acetate to CH_4. These reactions are often very prominent in reduced marine and estuarine sediments in shallow waters (Sørensen et al., 1979; Zeikus, 1977).

5.4   INORGANIC CONVERSIONS IN THE LITHOSPHERE AND HYDROSPHERE

Microbes as Geologic Agents   Some microbes in the biosphere act on inorganic matter as agents of concentration, dispersion, or fractionation. As **agents of concentration**, they cause localized accumulation of an inorganic substance by (1) intracellular deposition, as in the case of purple sulfur bacteria and other sulfur bacteria which deposit sulfur granules inside their cells upon oxidation of H_2S; (2) promoting precipitation of insoluble compounds external to the cell, as in most bacterial $CaCO_3$ precipitation; (3) adsorption to the cell surface, as in iron oxide deposition by certain sheathed bacteria; or (4) absorption and fixation in cells, as in the case of mercuric mercury uptake by some bacteria and fungi. The net effect of these localized precipitations is the manyfold local increase in concentration of an inorganic substance.

As **agents of dispersion**, microbes promote dissolution of insoluble mineral matter, for example, in the dissolution of $CaCO_3$ by respiratory CO_2, or in the biochemical reduction of ferric oxide or manganese dioxide to soluble compounds.

As **agents of fractionation**, microbes may act selectively on a mixture of inorganic compounds by promoting a selective chemical change of one or a few compounds of the mixture, causing a selective concentration or dispersion; for example, in the oxidation of arsenopyrite by *Thiobacillus ferrooxidans* (Ehrlich, 1964a) or in the preferential reduction of $Mn(IV)$ over $Fe(III)$ in ferromanganese nodules (Ehrlich et al., 1973). Microbes may also cause fractionation by attacking a particular stable isotope of an element in a compound in preference to another isotope of the same element, as in the reduction of $^{32}SO_4^{2-}$ in preference to $^{34}SO_4^{2-}$, the assimilation of $^{12}CO_2$ over $^{13}CO_2$, or the oxidation of ^{12}C organic matter in preference to ^{13}C organic matter by *Desulfovibrio desulfuricans* under conditions of slow growth (Doetsch and Cook, 1973). In general, the magnitude of these effects is large and may involve significant amounts of alteration in relatively short time. Studies so far lead to the impression that only a few mostly unrelated microorganisms have the capacity for isotope fractionation.

Mechanisms of Interaction of Microbes with Inorganic Compounds   Microbes may interact with inorganic compounds enzymatically (**direct interaction**) or nonenzymatically (indirect interaction). In enzymatic interactions, the microbes play an active role as catalysts, wheres in nonenzymatic interactions, they play a passive role. Examples of direct interaction include oxidations and reductions such as the oxidation or reduction of sulfur, iron, or manganese compounds. Direct interactions also include the breakdown of chelates of inorganic compounds and the hydrolysis of inorganic polymers such as phosphate esters and polyphosphates. Examples of indirect interactions include the deposition of inorganic compounds on a cell surface, such as the precipitation of manganic oxides on hyphae of fungi. Indirect interactions also include the reaction of metabolic end products of microbes with inorganic ions or compounds, leading to precipitation, solution, chelation, or structural alteration of insoluble inorganic compounds, as, for instance, in the formation of ferrous sulfide by interaction of microbially produced H_2S with ferrous iron.

5.5 SUMMARY

Microbes may be geologically active in lithification, mineral formation, mineral diagenesis, and sedimentation, but not in volcanism, tectonic activity, orogeny, or wind and water erosion. They may act as agents of concentration, dispersion, or fractionation of mineral matter. Their influence may be direct, through action of their enzymes, or indirect, through chemical action of their metabolic products, through passive concentration of insoluble substances on their cell surface, and through alteration of pH and E_h conditions in their environment. Their metabolic influence may involve anabolism or catabolism under aerobic or anaerobic conditions. Respiratory activity of prokaryotes may cause oxidation or reduction of certain inorganic compounds, resulting in their precipitation or solubilization. Some chemolithotrophic and mixotrophic bacteria can obtain useful energy from the oxidation of some inorganic substances, such as H_2, Fe(II), Mn(II), H_2S, S^0, and others. Photolithotrophic bacteria can use H_2S as a source of reducing power in the assimilation of CO_2 and, in the process, deposit sulfur. Anaerobically respiring organisms, which use certain oxidized substances as terminal electron acceptors, are important in the mineralization of organic matter in environments devoid of atmospheric oxygen. Mineralization of organic matter by microbes under aerobic conditions results in the formation of CO_2, H_2O, NO_3^-, SO_4^{2-}, PO_4^{3-}, and so on, while under anaerobic conditions it leads to the formation of CH_4, CO_2, NH_3, H_2S, and so on. Some organic matter is refractory to mineralization under anaerobic conditions and is microbially converted to humus. All microbial activities are greatly influenced by pH and E_h conditions in the environment.

6

Methods in Geomicrobiology

Geomicrobial phenomena can be studied in the field (in situ) or in the laboratory (in vitro). Field studies involve recognition and enumeration of geomicrobially active microorganisms, in situ measurement of the growth rate of geomicrobial agents, chemical and physical identification of reactants and products, rate measurements of the biogeochemical reaction, geochemical measurements as evidence of biological activity, and evaluation of the impact of different environmental factors on the geomicrobial process under study. Laboratory studies involve the isolation and characterization of the active agents in a field sample from a geomicrobially active site; recreation of the geomicrobial process under laboratory conditions using either an enriched or a pure culture from the field sample; characterization of the mechanism by which the active geomicrobial agent or agents act, including enzymatic study where appropriate; and an evaluation of environmental effect on the in vitro geomicrobial phenomenon.

6.1 FIELD STUDIES

Qualitative Field Observations; Detection of Geologically Active Microbes
Geomicrobially active microorganisms rarely occur as pure cultures in the field. They are frequently accompanied by other organisms, which may play no direct role in the geomicrobial process under study, although they may compete with the geomicrobially active agent for living space and nutrients and may even produce metabolites that stimulate or inhibit the geomicrobial agent to a degree. In the field, three types of microorganisms may be found associated with a geomicrobial process: (1) **indigenous organisms**, whose normal habitat is being examined, (2) **adventitious organisms**, which were introduced by chance into the habitat by natural circumstances and which may or may not grow in the new environment but do survive in it, and (3) **contaminants**, which were introduced during manipulation of the environment under study or in sampling. Distinctions among these groups are frequently difficult to make experimentally. A

81

criterion for identifying indigenous organisms may be their high frequency of occurrence in a given habitat and in similar habitats at different sites. A criterion for identifying adventitious organisms may be their inability to grow successfully in the habitat under study and their lower frequency of occurrence than in an adjacent habitat. Neither of these criteria are absolute, however. Identification of a contaminant may simply be based on knowledge about the organism that would make unlikely its natural existence in the habitat under consideration.

In situ observations of a geomicrobial phenomenon should include a study of the setting. In a terrestrial environment, the nature of rocks, soil, or sediment and the constituent minerals ought to be considered together with prevailing temperature, pH, oxidation-reduction potential (E_h), sunlight intensity, seasonal cycles, and the source and availability of moisture, oxygen, and nutrients. In an aqueous environment, water depth, oxygen availability, turbidity, light penetration, thermal stratification, pH and E_h, chemical composition of the water, nature of the sediment if part of the habitat, and nutrient availability should be evaluated.

Only rarely can visual observations of the microbial flora at a given location be made in place. They are possible when the microbes occur so massively as to be easily detectable, as, for instance, in the case of algal and bacterial mats in hot springs (e.g., Yellowstone Park, Montana; see Bauld and Brock, 1973), or in the case of lichen growth on rock surfaces. In most instances, observations of microbes in their natural habitat require special examination of a field sample. In soil, sediment, or rock samples, such observations may be by fluorescent microscopy in conjunction with staining with fluorescent dyes, or in conjunction with fluorescent antibody reaction if specific microbes are being sought (Casida, 1962, 1965, 1971; Eren and Pramer, 1966; Schmidt and Bankole, 1965). Such observations can also be made by scanning electron microscopy (LaRock and Ehrlich, 1975; Sieburth, 1975).

For detection by cultural methods, field samples may be aerobically plated on a variety of media with and without prior enrichment, or anaerobically grown in liquid and solid media (e.g., agar-shake cultures). Anaerobic culture may require special precautions relative to the exclusion of air from the culture during its manipulation. Such culture methods, however, are likely to fail in detecting a certain portion of the total flora in a sample because the growth requirements of some organisms in the field sample will not be met by the various media employed. In the case of soil or sediment samples, the buried slide method may avoid to some extent the disadvantage of offering only selective media for growth. In this method, a buried microscope slide furnishes a surface for growth of soil microbes at the expense of soil nutrients. Staining and microscopic examination of the slide after sufficient length of burial will reveal bacteria and other microbes attached to it, some of which would not have been seen in routine culturing. Figure 6.1 shows microbial growth on a slide that had been buried in soil.

The capillary technique of Perfil'ev is a more sophisticated version of the buried slide method (Perfil'ev and Gabe, 1969). It also relies on the natural environment for cultivation. In this technique, capillary tubes with optically flat sides are inserted into the soil, entrapping in their lumen soil solution and some tiny soil particles together with microbes. The lumen of the capillary, according to Perfil'ev, is representative of the capillary space in soil in which microbes grow. It may, however, fail to detect microbes growing on soil particles too large to be picked up in the capillary lumen. Soil particle surfaces are importnat microbial habitats. The content of the capillaries may or may not be perfused with appropriate solutions. The capillary tubes can be intermittently removed from soil and their content microscopically examined (Fig. 6.2). Using this technique, Perfil'ev and Gabe (1965) discovered several previously unknown bacteria in soil, including *Metallogenium, Kuznetsova,* and *Caulococcus.*

When enrichment is necessary, a method utilizing a two-dimensional steady-state diffusion gradient of nutrients in an agar medium may be utilized (Caldwell and Hirsch, 1973). It is especially useful for fastidious microorganisms.

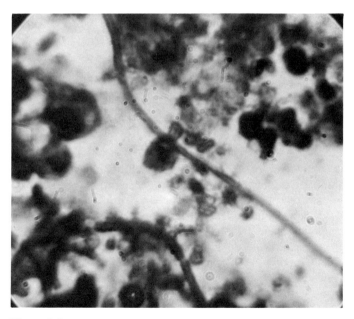

Figure 6.1   Demonstration of microbes in soil by the buried slide method (5,240X). This slide was stained with methylene blue and shows evidence of rod-shaped bacteria and a hypha.

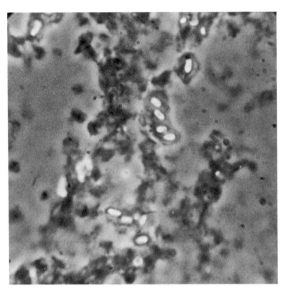

Figure 6.2   Microbial development in a capillary tube inserted into lake sediment contained in a beaker and incubated at ambient temperature (5,270X). The oval, refractile structures are bacterial spores.

In water samples, microbes, if present in sufficient numbers (10^3-10^5 per milliliter minimally, depending on size), may be observed in wet mounts by direct microscopic examination. If there are too few, direct visual observation will require prior concentration. One way of achieving such concentration, especially of bacteria, is by filtration through an appropriate filter membrane with pores small enough to hold back bacteria (0.45 μm). For visual observation, the cells on the membrane are then stained, and the membrane is made translucent with mineral oil or immersion oil and then examined microscopically (Fig. 6.3). Plankton may be concentrated by the classical sand filtration method (*Standard Methods for the Examination of Water and Wastewater,* 1965) prior to microscopic examination.

Microbes, especially bacteria and fungi in water, may also be detected by appropriate cultivation with or without prior concentration or enrichment. As in the case of particulate samples such as soil, cultivation may fail to detect all organisms present because of nutritional insufficiency of the culture media.

Quantitative Field Observations   The size of microbial populations in field samples may be determined visually and by culturing. In soil, sediment, and rock samples, visual estimation in situ of most microbes usually requires staining

with a fluorescent dye or fluorescent antibody followed by fluorescence microscopy (Trolldeiner, 1973). Cultural estimation of specific microbes in such
samples should, if possible, involve dislodging the organisms from the surface
of solid particles and suspending them in liquid medium. In some cases, this is
easily accomplished by simple agitation or by vortexing. It may also be
achieved by use of detergents, or by brief exposure to ultrasound (sonication)
of a suspension of soil, sediment, or rock particles in a suitable suspension
medium followed by settling of the particles while the microbial cells remain
in suspension. In other cases, however, the microbes may be so firmly attached
to the particles that agitation or other treatment will not dislodge them. In that
case, it may be necessary to plate suitable dilutions of unsettled suspension for
colony counts. Settling of the particles in the suspension medium may be
slowed in this case by increasing the viscosity of the suspension medium, as with
0.2% agar (Ehrlich et al., 1972). Counts obtained by plating suspensions of
particles with attached bacteria may be underestimates of the microbial population, however, since a certain fraction of particles are likely to harbor more than
one microbial cell on their surface which are likely to give rise to only a single
colony. An alternative approach, if the cells to be enumerated are too firmly
attached to particles, is to dissolve the particles in question. If the treatment is
mild enough, the freed cells, although killed, may remain intact and may thus be
countable by direct microscopic enumeration. If the dissolution process causes

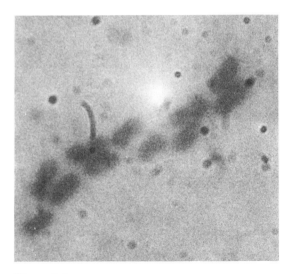

Figure 6.3   Bacteria deposited on a filter membrane (10,500X). Stained with
methylene blue.

cell destruction, an indirect estimation, such as measurement of cell nitrogen, may be performed. For this, the particles should be suspended in a nitrogen-free solution prior to treatment. The particles in the suspension are then subjected to solution followed by sulfuric acid digestion of the cells. The resultant NH_3-N is then measured. If no other source of NH_3-N was present in the sample, that which is recovered should represent cell nitrogen. In cases of mixed populations, however, cell nitrogen measurements cannot be readily converted to cell numbers, because different microbes do not necessarily contain similar concentrations of cell nitrogen. Even in pure cultures the nitrogen concentration per cell may vary depending on age and nutritional conditions.

Another indirect method of microbial enumeration involves the extraction of ATP from cells in soil, sediment, rock, or water samples and measuring it by a luminescence assay using firefly lantern extract (Holm-Hansen and Booth, 1966; Karl and LaRock, 1975; Karl, 1978). Since different kinds of microorganisms harbor different amounts of ATP, the ATP measurement, like that of NH_3-N, cannot be easily converted to cell number when dealing with a mixed population.

In aqueous samples, microbial enumeration may be made microscopically, usually after prior concentration. Bacteria and fungi are most frequently enumerated by cultural methods, using plating or filtration methods or the multiple tube dilution method (*Standard Methods for the Examination of Water and Wastewater*, 1965). Interpretation of the results of fungal enumeration by plating presents a problem in that the counts may have arisen from free spores, sporangia or conidia, or mycelial fragments. Usually, the fungal counts are expressed in terms of propagules.

Sample Collection   Special gear is often required for obtaining geomicrobial samples. Deeper-water samples may be obtained with a Van Dorn sampler, consisting of a piece of large-diameter plastic tubing, open at both ends and mounted in such a way on a cable or rope that it will descend vertically down a water column. The plastic tube can be closed at both ends with rubber closures operated by a spring mechanism that can be activated by a messenger (brass weight) which is allowed to slide down on the cable or rope after the sampler has reached a desired water depth. Deep-water samples can also be collected with a Niskin sampler, consisting of a collapsible, sterile plastic bag with a tube-like opening, mounted between two hinged metal plates. The sampler, with the bag in a collapsed state between the metal plates, is lowered on a cable or rope to a desired depth, where a spring mechanism of the sampler is activated by a messenger that causes the metal plates to open, expanding the bag, which now draws water into it.

Sediment samples may be obtained with dredging or coring devices. Lake sediment may be collected with an Ekman dredge (Fig. 6.4) or a Peterson

Figure 6.4 Ekman dredge. The brass messenger on the rope is 5 cm long.

dredge. Ocean sediment may be collected with a bucket dredge if the sediment surface only is to be sampled, or with a gravity corer (Fig. 6.5), a piston corer, or a box corer (Fig. 6.6) if the sediment column is to be sampled. Large mineral fragments or concretions on the sediment surface may be collected with a chain dredge or similar device (Fig. 6.7). The coring devices are rammed into the sediment to obtain a sample of the sediment column with minimal disturbance, whereas the dredges are dragged over the sediment surface so as to scrape it, thereby obtaining a highly disturbed sample.

Collection of rock samples may require chipping with a rock hammer or chisel. Collecting of soil samples may require the use of an auger, a type of hand drill.

Insofar as is possible, geomicrobial samples should be collected aseptically. This means that the critical surfaces of collecting gear should be sterile. Sterilization of metal surfaces can be achieved by thoroughly washing and then alcohol flaming. Plastic containers, if they can withstand it, may be autoclaved, or alternatively they may be sterilized with ethylene oxide. If a collection device cannot be sterilized, the sample it gathers should be subsampled to obtain the aliquot least likely to have been contaminated. If samples have to be manipulated by hand sterile surgical gloves should be worn.

Figure 6.5   Gravity corer. This is simply a hollow pipe containing a removable plastic liner and having a cutting edge at the lower end with a "core catcher" to retain the sediment core when the device is raised. A heavy lead weight at the top helps to ram the corer into the sediment when allowed to free-fall just above the sediment surface.

If samples cannot be examined immediately after collection, they should be stored so as to minimize microbial multiplication or death. Cooling the sample is usually the best way to preserve it temporarily in its native state, but the degree of cooling may be critical. Freezing may be destructive to the microbes. On the other hand, icing may not prevent the growth of psychrophiles or psychrotrophs.

Identification of Geomicrobially Active Agents in a Mixed Population in a Field Sample   In most instances, microscopic observation and enumeration of microbes in field samples reveal little of possible geomicrobial significance. Such assessment requires enrichment and the isolation in pure culture of as many of the different types of organisms present in a sample as seems reasonable. Each isolate must be tested for the particular geomicrobial activity under investigation. Enrichment and isolation from a mixed culture require selective culture conditions. If a microbial agent with a particular geomicrobial attribute is sought, the selective culture medium should have ingredients incorporated which favor the geomicrobial activity performed by the organism or organisms. Apart from nutrients, special pH, E_h, and temperature conditions may also have to be chosen to favor the selective growth of the geomicrobial agent. In attempting to identify the geomicrobially active microbe in a sample, it should be kept in mind that in some cases individual organisms may not be able to exhibit a particular geomicrobial activity by themselves but may have to be in combination with other specific organisms to be able to do so (synergism). At least three instances of such geomicrobial cooperation between microorganisms have been described in the case of microbial manganese oxidation: (1) the bacterium *Metallogenium symbioticum* in association with the fungus *Coniothyrium carpaticum* (Zavarzin, 1961; Dubinina, 1970; but see also Schweisfurth, 1969), (2) the bacteria *Corynebacterium* sp. and *Chromobacterium* sp. (Bromfield and Skerman, 1950; Bromfield, 1956), and (3) two strains of *Pseudomonas* (Zavarzin, 1962).

Figure 6.6   Box corer. After the frame hits bottom, the coring device is forced into the sediment mechanically. (Courtesy of Mark Sand.)

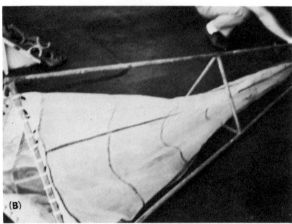

Figure 6.7 (A) Chain dredge. (B) Dredge for collecting manganese nodules from the ocean floor. A conical bag of nylon netting is attached to a pyramidal metal frame.

Indirect Identification of Past in situ Geomicrobial Activity in Terms of Isotope Fractionation   Microbial geochemical activity that occurred in the geologic past may under certain circumstances be identified indirectly in terms of isotope fractionation. Certain prokaryotic and eukaryotic microbes have been shown to distinguish between stable isotopes of elements such as sulfur, carbon, oxygen, and nitrogen. The microbes, especially under conditions of limited growth, prefer to metabolize substrates containing the lighter isotope of these elements (e.g., ^{32}S in preference to ^{34}S, ^{12}C in preference to ^{13}C, ^{16}O in preference to ^{18}O, and ^{14}N in preference to ^{15}N) (see Jones and Starkey, 1957; Emiliani et al., 1978; Wellman et al., 1968). Thus, products of metabolism will be enriched in lighter isotope when compared to the starting compound or to some reference standard that has not been subjected to isotope fractionation. In practice, isotope fractionation is measured by determining isotopic ratios (e.g., $^{34}S/^{32}S$, $^{13}C/^{12}C$, $^{18}O/^{16}O$, and $^{15}N/^{14}N$) by mass spectrometry and then calculating the amount of isotope enrichment from the relationship

$$\delta = \frac{\text{Isotope ratio of sample} - \text{Isotope ratio of standard}}{\text{Isotope ratio of standard}} \times 1{,}000$$

If the enrichment value δ is negative, it means that the sample tested was enriched in the lighter isotope relative to a reference standard, and if the value is positive, the sample tested was enriched in the heavier isotope relative to a reference standard. Thus, to determine, for instance, if a certain metal sulfide mineral deposit is of biogenic origin, various parts of a deposit are sampled and $\delta^{34}S$ values of the sulfide determined. If the values are generally negative (although the magnitude of the $\delta^{34}S$ value may vary from sample to sample and fall in the range -5 to -50 ‰), the deposit can be viewed as of biogenic origin because a chemical explanation for such ^{32}S enrichment under natural conditions is not likely. If the $\delta^{34}S$ values are positive and fall in a narrow range, the deposit is viewed as being of abiogenic origin.

Measurement of Ongoing, in situ Geomicrobial Activity by Use of Radioisotopic Tracers   Present-day geomicrobial activity may be measurable in situ. Such activity may be followed by use of radioisotopes. For instance, bacterial sulfate-reducing activity can be measured by adding $Na^{35}SO_4$ to a water, soil, or sediment sample of known sulfate content. After incubating in a closed vessel under in situ conditions (e.g., a water sample may be held in the water column at the depth from which it was taken), the sampe is analyzed for loss of $^{35}SO_4^{2-}$ and buildup of $^{35}S^{2-}$ by separating these two entities and measuring their quantity in terms of their radioactivity. A direct application of this method is that of Ivanov (1968). It allows estimation of the rate of sulfate reduction in the sample without having any knowledge of the number of physiologically active organisms. A modified method is that of Sand et al. (1975). It allows an estimation of sulfate-reducing activity in terms of the number of physiologically active bacteria in the sample as distinct from an estimation of the sum of physiologically active and inactive bacteria. The assay for the estimation of active bacteria can either be set up to measure the percentage of sulfate reduced in a fixed amount of time, which is proportional to the logarithm of active cell concentration, or it can be set up to measure the length of time required to remove a fixed amount of sulfate, which is related to the concentration of physiologically active sulfate-reducing bacteria in the sample. Ivanov's method can be adapted to measure the formation of sulfur and sulfates from sulfide by adding $^{35}S^{2-}$ to a sample and, after incubation in situ, separating $^{35}S^0$ and $^{35}SO_4^{2-}$ and measuring their quantity in terms of their radioactivity.

6.2 LABORATORY STUDIES

Laboratory Study of Geomicrobial Processes in Closed, Semiopen, and Open Systems   Present-day geomicrobial activity may also be studied in the laboratory by use of pure cultures or synergistic pairs of geomicrobially active organisms. The activity of growing organisms may be investigated in batch culture, in air-lift columns, in percolation columns, or in a chemostat. A batch culture represents a closed system in which an experiment is started with a finite amount of substrate that is continually depleted during growth of the organism with cell population and metabolic product buildup and possible changes in pH and E_h. Conditions within the culture are thus continually changing and become progressively less favorable. Batch experiments may be least representative of the natural process, which usually occurs in an open system with continual or intermittent replenishment of substrate and removal of at least some of the metabolic wastes. A culture in an air-lift column (Fig. 6.8) is a partially open system, in that the microbes grow and carry out their geomicrobial activity in a mineral charge of the column. They are fed with a continually recirculating feed solution from which nutrients are depleted and which removes metabolic products from the mineral charge in the column. Systems composed of percolation columns (Fig. 6.9) are even more open than those of air-lift columns. In this case, the microbes grow in the mineral charge of the column, but they are fed with a nutrient solution that is not recirculated. Thus, some fresh nutrient is always added, and wastes are removed in the effluent without recirculation, while pH, E_h, and temperature are held constant or nearly so. The chemostat idealizes an open culture system. It also does not imitate nature because culture conditions are too constant. It is a liquid culture system of constant volume in which nutrient supply and actively growing cell population as well as metabolic wastes can be held constant by introducing fresh nutrient solution at such a rate

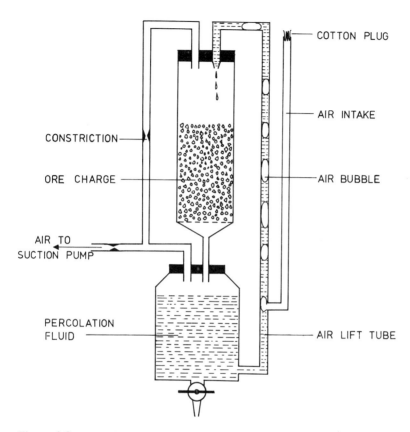

Figure 6.8 Air-lift column for ore leaching.

that the outflow removes cells and wastes at a rate that keeps the population and composition of the medium in the culture vessel constant (steady-state conditions). This can be expressed mathematically as

$$\frac{dx}{dt} = \mu x - Dx \tag{6.1}$$

where dx/dt is the rate of cell population change in the chemostat, μ the instantaneous growth rate constant, D the dilution rate, and x the cell concentration or cell number. Under steady-state conditions, dx/dt = 0 and $\mu x = Dx$ (i.e., instantaneous growth rate equals the dilution rate). Under conditions where $D > \mu$, the cell population in the chemostat will decrease and may ultimately be washed out. Conversely, if $\mu > D$, the cell population in the chemostat will

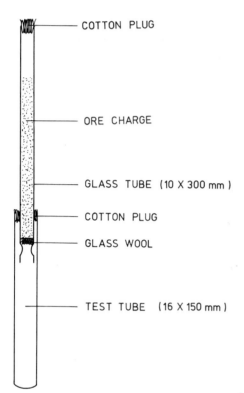

COTTON PLUG

ORE CHARGE

GLASS TUBE (10 X 300 mm)

COTTON PLUG

GLASS WOOL

TEST TUBE (16 X 150 mm)

Figure 6.9 Percolation column for ore leaching.

increase until a new steady state is reached, which is determined by the growth-limiting concentration of the substrate.

The steady state in the chemostat can also be expressed in terms of the rate of change of growth-limiting substrate concentration on the principle that the rate of change in substrate concentration is dependent on the rate of substrate addition to the chemostat, the rate of washout from the chemostat, and the rate of substrate consumption by the growing organism:

$$\frac{ds}{dt} = DS_{inflow} - DS_{outflow} - \mu(S_{inflow} - S_{outflow}) \tag{6.2}$$

where D is the dilution rate, S_{inflow} the substrate concentration entering the chemostat, $S_{outflow}$ the substrate concentration of unconsumed substrate, and μ the instantaneous growth-rate constant. At steady state, $ds/dt = 0$. The substrate consumed (i.e., $S_{inflow} - S_{outflow}$) is related to the cell mass produced according to the relationship

$$x = y(S_{inflow} - S_{outflow}) \tag{6.3}$$

where y is the growth-yield constant (mass of cells produced per mass of substrate consumed).

The chemostat can be used, for instance, to determine limiting substrate concentrations for growth of bacteria under simulated natural conditions. Thus, the limiting concentrations of lactate, glycerol, and glucose required for growth at different relative growth rates (D/μ_m) of *Achromobacter aquamarinus* (strain 208) and *Spirillum lunatum* (strain 102) in seawater have been determined by this method (Table 6.1) (Jannasch, 1967). The chemostat principle can also be applied to a study of growth rates of microbes in their natural environment by laboratory simulation (Jannasch, 1969) or directly in the microbial habitat. For instance, the algal population in an algal mat of a hot spring in Yellowstone Park, Montana, was found to be relatively constant, implying that the algal growth rate equaled its washout rate from the spring pool. If a portion of the algal mat is darkened by blocking access of sunlight, thereby stopping algal growth and multiplication in that part of the mat, the algal cells are washed out from it at a constant rate after a short lag (Fig. 6.10). The washout rate under these conditions equals the growth rate in the illuminated part of the mat. This follows from equation 6.1 when dx/dt = 0 (Brock and Brock, 1968). Similarly, the population of the sulfur-oxidizing thermophile *Sulfolobus acidocaldarius*, is in steady state in certain hot springs in Yellowstone Park, implying, as in the case of the algae, that the growth rate of the organism equals its washout rate from the spring. In this instance, the washout rate is measured by following the water turnover rate in terms of the dilution rate of a small, measured amount of NaCl added to the spring pool (Fig. 6.11). The dilution rate of the NaCl is

Table 6.1   Threshold concentrations of three growth-limiting substrates (mg liter^{-1}) in seawater at several relative growth rates of six strains of marine bacteria and the corresponding maximum growth rates (hr^{-1})

Strain	D/μ_m	Lactate	Glycerol	Glucose
208	0.5	0.5	1.0	0.5
	0.1	0.5	1.0	0.5
	0.005	1.0	5.0	1.0
μ_m:		0.15	0.20	0.34
102	0.3	0.5	no growth	0.5
	0.1	1.0		5.0
	0.05	1.0		10.0
μ_m:		0.45		0.25

Source: Excerpted from Table 2 in Jannasch (1967); with permission.

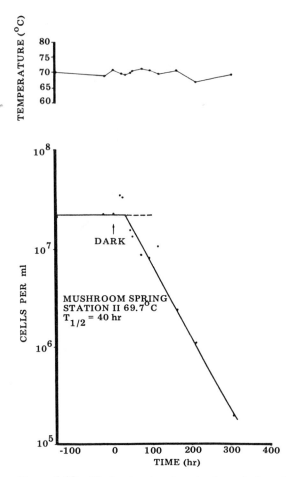

Figure 6.10 Washout rate of part of an algal mat at a station at Mushroom Spring in Yellowstone Park, Montana, after shading it experimentally. (From Bock and Brock, 1968; by permission.)

then a measure of the growth rate of *S. acidocaldarius* in the spring (Mosser et al., 1974).

Quantitative Study of Growth on Surfaces   The chemostat (or its principle of operation) is not applicable to all culture situations. In the study of geomicrobial phenomena, the key microbial activity often occurs on the surface of inorganic or organic solids. Indeed, the solid may be the growth-limiting substrate

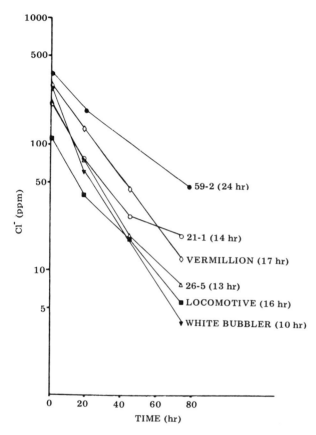

Figure 6.11 Chloride dilution in several small springs in Yellowstone Park, Montana. Estimated half-times for chloride dilution are given in parentheses. At site 21-1, chloride concentration had reached the natural background level by the final sampling time, and the dilution rate was estimated from the data from the first three sampling times. (From Mosser et al., 1974; by permission.)

upon which the organism acts. Under these conditions, the microbial population will increase geometrically on the solid surface, approximating the relationship

$$\log N = \log N_0 + \left(\frac{t}{g}\right) \log 2 \qquad (6.4)$$

where N is the final cell concentration per unit area, N_0 the initial cell concentration per unit area, t the time required for N_0 cells to multiply to N cells, and g the average doubling time (generation time). Once all available space on the surface has been occupied, the cell population on it will remain constant (provided that the surface area does not decrease significantly due to solid substrate consumption or dissolution), but the cell population in the liquid with which the solid is in contact (but which cannot support growth of the cells) will show an arithmetic increase in cell numbers according to the relationship

$$N_{final, liquid} = kN_{initial, liquid} \qquad (6.5)$$

because for every doubling on the surface of the cell-saturated solid, one daughter cell will be displaced into the liquid medium for lack of space on the solid. As long as there is no significant change in surface area of the solid phase, this model system applies, for instance, to the growth of autotrophic thiobacteria using water-insoluble, elemental sulfur or appropriate metal sulfides as their energy source in a mineral salts solution that satisfies all other nutrient requirements except the energy source (see Chapters 14 and 15).

Test for Distinguishing Between Enzymatic and Nonenzymatic Geomicrobial Activity   To determine if a geomicrobial transformation is an enzymatic or a nonenzymatic process, attempts to reproduce the phenomenon with cell-free extract should be tried. If active catalysis is observed, identification of the enzyme system involved should be undertaken. Spent culture medium, free of cells and enzyme activity, should also be tested for activity. If activity with spent medium is observed, operation of a nonenzymatic process may be inferred.

Ideally, the products of geomicrobial transformation, if they are precipitates, should be studied not only with respect to chemical composition but also with respect to mineralogical properties through one or more of the following techniques: electron microscopic examination, infrared spectroscopy, X-ray diffraction, electron microprobe examination, and crystallographic study. Similar studies should ideally be undertaken on the substrate if it is an insoluble mineral or mineral complex, to be able to detect any mineralogical changes it may undergo with time.

Studies of geomicrobial phenomena require ingenuity in the application of standard microbiological, chemical, and physical techniques, and often require collaboration among microbiologists, geochemists, mineralogists, and other specialists to unravel the problem.

6.3 SUMMARY

Geomicrobial phenomena can be studied in the field and in the laboratory. Direct observation may involve microscopic examination and chemical and physical measurements. Laboratory study may involve an artificial reconstruction of a geomicrobial process. Special methods have been devised for sampling, for direct observation, and for laboratory manipulation. The latter two categories include the use of fluorescence microscopy, radioactive tracers, and mass spectrometry for observing microorganisms in situ, for measuring process rates, and for measuring microbial isotope fractionation. The chemostat principle has been applied in the field to measure natural growth rates. It has also been used under simulated conditions for determining limiting substrate concentrations. Growth on particle surfaces may require special experimental approaches.

7

Microbial Formation and Decomposition of Carbonates

Carbonates, especially calcium carbonate and calcium-magnesium carbonates, occur extensively on the earth's surface. Assuming a total carbon concentration of 2,982.34 g cm^{-2} of earth surface, carbonates (as calcite and dolomite) account for 78.5% of the total; HCO_3^-, CO_3^{2-}, and CO_2 in the oceans account for 0.25%. Living and dead organic matter account for only 0.06% (data from Degens, 1969).* Thus, carbonates are an important reservoir of carbon on the earth's surface. Although some of the calcium carbonate is the result of magmatic and metamorphic activity (e.g. Bonatti, 1966; Skirrow, 1975), much is directly or indirectly the result of biological activity. Calcium carbonate is deposited extracellularly under various conditions by some bacteria, including cyanobacteria, and fungi (Bavendamm, 1932; Monty, 1972; Krumbein, 1974, 1974, 1979), and intracellularly by the bacterium *Archromatium oxaliferum* (*Bergey's Manual,* 1974). It is also laid down in surface structures of some algae, including certain green, brown, and red algae, and chrysophytes such as coccolithophores (Lewin, 1965a), and as a test or shell by some protozoans (foraminifera). In addition, it is incorporated into skeletal support structures of certain sponges and invertebrates such as coelenterates (corals), echinoderms, bryozoans, brachiopods, mollusks, and arthropods, in which it is associated with the chitinous exoskeleton. In all these cases, calcium and some magnesium are combined with carbonate ions of biogenic origin (Lewin, 1965a). Figure 7.1 illustrates a massive biogenic carbonate deposit, the White Cliffs of Dover, England.

7.1 CARBONATE DEPOSITION

Why Carbonates Are Readily Precipitated   Carbonate compounds are relatively insoluble. Table 7.1 lists the solubility constants for several different, geologically important carbonates. Because of the relative insolubility of the carbonates,

*Shales and sandstones (21.2%) and atmospheric CO_2 (0.25%) account for the remaining carbon.

Figure 7.1   White Cliffs of Dover, England; a foraminiferal chalk deposit. (Courtesy of the British Tourist Authority, 680 Fifth Avenue, New York, N.Y. 10019.)

they are readily precipitated from solution at relatively low carbonate and counterion concentrations. The following will illustrate the fact.

In a solution containing $10^{-4.16}$ M Ca^{2+},* the calcium will be precipitated by CO_3^{2-} at a concentration in excess of $10^{-4.16}$ M, since the product of the concentrations of the two ions will exceed the solubility product of $CaCO_3$, which is

$$[Ca^{2+}][CO_3^{2-}] \rightleftharpoons K_{sol} = 10^{-8.32} \tag{7.1}$$

In general, Ca^{2+} will be precipitated as $CaCO_3$ when the carbonate-ion concentration is in excess of the ratio $10^{-8.32}/[Ca^{2+}]$ (see equation 7.1).

Carbonate ion is an unbuffered aqueous solution undergoes hydrolysis. This process causes a solution of 0.1 M Na_2CO_3 to develop a hydroxyl-ion concentration of 0.004 M. The following reactions explain this phenomenon.

---

*The ion concentrations here really refer to ion activities.

Table 7.1   Solubility Products of Some Carbonates

Compound	Solubility constant (K_{sol})
$CaCO_3$	$10^{-8.32}$
$MgCO_3 \cdot 3H_2O$	10^{-5}
$MgCO_3$	$10^{-3.4}$
$CaMg(CO_3)_2$	$10^{-19.6}$

$$Na_2CO_3 \leftrightharpoons 2Na^+ + CO_3^{2-} \tag{7.2}$$

$$CO_3^{2-} + H_2O \leftrightharpoons HCO_3^- + OH^- \tag{7.3}$$

$$HCO_3^- + H_2O \leftrightharpoons H_2CO_3 + OH^- \tag{7.4}$$

Of these reactions, the third can be considered negligible.
Since bicarbonate dissociates according to the reaction

$$HCO_3^- = CO_3^{2-} + H^+ \tag{7.5}$$

whose dissociation constant (K_2) is

$$\frac{[CO_3^{2-}][H^+]}{[HCO_3^-]} = 10^{-10.33} \tag{7.6}$$

and since the ionization constant for water (K_w) at $25°C$ is

$$[H^+][OH^-] = 10^{-14} \tag{7.7}$$

the equilibrium constant (K_{equil}) for the hydrolysis of CO_3^{2-} (reaction 7.3) is

$$\frac{[HCO_3^-][OH^-]}{[CO_3^{2-}]} = \frac{10^{-14}}{10^{-10.33}} = 10^{-3.67} \tag{7.8}$$

At pH 7.0, where the hydroxyl-ion concentration is 10^{-7} (see equation 7.7), the ratio of bicarbonate to carbonate is therefore

$$\frac{[HCO_3^-]}{[CO_3^{2-}]} = \frac{10^{-3.67}}{10^{-7}} = 10^{3.33} \tag{7.9}$$

This means that at pH 7.0, the bicarbonate concentration is $10^{3.33}$ times the carbonate concentration in an aqueous solution, assuming that an equilibrium exists between the CO_2 of the atmosphere in contact with the solution and the CO_2 in the solution.

From equation 7.1 we can predict that at a freshwater concentration of 0.3 g of Ca^{2+} per liter of solution (i.e., at a $10^{-3.14}$ M Ca^{2+} concentration), an excess of $10^{-5.18}$ M carbonate ion would be required to precipitate the calcium as calcium carbonate ($CaCO_3$) because from equation 7.1 it follows that

$$[CO_3^{2-}] = \frac{10^{-8.32}}{10^{-3.14}} = 10^{-5.18} \qquad (7.10)$$

Now, at pH 7.0, $10^{-5.18}$ M carbonate would be in equilibrium with $(10^{-5.18}) \times (10^{3.33})$ or $10^{-1.85}$ M HCO_3^- according to equation 7.9. This amount of bicarbonate plus carbonate is equivalent to approximately 0.6 g of CO_2 per liter of solution.

Similarly, from equation 7.1, we can predict that at the calcium-ion concentration in normal seawater, which is 10^{-2} M, a carbonate concentration in excess of $10^{-6.32}$ M would be required to precipitate it. At pH 8, this amount of carbonate ion would be expected to be in equilibrium with approximately $10^{-3.92}$ M bicarbonate ion, which is equivalent to about 0.0044 g of CO_2 per liter of solution. Assuming the combined concentration of carbonate and bicarbonate ions in seawater to be 2.8×10^4 μg of carbon per liter (*Marine Chemistry*, 1971), we can calculate that the carbonate concentration at pH 8.0 must be about $10^{-4.97}$ M and the bicarbonate concentration about $10^{-2.64}$ M. Since the product of the carbonate concentration ($10^{-4.97}$ M) and that of the calcium concentration (10^{-2} M) is $10^{-6.97}$, which is greater than the solubility product for $CaCO_3$ ($10^{-8.32}$), seawater is saturated with respect to calcium carbonate. In reality, this appears to be true only for surface marine waters (Schmalz, 1972). (Mg^{2+} ion in seawater is not readily precipitated as $MgCO_3$ because of the relatively high solubility of $MgCO_3$.)

A quantity of 0.6 g of CO_2 can be derived from the complete oxidation of 0.41 g of glucose according to the equation

$$C_6H_{12}O_6 + CO_2 \rightarrow 6CO_2 + 6H_2o \qquad (7.11)$$

Similarly, 0.0044 g of CO_2 can be derived from the complete oxidation of 0.003 g of glucose. Such quantities of glucose are readily oxidized by appropriate populations of bacteria or fungi in a relatively short time.

Conditions of Microbial Carbonate Precipitaton   As already mentioned, some bacteria, including cyanobacteria, and fungi precipitate $CaCO_3$ or other insoluble carbonates extracellularly under various conditions. Let us now define the conditions under which this can take place. They include:

1. Aerobic and anaerobic oxidation of carbon compounds consisting of carbon and hydrogen with or without oxygen, for example carbohydrates, organic acids, and hydrocarbons, leading to the formation of CO_2 in a *well-buffered neutral or alkaline environment* containing adequate amounts of calcium, magnesium, or other appropriate cations. Under these conditions at least some of the CO_2 will be transformed into carbonate, which will then precipitate with appropriate cations

$$CO_2 + H_2O \rightleftharpoons H_2CO_3 \qquad\qquad (7.12)$$

$$H_2CO_3 + OH^- \rightleftharpoons HCO_3^- + H_2O \qquad\qquad (7.13)$$

$$HCO_3^- + OH^- \rightleftharpoons CO_3^{2-} + H_2O \qquad\qquad (7.14)$$

An example of calcium carbonate precipitation under these conditions is the formation by bacteria and fungi of aragonite—a crystal form of $CaCO_3$—and of other calcium carbonates in seawater media containing organic matter at concentrations of 0.01-0.1% (Krumbein, 1974). The organic matter consisted in different experiments of glucose, sodium, acetate, or sodium lactate. The aragonite precipitated on the surface of the bacteria or fungi after 36 hr of incubation. Between 36 and 90 hr, the cells in the $CaCO_3$ precipitate were still viable (although deformed), but after 4 to 7 days they were nonviable.

2. Aerobic or anaerobic oxidation of organic nitrogen compounds with the release of NH_3 and CO_2 in unbuffered environments containing sufficient amounts of calcium, magnesium, or other appropriate cations. The NH_3 is formed especially by bacteria in the deamination of amines, amino acids, purines, pyrimidines, and other nitrogen-containing compounds. In water, the NH_3 hydrolyzes to NH_4OH, which dissociates partially to NH_4^+ and OH^-, thereby raising the pH of the environment to the point where at least some of the CO_2 produced is transformed into carbonate. Examples of $CaCO_3$ precipitation under these conditions are the formation of aragonite and other calcium carbonates by bacteria and fungi in seawater media containing nutrients such as asparagine or peptone in a concentration of 0.01 or 0.1%, or homogenized cyanobacteria, as observed by Krumbein (1974). Other examples are the precipitation of calcium carbonate by various species of *Micrococcus* and a gram-negative rod in peptone media made up in natural seawater and in Lyman's artificial seawater (Shinano and Sakai, 1969; Shinano, 1972a,b). The organisms in this case came from inland seas of the North Pacific and from the western Indian Ocean. Lithification of beachrock along the shores of the Gulf of Aqaba (Sinai) is an example of in situ formation of $CaCO_3$ by heterotrophic bacteria in their mineralization of cyanobacterial remains (Krumbein, 1979).

3. The reduction of $CaSO_4$ to CaS by sulfate-reducing bacteria such as *Desulfovibrio* spp. and *Desulfotomaculum* spp. using organic carbon, indicated by the formula (CH_2O) in equation 7.15, as their source of reducing power. The CaS formed by these organsims hydrolyzes readily to H_2S, which has a small dissociation constant $(K_1 = 11 \times 10^{-7}; K_2 = 1 \times 10^{-15})$. The Ca^{2+} then reacts with CO_3^{2-} derived from the CO_2 produced in the oxidation of the organic matter by the sulfate-reducing bacteria. The following equations summarize the chemical sequence of reactions:

$$CaSO_4 + 2(CH_2O) \xrightarrow[\text{reducers}]{\text{sulfate}} CaS + 2CO_2 + 2H_2O \tag{7.15}$$

$$CaS + H_2O \longrightarrow Ca(OH)_2 + H_2S \tag{7.16}$$

$$CO_2 + H_2O \longrightarrow H_2CO_3 \tag{7.17}$$

$$Ca(OH)_2 + H_2CO_3 \longrightarrow CaCO_3 + 2H_2O \tag{7.18}$$

In the foregoing reactions it should be noted that 2 mol of CO_2 are formed for every mole of SO_4^{2-} reduced. Yet, only 1 mol of CO_2 is required to precipitate the Ca ions. Hence, $CaCO_3$ precipitation under these conditions depends either on the loss of CO_2 from the environment or on the presence of a suitable buffer system, or on the development of alkaline conditions. Evidence for $CaCO_3$ deposition linked to bacterial sulfate reduction is found in the work of Abd-el-Malek and Rizk (1963a). They demonstrated the formation of $CaCO_3$ during bacterial sulfate reduction in laboratory experiments using fertile clay-loam soil enriched with starch and sulfate, and in sandy soil enriched with sulfate and plant matter. Other evidence of microbial carbonate formation during sulfate reduction is found in the work of Ashirov and Sazonova (1962) and Roemer and Schwartz (1965). Ashirov and Sazonova showed that secondary calcite was deposited when an enrichment of sulfate-reducing bacteria was grown in quartz sand bathed in Shturm's medium: $(NH_4)_2 SO_4$, 4 g; $NaHPO_4$, 3.5 g; KH_2PO_4, 1.5 g; $CaSO_4$, 0.5 g; $MgSO_4 \cdot 7H_2O$, 1.0 g; NaCl, 20 g; $(NH_4)_2 FeSO_4 \cdot 6H_2O$, 0.5 g; Na_2S, 0.030 g; $NaHCO_3$, 0.5 g; distilled water, 1 liter. The hydrogen donor was gaseous hydrogen, calcium lactate plus acetate (the acetate probably acted as a carbon source), or petroleum. Petroleum constituents were probably first broken down to usable hydrogen donors for sulfate reduction by other organisms in the culture enrichment which the investigators used for inoculation before the sulfate reducers carried out their activity. The results from these experiments have lent support to the notion that incidents of sealing of some oil deposits by $CaCO_3$ may be due to the activity of sulfate-reducing bacteria at the petroleum-water interface of an oil reservoir.

Roemer and Schwartz (1965) showed that sulfate reducers were able to form calcite ($CaCO_3$) from gypsum ($CaSO_4 \cdot 2H_2O$) and anhydrite ($CaSO_4$); strontianite ($SrCO_3$) from celestite ($SrSO_4$); and witherite ($BaO \cdot CO_2$) from barite ($BaSO_4$). They found, however, that the sulfate reducers formed aluminum hydroxide from aluminum sulfate.

Still another example of calcium carbonate formation as a result of bacterial sulfate reduction is the deposition of secondary calcite in cap rock of salt domes, which has been inferred from a study of $^{13}C/^{12}C$ ratios (Feely and Kulp, 1957).

4. The hydrolysis of urea leading to the formation of $(NH_4)CO_3$

$$NH_2(CO)NH_2 + 2H_2O \rightarrow (NH_4)_2CO_3 \qquad (7.19)$$

This reaction causes precipitation of Ca, Mg, or other appropriate cations when present at appropriate concentrations. Urea is an excretory product of ureotelic animals, including adult amphibians and mammals. This hydrolytic reaction was first observed in experiments performed by Murray and Irvine (see Bavendamm, 1932). They believed it to be important in the marine environment. However, they did not implicate bacteria in urea hydrolysis. Steinmann (1899, 1901), on the other hand, working independently, did implicate bacteria in urea hydrolysis. Bavendamm (1932) observed extensive $CaCO_3$ precipitation by urea-hydrolyzing bacteria from the Bahamas. As it is now perceived, this is the least important mechanism of microbial carbonate deposition in nature because urea is not a widely disseminated compound.

5. Removal of CO_2 from a bicarbonate-containing solution. Such removal will cause an increase in carbonate-ion concentration according to the relationship:

$$2HCO_3^- \rightleftharpoons CO_2\uparrow + H_2O + CO_3^{2-} \qquad (7.20)$$

An important process of CO_2 removal is photosynthesis, in which CO_2 is assimilated as the chief source of carbon. However, some chemolithotrophs, as long as they do not generate acids in the oxidation of their inorganic energy source, may play a role because such organisms also rely on CO_2 as their carbon source. Examples of microbial organisms that precipitate $CaCO_3$ around them as a result of their photosynthetic activity include certain filamentous cyanobacteria associated with stromatolites (Monty, 1972; Golubic, 1973). Other examples are some of the green, brown, and red algae, and some of the chrysophytes, which are all known to deposit calcium carbonate in their walls (e.g., Lewin, 1965a; Friedman et al., 1972). Calcium carbonate formation by the latter algae is, however, not always directly related to photosynthetic activity. CO_2 removal is probably one of the most important mechanisms of biogenic $CaCO_3$ formation in the open environment.

Some Early Studies of Microbial Carbonate Precipitation   G. A. Nadson in 1903 (see Nadson, 1928) presented the first extensive evidence that bacteria could precipitate $CaCO_3$. Nadson studied the process in Lake Veisovoe in Kharkov, southern Russia. He described this lake as shallow with a funnel-like deepening to 18 m near its center and as resembling the Black Sea physico-chemically and biologically. He found the lake bottom to be covered by black sediment with a slight admixture of calcium carbonate, and he noted that the lake water contained $CaSO_4$, with which it was saturated at the bottom. He found that in winter the lake water was clear to a depth of 15 m, and from there to the bottom it was turbid, owing to a suspension of elemental sulfur (S^0) and $CaCO_3$. The clear water revealed a varied flora and fauna and a microbial population resembling that of the Black Sea.

When Nadson incubated black sediment from Lake Veisovoe covered with water from the lake in a test tube, he noted the appearance over time of crystalline $CaCO_3$ in the water above the sediment and the development of a rust color due to oxidized iron on the sediment surface itself (Fig. 7.2). Sterilized mud did not undergo these changes unless inoculated with a small portion of unsterilized mud. When access of air was blocked after $CaCO_3$ and ferric oxide development and the test tube reincubated, the reactions were reversed. The $CaCO_3$ dissolved and the sediment again turned black (Fig. 7.2). He isolated a number of organisms from the black lake sediment: *Proteus vulgaris, Bacillus mycoides, B. salinus* n.sp., *Bacterium albo-luteum* n.sp., *Actinomyces albus* Gasper, *A. verrucosis* n.sp., and *A. roseolus* n.sp.

When in a separate experiment he incubated a pure culture of *P. vulgaris* in a test tube containing sterilized, dried lake sediment and a 2% peptone solution prepared in distilled water, he noted the development of a pellicle of silicic acid which yellowed in time and then became brown and opaque from ferric hyroxide deposition. At the same time he noted that $CaCO_3$ was deposited in the pellicle. None of these changes occurred in sterile uninoculated medium.

In further experiments, Nadson showed that *B. mycoides* precipitated $CaCO_3$ in nutrient broth, nutrient agar, and gelatin medium. He did not identify the source of the calcium in these instances.

In still other experiments, Nadson showed that bacterial decomposition of dead algae and invertebrates in seawater led to $CaCO_3$ precipitation. In the decomposition of the alga *Alcyonidium*, a bacterium developed which Nadson called *Bacterium helgolandium*. In pure culture in 1% peptone-seawater broth, this organism formed ooliths (i.e., shell-like deposits around normal and involuted cells).

In a last set of two experiments, Nadson claimed to have observed the precipitation of magnesian calcite or dolomite. In one experiment, bacterial activity over 3½ years of incubation produced dolomite crystals in a culture of black mud from a salt lake near Kharkov. In a second experiment, a culture of

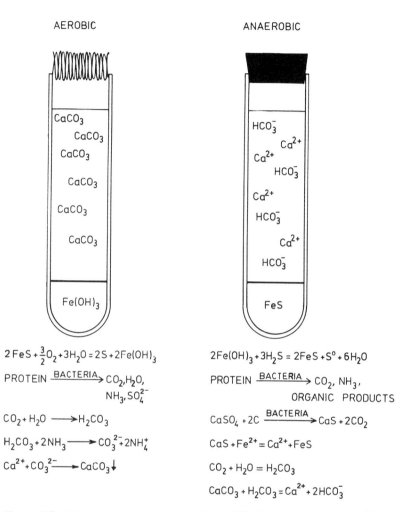

Figure 7.2 Diagrammatic representation of Nadson's experiments with mud from Lake Veisovoe. Contents of the tubes represent the final chemical state after microbial development. Chemical equations describe important reactions leading to the final chemical state.

P. vulgaris growing in sterilized mud from Lake Veisovoe which was mixed with 2% peptone and seawater and incubated for 1½ years also produced dolomite crystals.

Nadson's experiments showed that $CaCO_3$ precipitation was not brought about by any special group of bacteria, but depended on appropriate conditions, such as those previously listed in this chapter, for extracellular $CaCO_3$ precipitation to occur. Bavendamm (1932) has pointed out that although Nadson's studies and those of some others indicated that the ability to precipitate $CaCO_3$ was not a special property of any one group of bacteria, these observations were apparently unknown to other researchers, such as Drew, who implicated denitrifying bacteria, especially *Bacterium calcis,* as the chief precipitators of $CaCO_3$ in the tropical seas around the Bahamas (Drew, 1911, 1914). He found denitrifyers in significant numbers in these waters and demonstrated their ability to precipitate $CaCO_3$ in artificial culture media in the laboratory. He believed to have proven through these experiments that denitryfying bacteria were responsible for the extensive structureless $CaCO_3$ deposits in the marine environment. Although many geologists of his day apparently accepted this interpretation without question, it is not accepted today. There is still not complete consensus about the mechanism of calcium carbonate precipitation in the seas around the Bahamas, but an inorganic mechanism seems to be favored by geochemists (Skirrow, 1975). They feel that a finding of supersaturation with $CaCO_3$ and an abundant presence of nuclei for $CaCO_3$ deposition explain $CaCO_3$ precipitation in these waters in the simplest way. However, others, such as Lowenstam and Epstein (1957), favor algal involvement based on a study of $^{18}O/^{16}O$ ratios of the precipitated carbonates.

Lipman (1929) rejected Drew's conclusion on the basis of some studies of Pacific waters off the island of Tutuila (American Samoa) and the Tortugas. In his work, Lipman confined himself mainly to an examination of water samples and only a few sediment samples. Using various media, he was able to demonstrate $CaCO_3$ precipitation in the laboratory by a variety of bacteria, not just *B. calcis,* as Drew had suggested. However, since the number of viable organisms in the water samples and in the few sediment samples examined by him seemed small, Lipman felt that these organisms were not important in $CaCO_3$ precipitation in the sea. He probably should have examined the sediments more extensively.

Bavendamm reintroduced Nadson's concept of microbial $CaCO_3$ precipitation as an important geochemical process that is not a specific activity of a limited group of bacteria. As a result of his microbiological investigations, he concluded that this precipitation occurs chiefly in sediments of shallow bays, lagoons, and swamps. He isolated and enumerated heterotrophic and autotrophic bacteria, including sulfur bacteria, photosynthetic bacteria, agar-hydrolyzing bacteria, cellulose-hydrolyzing bacteria, urea-hydrolyzing bacteria, nitrogen-fixing bacteria, and sulfate-reducing bacteria, all of which he implicated in an ability to precipitate $CaCO_3$. He rejected the idea of significant participation of cyanobacteria in $CaCO_3$ precipitation. As we now know, however, cyanobacteria may cause significant $CaCO_3$ deposition.

In general, morphological and physiological studies have shown that whereas bacteria and cyanobacteria cause precipitation of $CaCO_3$ mostly in the immediate environment around their cells, some eukaryotic algae and protozoans lay down $CaCO_3$ as cell-support structures on their cells. Examples of such eukarotic organisms are green algae (Chlorphyceae) such as *Acetabularia, Chara,* and *Nitella*; coccolithophores (Fig. 7.3) such as *Scyphosphaera, Rhabdosphaera,* and *Calciococcus*; red algae such as *Lithothamnion, Porolithon* (Fig. 7.4), and

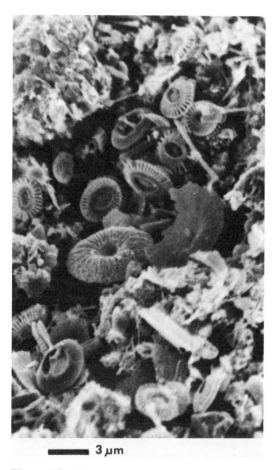

3 μm

Figure 7.3 Coccoliths, the calcerous skeletons of an alga belonging to the class Chrysophyceae. These specimens were found residing on the surface of a ferro-manganese nodule from Blake Plateau off the Atlantic coast of the United States. (From LaRock and Ehrlich, 1975; by permission.)

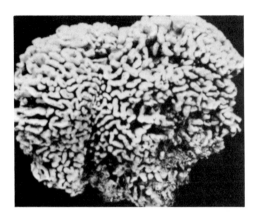

Figure 7.4   *Porolithon gardineri* f., a calcareous alga belonging to the Rhodo-phyceae (0.6X). Specimen from the reef near Rumit Island. (From W. R. Taylor, 1950. Copyright (1950) by The University of Michigan Press.)

Gonolithon; and foraminifera such as *Heterostegina* and *Globigerina* (Fig. 7.5). The details of the biochemical mechanism by which they deposit $CaCO_3$ are largely unknown, although it has been assumed that they derive the carbonate externally as a result of photosynthetic activity or internally as a result of their respiratory activity. But this hardly explains the morphogenesis of such intricate structures as those found associated with foraminifera or coccolithophores, or the crystalline wall deposits of some other algae.

Skeletal Carbonate Formation   A possible clue to the biochemical mechanism of $CaCO_3$ deposition in cell structures, especially of protozoans (heterotrophs), may be found in an observation of Berner (1968). He noted that during bacterial decomposition of butterfish and smelts in seawater in a sealed jar, calcium was precipitated not as $CaCO_3$ but as calcium soaps or adipocere (calcium salts of fatty acids) in spite of the presence of HCO_3^- and CO_3^{2-} species and an alkaline pH in the reaction mixture. The prevailing fatty acid concentration favored calcium soap formation over calcium carbonate formation. Berner suggested that in nature such soaps could later be transformed into $CaCO_3$. It seems possible that in some living calcerous organisms, calcium intended for calcium

Fig. 7.5   Foraminifera. (A) A living foraminiferan specimen, *Heterostegina depressa*, from laboratory cultivation. Note the multichambered test and the fine protoplasmic threads projecting from the test. Test diameter, 3 mm. (Courtesy of R. Röttger. See also Röttger, 1976). (B) Foraminiferan test (arrow) in Pacific sediment: *Globigerina* (?) (2430 X).

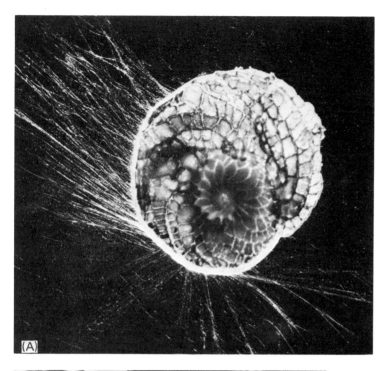

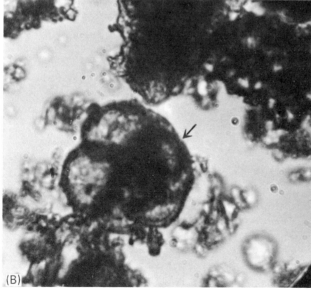

Figure 7.5

carbonate structures is first localized by formation of an organic calcium salt at the site of calcium deposition (the plasma membrane?) and is then converted to $CaCO_3$ in the presence of carbonic anhydrase, an enzyme that promotes the conversion of dissolved metabolic CO_2 to bicarbonate and carbonate. *Hymenomonas,* a coccolithophorid, has been said to utilize "a hydroxyproline-proline-rich peptide and sulfated polysaccharide moieties" in $CaCO_3$ deposition (Isenberg and Lavine, 1973). Microscopically, vesicles derived from the Golgi apparatus of the cell and intracellular fibers produced by them play a role in the mineral deposition process in this organism. Degens (1979) has related this model to calcium transport mechanisms in cells.

A related model for structural carbonate deposition may be constructed from observations with a marine pseudomonad, strain MB-1, which adsorbs calcium and magnesium ions to its cell surface (cell wall-membrane complex) (Greenfield, 1963). Living or dead cells have this capacity. The dead cells adsorb calcium ions more extensively than they adsorb magnesium ions. Carbonate in the medium, mostly derived from respiratory CO_2 which is converted to carbonate by ammonia produced from organic nitrogen compounds, results in aragonite formation. The bacterial cells coated in aragonite then serve as a nucleus for further calcium carbonate precipitation.

Carbonate Deposition by Cyanobacteria   Carbonate deposition by cyanobacteria has been described (Golubic, 1973). In this process, a distinction must be made between cyanobacteria that entrap and agglutinate preformed calcium carbonate in their thalli and those that precipitate calcium carbonate in their thalli as a result of their photosynhetic activity. The preformed calcium carbonate used in the entrapment and agglutination process was formed at a site other than the site of deposition, whereas the calcium carbonate deposited as a result of photosynthetic activity of the cyanobacteria is being formed at the site of its deposition. Calcium carbonate associated with stromatolite structures (special types of cyanobacterial mats) is an example of deposition by entrapment and agglutination or of deposition through cyanobacterial photosynthesis. Calcium carbonate associated with travertine, lacustrine carbonate crusts, and nodules is an example of deposition resulting from photosynthetic activity of cyanobacteria in freshwater environments. Travertine, a porous limestone, is formed from rapid calcium carbonate precipitation due to cyanobacterial photosynthesis in waterfalls and streambeds of fast-flowing rivers, which tends to bury the cyanobacteria. By outward growth or movement, the cyanobacteria avoid being trapped and contribute to the porosity of the deposit (Golubic, 1973). Lacustrine carbonate crusts are formed by benthic cyanobacteria attached to rocks or sediment, which cause calcium carbonate to be deposited through their photosynthetic activity in the shallow portions of lakes with carbonate-saturated waters (Golubic, 1973). Calcareous nodules also arise from rounded rocks and pebbles or shells to which calcium carbonate-depositing cyanobacteria are attached and which are rolled by water currents, thus exposing different parts of their surface to sunlight at different times and promoting photosynthetic activity and calcium carbonate precipitation by the attached cyanobacteria (Golubic, 1973). High-magnesium calcite precipitation by sheaths of certain cyanobacteria such as *Scytonema,* which may be related to the ability of the sheaths to concentrate magnesium three to five times over its concentration in seawater, has also been reported (Monty, 1967; see also the discussion by Golubic, 1973).

Calcification by Eukaryotic Algae   Calcification by eukaryotic algae seems to be the result of calcium carbonate deposition as calcite or aragonite crystals (Lewin, 1965a). Gelatinous or mucilaginous substances in association with the cell walls of these algae may be involved in the deposition and organization of the crystalline mineral. In some algae the CO_2 of the deposited $CaCO_3$ derives from CO_2 dissolved in seawater and is converted to carbonate during photosynthesis, whereas in other algae the CO_2 of the deposited carbonate has a metabolic origin not related to photosynthesis (Lewin, 1965a). In the former case the algal carbonate is enriched in ^{13}C relative to seawater CO_2, whereas in the latter case the algal carbonate is enriched in ^{12}C relative to seawater CO_2. Enrichment in ^{12}C signifies a metabolic rather than a purely chemical process.

Sodium Carbonate Deposition in the Wadi Natrûn   Carbonate may occur in solid phases not only as calcium and magnesium salts but also as a sodium salt (natron, $Na_2CO_3 \cdot 10H_2O$). In at least one instance, the Wadi Natrûn in the Libyan Desert, such a deposit has been clearly associated with the activity of sulfate-reducing bacteria (Abd-el-Malek and Rizk, 1963b). The wadi (a channel of a watercourse that is dried up except during rainfall; an arroyo) in this case contains a chain of small lakes, 23 m below sea level. The smallest of the lakes dries up almost completely during the summer, and the larger ones dry out partially at that time. Natron is in solution in the water of all the lakes and in solid form on the bottom of some of the lakes. Its origin has been explained as follows. The water feeding the lakes is supplied partly by springs and partly by streamlets which probably derive their water from the nearby Rosetta branch of the Nile. On its way to the lakes, the surface water passes through cypress swamps. Sulfate and carbonate have been found in high concentration in the lakes (189-204 mEq of carbonate per liter and 324-1107 mEq of sulfate per liter) and at low concentration in the cypress swamp (0 mEq to traces of carbonate per liter and 2-13 mEq of sulfate per liter). Bicarbonates have been reported to occur in significant amounts in lakes and swamps (22-294 and 11-61 mEq liter^{-1}, respectively). Soluble sulfides have been found to predominate in the lakes and swamps (7-13 and 1-4 mEq liter^{-1}, respectively). Considerable numbers of sulfate-reducing bacteria (1×10^6 to 8×10^6 per milliliter) have

Table 7.2 Chemical and bacteriological analyses of water and soil samples from Wadi Natrûn

Type and source of sample	pH range	Milliequivalents[a] of:					Total soluble salts (g liter^{-1})	Organic matter (%)	Viable counts of sulfate reducers[b] (10^6 ml^{-1})
		$CO_{3}{}^{2-}$	$HCO_{3}{}^{-}$	$SO_{4}{}^{2-}$	Cl^{-}	$S_{2}{}^{-}$			
Water									
Artesian wells	7.4–7.8	0	2–5	9–13	18–30	0.2–0.5	2–3	c	d
Burdi swamps	6.8–7.2	0-trace	11–61	2–13	1–7	1–4	1–8	c	1–5
Lakes	9.5–10.1	189–240	22–294	324–1107	107–210	7–13	180–239	c	d
Soil									
Newly cultivated uplands	7.0–7.6	0	1–2	14–18	11–19	c	c	0.2–0.5	d
About 150 m from lakes	7.2–7.5	0	2–3	12–23	1–6	c	c	0.1–5.2	5–8
Swamps	7.4–7.8	Trace	3–11	4–7	3–8	c	c	3.4–7.8	0.8–2

[a] Milliequivalents per liter of water and per 100 g of soil.
[b] Counts per milliliter of water and per gram of soil.
[c] Not determined.
[d] Not detected.
Source: Abd-el-Malek and Rizk (1963); by permission.

been detected in the swamps and in the soil at a distance of 150 m from the lakes, but not elsewhere (see Table 7.2 for more detailed data). Sulfate reduction was inferred to occur chiefly in the cypress swamps because of the significant presence of sulfate reducers and the readily available organic nutrient supply at those sites. The sulfate reduction leads to the production of bicarbonate, as follows:

$$SO_4{}^{2-} + 2(CH_2O) \rightleftharpoons H_2S + 2CO_2 + 2OH^- \qquad (7.21)$$

$$2CO_2 + 2H_2O + 2OH^- \rightleftharpoons 2HCO_3{}^- + 2H_2O \qquad (7.22)$$

Most of the soluble products of sulfate reduction, $HCO_3{}^-$ and S^{2-}, are washed into the lakes where they become concentrated and partially precipitated as carbonate and sulfide salts on evaporation. Some of the sulfide produced in the swamps, however, combines with iron to form ferrous sulfide, which imparts a characteristic black color to the swamps. The carbonate in the lakes results from the loss of CO_2 from the water to the atmosphere, which turns some of the bicarbonte into carbonate (equation 7.20).

Managanese and Iron Carbonates   Carbonate may also occur in combination with manganese and iron in some deposits in nature. At least some of these deposits have been ascribed directly to microbial activity. Formation of rhodochrosite ($MnCO_3$) and siderite ($FeCO_3$) has been reported ot occur in Lake Punnus-Yarvi on the Karelian peninsula in the U.S.S.R. (Sokolova-Dubinina and Deryugina, 1967a). This lake has a length of 7 km. Its greatest width is 1.5 km and its greatest depth is 14 m. It is only slightly stratified thermally. The oxygen concentration in its surface water is 11.8-12.1 mg liter^{-1} and in its bottom waters 5.7-6.6 mg liter^{-1}. The pH of its water is slightly acid (pH 6.3-6.6). The Mn^{2+} concentration ranges from 0.09 mg liter^{-1} in its surface waters to 0.02-0.2 mg liter^{-1} in its bottom waters (1.4 mg liter^{-1} in winter). The lake is fed by the Suantaka-Ioki and Rennel rivers and by 24 small streams that drain surrounding swampland. The lake, in turn, feeds into the Punnus-Yarvi River. It is estimated that 48% of the water in the lake is exchanged each year. The manganese and iron in the lake are derived from surface and underground drainage from Cambrian and Quaternary glacial deposits, carrying 0.2-0.8 mg of manganese per liter and 0.4-2 mg of iron per liter. The oxidized forms of manganese and iron are incorporated into silt, where they are reduced and subsequently concentrated by upward migration to the sediment-water interface and reoxidation into lake ore. This occurs mostly in Punnus-Ioki Bay, which has oxide deposits on the sediment at water depths down to 5-7 m. The oxide layer is 5-7 cm thick.

   All sediment and ore samples taken from the lake (mainly from Punnus-Ioki Bay) contained manganese-reducing bacteria. They were concentrated chiefly in the upper sediment layer. They included an unidentified, nonsporeforming bacillus in addition to *Bacillus circulans* and *B. polymyxa*. Sulfate-reducing bacteria, limited in numbers by the lack of extensive accumulation of organic matter, were also recovered from the ore. They were associated with hydro-troilite (FeS · nH$_2$O). Carbonates of calcium and manganese at most stations in the lake and bay have been reported of low occurrence (0.1%, calculated on the basis of CO$_2$). However, in a limited area near the center of Punnus-Ioki Bay, ore contains as much as 4.7% calcite, 5.96% siderite, and 4.99% rhodo-chrosite, together with 15.81% hydrogoethite and 38.9% barium psilomelanes and wads (complex oxides of manganese). The relatively localized concentration of carbonates has been related to the localized availability of organic matter, the ultimate source of CO$_2$-CO$_3{}^{2-}$ and the cause of the essential low E$_h$. It has been noted that the decaying remains of the plant life on the lakeshore accumulated in sufficient quantities in the Punnus-Ioki Bay area only where extensive carbonate ores were found. The low E$_h$ at this spot was attributed to the intense activities of heterotrophs attacking this organic matter. The weak sulfate-reducing activity at this location may explain the low iron and manganese sulfide formation and the significant carbonate formation.

Siderite (ferrous carbonate) beds in the Yorkshire Lias in England are thought to have resulted from Fe$_2$O$_3$ reduction and subsequent reaction with HCO$_3{}^-$. This can only be explained by the exclusion of sulfate by an overlying clay layer, which blocked the entry of sulfate to the site of siderite deposition. If sulfate had entered the site, it would have been bacterially reduced to sulfide and led to the formation of iron sulfide instead of siderite (see Sellwood, 1971).

7.2   CARBONATE DEGRADATION

Mechanism of Action   Carbonates in nature may be readily degraded as a direct or indirect result of biological activity, especially microbiological activity. The chemical basis for this decomposition is the instability of carbonates in acid solution. For instance,

$$CaCO_3 + H^+ \rightarrow Ca^{2+} + HCO_3{}^- \tag{7.23}$$

$$HCO_3{}^- + H^+ \rightarrow H_2CO_3 \tag{7.24}$$

$$H_2CO_3 \rightarrow H_2O + CO_2 \tag{7.25}$$

Since Ca(HCO$_3$)$_2$ is very soluble compared to CaCO$_3$, the CaCO$_3$ begins to dissolve even weakly in acid solution. In more strongly acid solution, CaCO$_3$ dissolves more rapidly because, as equation 7.25 shows, the carbonate is like to be lost from solution as CO_2. Therefore, from a biochemical standpoint, any organism that generates acid metabolic wastes is capable of dissolving insoluble carbonates. Even the mere metabolic generation of CO_2 during respiration may have this effect, because

$$CO_2 + H_2O \rightarrow H_2CO_3 \qquad (7.26)$$

and

$$H_2CO_3 + CaCO_3 \rightarrow Ca^{2+} + 2HCO_3^- \qquad (7.27)$$

Thus, it is not surprising that various kinds of CO_2 and acid-producing microbes have been implicated in the breakdown of insoluble carbonates in nature.

Limestone Degradation by Externally Located Microbes   An early report of a breakdown of lime in the cement of reservoir walls and docksides attributed in part to bacterial activity was made by Stutzer and Hartleb in 1899. However, extensive investigation into microbial decay of limestone was first undertaken by Paine et al. (1933). These workers showed that both sound and decaying limestones usually carried a significant bacterial flora, the numbers ranging from 0 to over 18×10^6 per gram. The size of the population seemed to depend in part on the environment around the stone. As one might expect, the surface of the limestone was generally more densely populated than was the interior of the stone. The authors suggested that in many limestones, the bacteria occur unevenly distributed, inhabiting pockets or interstices in the limestone structures. The kinds of bacteria found in limestones examined included gram-variable and gram-negative rods and cocci. Sporeformers, such as *Bacillus mycoides, B. megaterium,* and *B. mesentericus,* appeared to have been rare or absent. In an experiment to estimate the rate of limestone decay through bacterial action under controlled conditions in a special apparatus in which evolved CO_2 was trapped in barium hydroxide solution and measured, 0.18 mg of CO_2 per hour per 350 mg of stone was detected in one case, and 59 mg of CO_2 per hour per 350 mg of stone in another. In the latter case it was calculated from the data that 28 g of CO_2 would have been evolved from 1 kg of stone in 1 year. As expected, the rate of CO_2 evolution from decaying stone was greater than from sound stone. Although in these experiments organic acids and CO_2 from heterotrophic metabolism of organic matter were the cause of the observed limestone decay, autotrophic nitrifying and sulfur-oxidizing bacteria were also shown to be able to promote limestone decay through the production of nitric and sulfuric acids, respectively. Nitrifying and sulfur-oxidizing bacteria were actually detected in some limestone samples by the authors.

A much later study of German limestones found variable numbers of fungi, algae, and ammonifying, nitrifying, sulfur- and H_2S-oxidizing, sulfate-reducing, and organic sulfur-mineralizing bacteria on their surface (Krumbein, 1968).

Figure 7.6 shows the effect of such organisms on structural limestone. The number of organisms found depended on the type of stone, the length of elapsed time since the collection of the stone from a natural site, the surface structure of the stone (i.e., the degree of weathering), the cleanliness of the stone, and the climatic and microclimatic conditions prevailing at the collection site. In the case of strongly weathered stone, the bacteria had sometimes penetrated the stone to a depth of 10 cm. Ammonifying bacteria were generally most numerous. Nitrogen-fixing bacteria were few and denitrifyers were absent. The number of the organisms was not directly related to the presence of lichens or algae. The greatest number of ammonifiers were found on freshly collected and weathered stone. This was related to the pH of the stone surface (pH 8.1-8.3 in water extract). Sulfur oxidizers were more numerous in city environments,

Figure 7.6   Limestone wall of a stable in Würzburg, West Germany. The stable was destroyed in a bombing raid during World War II. Bacterial testing of the wall revealed high numbers of heterotrophic ammonifying bacteria supporting up to 10,000 nitrifying autotrophic bacteria per gram of stone. The nitric acid that the latter organisms produce has led to extensive weathering of the lime-stone and mortar of the wall. (From Krumbein, 1973; by permission.)

where the atmosphere contains more oxidizable sulfur compounds than in the countryside. The concentration of nitrifiers on limestone could not be correlated with city and country atmospheres. The number of ammonifiers on limestone was also found to be greater in stone exposed to city air than that exposed to country air. This can be explained on the basis that city air contains more organic carbon that can serve as nutrient to those bacteria than does country air. Laboratory experiments confirmed the weathering activity of the limestone by its natural flora. The ammonifiers in these observations were less directly responsible for the weathering of the limestone than they were in generating the ammonia from which the nitrifying bacteria could form the nitric acid which then corroded the limestone.

In another important study of the decomposition of carbonates by microorganisms, numerous bacteria and fungi were isolated from a number of limestone samples (Wagner and Schwartz, 1965). The active organisms appeared to weather the stone through production of oxalic and gluconic acids. The investigators also noted the presence of nitrifying bacteria and thiobacilli in their samples and suggested that the corresponding mineral acids produced by these organisms also contributed to the weathering of the stone.

Limestone Degradation by Endolithic Algae and Fungi Endolithic algae and fungi bore into limestone (e.g., coral reefs, beach rock, etc.) and live in the resultant cavities (Fig. 7.7) (Golubic et al.,1975). Active algae include single-celled and filamentous prokaryotes, the cyanobacteria, as well as eukaryotic algae, especially green algae, and some brown and red algae (Golubic, 1969). The mechanism by which the algae bore into the limestone is not known. Some filamentous boring cyanobacteria feature a terminal cell that is directly responsible for the boring action, presumably dissolution of the calcium carbonate (Golubic, 1969). Different types of boring microorganisms form tunnels of characteristic morphology (Golubic et al., 1975). In a pure mineral such as iceland spar, boring follows the planes of crystal twinning, diagonal to the main cleavage planes (Golubic et al., 1975). The depth to which algae bore into limestone is limited by light penetration into the rock, since they need light for photosynthesis. Boring cyanobacteria may have an unusually high concentration of phycocyanin to compensate for the low light intensity in the limestone. As would be expected, boring fungi are not limited by light availability. In some extreme environments, such as antarctic dry valleys, the cavities in limestones seem to provide the major habitat for microorganisms such as cyanobacteria (Friedman and Ocampo, 1976). Some organisms may, however, be **chasmolithic** rather than endolithic. Chasmolithic means inhabiting preexistent pores or cavities, whereas endolithic, in the sense used here, means inhabiting tunnels or cavities produced by boring.

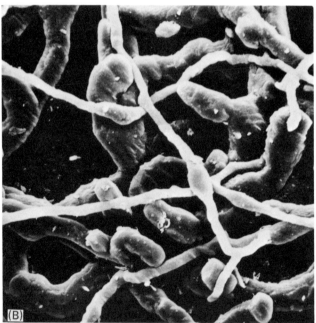

Figure 7.7

7.3 SUMMARY

Carbon dioxide is trapped on earth mainly as calcium and magnesium carbonates, but also to a much lesser extent as iron, manganese, and sodium carbonates. In many cases, these carbonates are of biogenic origin. Calcium or calcium-magnesium carbonates may be precipitated by bacteria, cyanobacteria, and fungi by (1) aerobic or anaerobic oxidation of carbon compounds devoid of nitrogen in neutral or alkaline environments with a supply of calcium or magnesium, (2) aerobic or anerobic oxidation of organic nitrogen compounds in unbuffered environments with a supply of calcium of magnesium, (3) $CaSO_4$ reduction by sulfate-reducing bacteria, (4) the hydrolysis of urea, or (5) photosynthesis.

The complete mechanism of calcium carbonate precipitation by living organisms is not understood in all cases, although it always depends on metabolic CO_2 production or consumption and on the relative insolubility of calcium carbonate. Bacteria and fungi may deposit it extracellularly. Bacterial cells may also serve as nuclei around which calcium carbonate is laid down. Some algae and protozoans form structural carbonate. Its deposition may involve local fixation of calcium prior to its reaction with metabolically produced carbonate.

Microbial calcium carbonate precipitation has been observed in soil and in freshwater and marine environments. Sodium carbonate (natron) deposition associated with microbial sulfate-reducing activity has been noted in the hot climate of the Wadi Natrûn in the Libyan Desert. Ferrous carbonate deposition associated with microbial activity has been noted in some special coastal marine environments. Manganese carbonate deposition associated with microbial activity has been noted in a freshwater lake.

Insoluble carbonates may also be broken down by microbial attack. This is usually the result of organic and inorganic acid formation by the microbes. A variety of bacteria, fungi, and even algae have been implicated. Such activity is particularly evident on limestone used in building construction, but it is also evident in natural limestone formations, such as coral reefs, where limestone-boring algae and fungi are active in the breakdown process.

Figure 7.7 Microorganisms that bore into limestone. (A) Limestone sample experimentally recolonized by the cyanobacterium *Hyella balani* Lehman (234X). The exposed tunnels are the result of boring by the cyanobacterium. (B) Casts of the borings of the green alga *Eugomontia sacculata* Kormann (larger filaments) and the fungus *Ostracoblabe implexa* Bornet and Flahault in calcite spar (2,000X). The casts were made by infiltrating a sample of fixed and dehydrated bored mineral with synthetic resin and then etching sections of the embedded material (e.g., with dilute mineral acid) to expose the casts of the organisms. (Courtesy of S. Golubic.)

8

The Geomicrobiology of Silica and Silicates

The element silicon is one of the most abundant in the earth's crust, ranking second only to oxygen. Its estimated crustal abundance is 27.7% (wt/wt) whereas that of oxygen is 46.6% (wt/wt) (Mitchell, 1955). It occurs generally in the form of silicates and silicon dioxide (silica) and is found in primary and secondary minerals. It can be viewed as an important part of the backbone of rock mineral structure. In silicates, silicon is usually surrounded by four oxygen atoms in tetrahedral fashion (see Kretz, 1972). When in solution at pH 2-9, silica exists in the form of undissociated but dissolved monosilicic acid (H_4SiO_4), whereas at pH 9, silicate ion appears (see Hall, 1972). Silica can be viewed as an anhydride of silicic acid:

$$SiO_2 + 2H_2O \rightarrow H_4SiO_4 \tag{8.1}$$

Silicic acid dissociates to form silicate (see Anderson, 1972):

$$H_4SiO_4 \rightleftharpoons H^+ + H_3SiO_4^- \qquad (K_1 = 10^{-9.5}) \tag{8.2}$$

$$H_3SiO_4^- \rightleftharpoons H^+ + H_2SiO_4^{2-} \qquad (K_2 = 10^{-12.7}) \tag{8.3}$$

Colloidal forms of silica seem to exist at high concentrations or at saturation levels only locally and are favored by acid conditions (Hall, 1972). Common silicate minerals include quartz (SiO_2), olivine [$(Mg,Fe)_2SiO_4$], orthopyroxene ($Mg,FeSiO_3$), biotite [$K(Mg,Fe)_3 AlSi_3O_{10}(OH)_2$], feldspar ($KAlSiO_3$), plagioclase [$(NaSi,CaAl)AlSi_2O_8$], kaolinite [$(Al_4Si_4O_{10}(OH)_8$], and others. The concentration of silicon in various components of the earth's surface is listed in Table 8.1.

Silicon is taken up and thus concentrated in significant quantities by certain forms of life. These include microbial forms, such as diatoms and other chrysophytes; silicoflagellates and some xanthophytes; radiolarians and actinopods; plant forms, such as horsetails, ferns, grasses, and some flowers and trees; and

125

Table 8.1 Abundance of Silicon on the Earth's Surface

Phases	Concentration	Reference
Igneous rock	281,000 ppm	Bowen (1966)
Shales	73,000 ppm	Bowen (1966)
Sandstones	368,000 ppm	Bowen (1966)
Limestones	24,000 ppm	Bowen (1966)
Soils	300,000 ppm	Bowen (1966)
Seawater	3×10^3 μg liter^{-1}	*Marine Chemistry* (1971)
Fresh Water	6.5 ppm (2-12 ppm)	Bowen (1966)

some animal forms, such as sponges, insects, and even vertebrates. Some bacteria (Heinen, 1960) and fungi (Heinen, 1960; Holzapfel and Engel, 1954a,b) have also been reported to take up silicon to a limited extent. According to Bowen (1966), diatoms may contain from 1,500 to 20,000 ppm, land plants from 200 to 5,000 ppm, and marine animals from 120 to 6,000 ppm of silicon. In some microorganisms, such as diatoms, silicon is not only important structurally, but it also seems to play a metabolic role in the synthesis of chlorophyll (Werner, 1966, 1967), the synthesis of DNA (Darley, 1969; Darley and Volcani, 1969), and the synthesis of DNA polymerase and thymidylate kinase (Sullivan, 1971; Sullivan and Volcani, 1973).

Silica and silicates form an important buffer system in the oceans (Garrels, 1965), together with carbon dioxide. The CO_2/CO_3^{2-} system, because of its rapid rate of reaction, is a short-range buffer, whereas silica and silicates, owing to their slow rate of reaction, are part of a long-range buffer system (Sillén, 1967).

The silicates of clay also perform a buffering function in mineral soils. This is because of their ion-exchange capacity, their electronegative charge, and their adsorption powers. Their ion-exchange capacity and adsorption powers, moreover, make them important reservoirs of cations and organic molecules. Montmorillonite exhibits the greatest ion-exchange capacity, illites exhibit less, and kaolinites least (Dommergues and Mangenot, 1970, p. 469).

8.1 BIOLOGICAL EFFECTS OF CLAYS

Some clays may stimulate or inhibit microbial metabolism, depending on conditions (see the discussion by Marshall, 1971). Stimulatory effects on bacteria are explained by the varying ability of different clays to serve as pH buffers and as a source of inorganic nutrients through their cation-exchange capacity. The surface area of clay particles may also be somehow stimulatory for bacteria, probably because of their adsorptive capacity (Stotzky and Rem, 1966; Stotzky, 1966a,b). Inhibitory effects on fungi are explained by the enhancement of viscosity by clays, which interferes with oxygen diffusion (Stotzky and Rem, 1967). Owing to their adsorptive capacity, clays also have an affinity for enzymes secreted by cells or liberated by them on autolysis after death (Skujins, 1967). Upon adsorption, the enzymes exhibit modified properties, such as changed potassium, pH optimum, and redox requirements. Clays by themselves may have played a significant role in the evolution of life on earth (Calvin, (1975).

8.2   SILICON UPTAKE BY BACTERIA, FUNGI, AND DIATOMS

Bacteria   A soil bacterium B_2 (Heinen, 1960) and a strain of *Proteus mirabilis* (Heinen, 1968) have been found to take up limited amounts of silicon when furnished in the form of silica gel, quartz, or sodium silicate (Heinen, 1960). Sodium silicate was assimilated most easily, quartz least easily. The silicon seemed to substitute partially for phosphorus in phosphorus-deficient media (Fig. 8.1). This substitution reaction has been found to be reversible (Heinen, 1962). The silicon taken up by the bacteria appeared to be organically bound in a metabolizable form (Heinen, 1962). Sulfide, sulfite, and sulfate were found to affect phosphate-silicate exchange in different ways, depending on

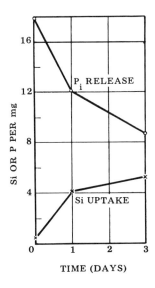

Figure 8.1 Relationship between Si uptake and P_i release during incubation of resting cells of strain B_2 in silicate solution (80 μg of Si per milliliter). (From Heinen, 1962; by permission.)

concentration, whereas KCl and NaCl were without effect. NH_4Cl, NH_4NO_3, and $NaNO_3$ stimulated the formation of adaptive enzymes involved in the phosphate-silicate exchange (Heinen, 1963a). The presence of carbohydrates such as glucose, fructose, or sucrose, and amino acids such as alanine, cysteine, glutamine, methionine, asparagine, and citrulline, as well as the metabolic intermediates pyruvate, succinate, and citrate, stimulated uptake, whereas acetate, lactate, phenylalanine, peptone, and wheat germ oil inhibited it. Glucose at an initial concentration of 1.2 mg per milliliter of medium stimulated silicon uptake maximally (Fig. 8.2). Higher concentrations of glucose caused the formation of particles of protein, carbohydrates, and silicon outside the cell. $CdCl_2$ inhibited the stimulatory effect of glucose on silicon uptake, but 2,4-dinitrophenol was without effect. The simultaneous presence of $NaNO_3$ and KH_2PO_4 lowered the stimulatory effect of glucose but did not eliminate it (Heinen, 1963b). The silcon that is fixed in bacteria is readily displaced by phosphate in the absence of external glucose. In cells that were preincubated in a glucose-silicate solution, only a small portion of the silicon was released by glucose-phosphate, but all of the silicon was releasable on incubation in glucose-carbonate solution. C–Si bonds may exist in the cells (Heinen, 1963c).

Fungi   The fungus *Aspergillus niger* seems to behave similarly to bacteria (Holzapfel and Engel, 1954a,b). The lower mycelium of a culture of this fungus growing on a solid medium takes up silicon in the form of inorganic silicate after an induction period of 5-7 days. When the silicon is in the form of galactose- or glucose-quartz complexes, the fungus takes it up in 12-18 hours.

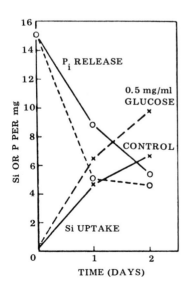

Figure 8.2   Influence of glucose on silicate uptake (x———x, without glucose, control; x– – –x, with glucose), and phosphate release (o———o, without glucose, control; o– – –o with glucose). (From Heinen, 1963b; by permission.)

Evidently, inorganic silicates have to be converted to organic complexes by the fungus during prolonged induction before they can be taken up.

Diatoms   Among the eukaryotes that take up silicon, the diatoms have been the most extensively studied with respect to this property (Fig. 8.3) (Lewin, 1965b). Their silicon uptake ability affects redistribution of silica between fresh and marine waters. In the Amazon River estuary, for instance, they remove 25% of the dissolved silica of the river water. Their frustules are not swept oceanward upon their death, but are transported coastward and incorporated into dunes, mud, and sand bars (Milliman and Boyle, 1975).

Diatoms are unicellular algae, enclosed in a wall of silica consisting of two valves in a pillbox arrangement. The valves are usually perforated plates which may have thickened ribs and may be of pennate or centric geometry. In cell division, each daughter cell receives one of the two valves and synthesizes the other de novo to fit into the one already present. To prevent the excessive diminution of daughter diatoms which receive the smaller of the two valves after each cell division, a special reproductive step, called **auxospore formation**, returns the organism to maximum size. Auxospore formation is a sexual reproductive process in which the cells escape from their frustules and increase in size in their zygote membrane, which may become weakly silicified. After a time, the protoplast in the zygote membrane contracts and forms the typical frustules of the parent cell (Lewin, 1965b).

The silica walls of the diatoms consist of hydrated, amorphous silica, a polymerized silicic acid (Lewin, 1965b). In the walls of marine diatoms, 96.5% SiO_2 and only 1.5% Al_2O_3 or Fe_2O_3 and 1.9% water have been found (Rogall, 1939). In clean, dried frustules of freshwater *Navicula pelliculosa,* 9.6% water has been found (Lewin, 1957). Thin parts of diatom frustules under the electron microscope reveal a foamlike substructure, suggesting silica gel (Helmcke, 1954), which may account for the adsorptive power of such frustules. The silica gel appears to be arranged in small spherical particles about $22\mu m$ in diameter. Despite a low degree of solubility of amorphous silica at the pH of most natural waters, frustules of living diatoms do not dissolve readily (Lewin, 1965b). Yet it has been reported that at pH 8, 5% of the silica in the cell walls of *Thalassiosira nana* and *Nitschia linearis* dissolves, and at pH 10, 20% of the silica in the frustules of *N. linearis* and all of the silica in the frustules of *T. nana* dissolves (Jφrgensen, 1955). After the death of diatoms, their frustules may dehydrate to form anhydrous SiO_2, which is much less soluble in alkali and may account for the accumulation of diatomaceous ooze.

Silicic acid uptake and incorporation by diatoms can be easily measured in terms of radioactive [68Ge]germanic acid uptake and incorporation (Azam et al., 1973; Azam, 1974). At low concentration, germanium, which is chemically similar to silicon, is apparently incorporated together with silicon into the silicic acid polymer of the frustule. At higher concentration, germanium is toxic to diatoms.

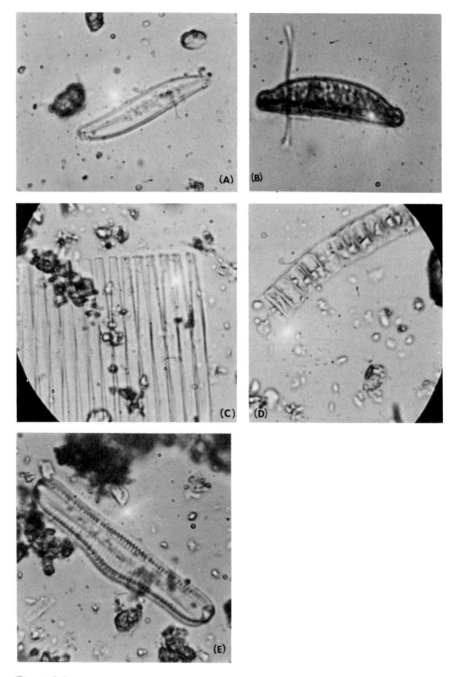

Figure 8.3

How silica is deposited in the wall structure of diatoms and arranged in the characteristically intricate geometric designs is still poorly understood. Important aspects have, however, been revealed (Lewin, 1965b). Diatoms take up silicon as contained in orthosilicate. More highly polymerized forms of silica are apparently not available to them. Organic silicates are not available to them either, nor can Ge, C, Sn, Pb, P, As, P, B, Al, Mg, or Fe replace silicon extensively, if at all (Lewin, 1965b). The concentration of silicon in a diatom depends to a degree on its concentration in the growth medium and on the rate of cell division (the faster the cells divide, the thinner their frustules). Silicon is essential for cell division, but resting cells in a medium in which silica is not at a limiting concentration continue to take up silica (Lewin, 1965b). Synchronously growing cells of *Navicula pelliculosa* take up silica at a constant rate during the cell-division cycle (Lewin, 1965b). Silica uptake appears dependent on energy-yielding processes (Lewin, 1965b; Azam et al., 1974; Azam and Volcani, 1974) and seems to involve intracellular receptor sites (Blank and Sullivan, 1979). Uncoupling of oxidative phosphorylation stops silica uptake by *N. pelliculosa* and *Nitschia angularis*. Starved cells of *N. pelliculosa* show an enhanced silicon uptake rate when fed glucose or lactate in the dark, or when returned to the light, where they can photosynthesize (Healy et al., 1967). Respiratory inhibitors prevent Si and Ge uptake by *Nitschia alba* (Azam et al., 1974; Azam and Volcani, 1974). It has also been noted that total uptake of phosphorus and carbon is decreased during silica starvation of *N. pelliculosa*. Upon restoration of silica to the medium, the total uptake of phosphorus and carbon is again increased (Coombs and Volcani, 1968). Sulfhydryl groups appear to be involved in silica uptake (Lewin, 1965b). The cell membrane probably plays a fundamental role in the laying down of new silica to form a frustule (Lewin, 1965b), but the details of this process are not yet understood.

8.3   SILICA AND SILICATE SOLUBILIZATION BY BACTERIA AND FUNGI

Some bacteria and fungi seem to play an important role in the mobilization of silica and silicates in nature. Part of their involvement is in the weathering of various silicates, including aluminosilicates. Their action is indirect, either through the production of chelates or the production of acid (mineral or organic acid), or through the production of alkali (e.g., the production of ammonia or

---

Figure 8.3   Diatoms. (A) *Gyrosigma*; from fresh water (1944X). (B) *Cymbella*; from fresh water (1944X). (C) *Fragellaria*; from fresh water (1864X). (D) Ribbon of diatoms; from fresh water (1990X). (E) Marine diatom frustule; from Pacific Ocean sediment (1944X).

amines). The effect of chelation in silicate dissolution was shown in laboratory experiments with a soil strain of *Pseudomonas* that produced 2-ketogluconic acid from glucose which dissolved synthetic silicates of calcium, zinc, and magnesium and the minerals wollastonite ($CaSiO_3$), apophyllite [$KCa_4Si_8O_{20}$-$(F,OH)\cdot 8H_2O$], and olivine [$(Mg,Fe)_2SiO_4$] (Webley et al., 1960). The demonstration consisted of culturing the organism for 4 days at 25°C on separate agar media, each one containing 0.25% (wt/vol) of one of the synthetic or natural silicates. A clear zone was observed around the bacterial colonies when the silicate was dissolved (Fig. 8.4). A similar silicate-dissolving action with a gram-negative bacterium, strain D_{11}, which resembled *Erwinia* spp., and with *Bacterium herbicola* or with some *Pseudomonas* strains, all of which were able to produce 2-ketogluconate from glucose, has also been shown (Duff et al., 1963). The action of these bacteria was tested in glucose-containing basal medium: KH_2PO_4, 0.54 g; $MgSO_4 \cdot 7H_2O$, 0.25 g; $(NH_4)_2SO_4$, 0.75 g; $FeCl_3$, trace; Difco yeast extract, 2 g; glucose 40 g; distilled water, 1 liter; and 5-500 mg of pulverized mineral per 5-10 ml of medium. It was found that the

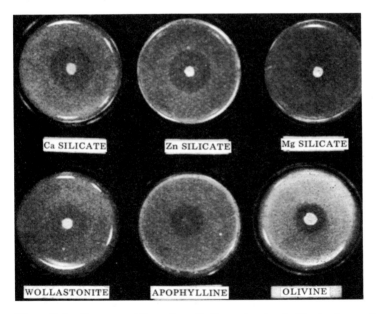

Figure 8.4 Decomposition of synthetic and natural silicates by a soil bacterium. The basal medium was that used by Bunt and Rovira (1955) but containing 2% glucose (wt/vol). The final concentration of the silicates was 0.25% (wt/vol). After inoculation, the plates were incubated for 4 days at 25°C. (From Webley et al., 1960; by permission.)

dissolution of silicates resulted from the complexation of their cationic components. The 2-ketogluconate complexes were apparently more stable than the silicates. For example,

$$CaSiO_3 \rightleftharpoons Ca^{2+} + SiO_3^{2-} \qquad (8.4)$$

$$Ca^{2+} + 2\text{-ketogluconate} \rightleftharpoons Ca(2\text{-ketogluconate}) \qquad (8.5)$$

The structure of 2-ketogluconic acid is

$$
HO-\overset{\displaystyle H}{\underset{\displaystyle H}{C}}-\overset{\displaystyle H}{\underset{\displaystyle OH}{C}}-\overset{\displaystyle H}{\underset{\displaystyle OH}{C}}-\overset{\displaystyle HO}{\underset{\displaystyle H}{C}}-\overset{\displaystyle}{\underset{\displaystyle O}{C}}-\overset{\displaystyle O}{\underset{\displaystyle OH}{C}}
$$

The silica in these experiments was liberated, or released and subsequently transformed, in three forms: (1) as low-molecular-weight or ammonium molybdate-reactive silica (monomeric?), (2) as a colloidal-polymeric silica of higher molecular weight which did not react with ammonium molybdate except after brief reaction with dilute hydrofluoric acid, and (3) as an amorphous form which could be removed from solution by centrifugation and dissolved in cold 5% aqueous sodium carbonate. Polymerized silica can be transformed into monomeric silica, as has been shown more recently in studies with *Proteus mirabilis* and *B. caldolyticus* (Lauwers and Heinen, 1974). The *Proteus* culture is able to assimiliate some of the monomeric silica. The mechanism of depolymerization has not been elucidated.

The effect of the action of acids in solubilizing silicates has been noted in various studies. Waksman and Starkey (1931) cited the action of CO_2 on orthoclase:

$$2KAlSi_3O_8 + 2H_2O + CO_2 \longrightarrow H_4Al_2Si_2O_9 + K_2CO_3 + 4SiO_2 \qquad (8.6)$$
orthoclase   kaolinite

The CO_2 is, of course, very likely to be a product of respiration or fermentation.

A practical illustration of silicate dissolution by acid formed by bacteria is the erosion of concrete sewer pipes (Parker, 1947). Concrete is formed from a mixture of cement (heated limestone, clay, and gypsum) and sand. On setting, the cement includes the compounds Ca_2SiO_5 and $Ca_3(AlO_4)_5$, which hold the sand in their matrix. The erosional process of concrete sewers can be explained as the result of several biochemical reactions on the calcium silicate-calcium aluminate matrix (Parker, 1947):

1. H_2S, resulting from the mineralization of organic sulfur compounds by various bacteria and from bacterial sulfate reduction, reacts with the calcium compounds in concrete, leading to the formation of CaS.

2. The reaction of CaS with CO_2 from the atmosphere to form $CaCO_3$ and H_2S accompanied by a drop in pH from 11-12 to 8.4. H_2S is transformed to thiosulfuric acid ($H_2S_2O_3$) and polythionic acids, causing a further drop in pH to about 7.5. These transformations are chemical.

3. Next, two groups of thiosulfate oxidizers act on the thiosulfate. One group consist of slow oxidizers which form polythionates and sulfates but not sulfur and which can tolerate H_2S when $CaCO_3$ is present but cannot grow on elemental sulfur or sulfides. Its pH optimum for growth is 7-10. It tolerates a pH range from 2 to 10.5. The second group consists of a *Thiobacillus thioparus*-like organism which can oxidize thiosulfate and polythionates to sulfuric acid and elemental sulfur. It exists over a pH range of 3.5-8.5. It starts its activity at the high end of its pH range and lowers it to pH 5 as it forms sulfuric acid.

4. Finally, *Thiobacillus concretivorus* takes over below pH 5. This organism oxidizes sulfur formed biochemically and chemically from other reduced sulfur compounds to sulfuric acid, which attacks the concrete and slowly solubilizes its silicates and aluminates.

Besides bacteria and actinomycetes, fungi such as *Botrytis, Mucor, Penicillium,* and *Trichoderma* from rock surfaces and weathered stone are able to solubilize Ca, Mg, and Zn silicates (Webley et al., 1963). The fungi cause dissolution of silicates through the production of oxalic and citric acids in their metabolism of glucose. The fungus *Aspergillus niger* M has been shown to release silicate from apophyllite, olivine, saponite, vermiculite, and wollastonite by oxalic acid produced on a glucose-salts-yeast extract agar (Henderson and Duff, 1963). Augite, garnet, heulandite, hornblende, hypersthene, illite, kaolinite, labradorite, orthoclase, and talc were not visibly affected by this fungus. In general, it has been found that feldspars are resistant to attack, whereas zeolites are susceptible (Henderson and Duff, 1963). It has also been found that the proportion of metal components, such as aluminum, to silica in solution after attack on a mineral are often not in the ratio of their occurrence in the original mineral (Henderson and Duff, 1963). Whether this is due to analytical limitations or to differential solvent action has not been ascertained.

The attack upon quartz, plagioclase, and nepheline by *Penicillium notatum* and *Pseudomonas* sp. with release of silica has also been studied (Aristovskaya and Kutuzova, 1968; Kutuzova, 1969). Of these, nepheline ($Na_3 KAl_4Si_4O_{16}$) was found to be most readily attacked by the fungus and the bacterium when growing in a glucose-salts medium. The action was attributed to acid production. The *Pseudomonas* was slower in its action than the *Penicillium*. Kutuzova believed that silica liberation from aluminosilicates (such as plagioclase?) was not the result of direct action of acid metabolic products, but rather the result of removal through complexation of alkali and alkaline earth metals from the crystal lattice, causing a reorganization of the mineral structure with the release of some silica. Silica release has also been noted in the action of *Penicillium simplicissimus* growing in a glucose-mineral salts medium in the presence of basalt, granite, granodiorite, rhyolite, andesite, peridotite, dunite, and quartzite (Silverman and Munoz, 1970). Metabolically produced citric acid was the agent of dissolution.

An example of the action of an alkaline substance in the dissolution of silica is silica release from nepheline, plagioclase, and quartz by *Sarcina ureae* growing in a peptone-urea broth (Aristovskaya and Kutuzova, 1968; Kutuzova, 1969). Here the ammonia, by generating alkalinity in solution, was the dissolving agent.

8.4 MICROBIAL WEATHERING OF SILICA AND SILICATES IN NATURE

The silica-liberating reactions by acids and complexing agents in the foregoing laboratory studies were all carried out at higher glucose concentrations than might be expected on the average in nature. This, of course, has raised the question of whether these reactions actually do occur under natural conditions. Direct evidence that lichens, a symbiotic association between fungi and algae, are very active in rock weathering has been cited [see, however, the disclaimers by Ahmadjian (1967) and Hale (1967)]. Direct evidence for similar action by other fungi and bacteria is mostly lacking. Nevertheless, it can be argued that in microenvironments, appropriately high concentrations of utilizable carbohydrates, nitrogenous compounds, and other needed nutrients are likely to occur in the form of excretory products and the dead remains of organisms. Their degradation should lead to the production of the metabolic products that cause silica removal from rock minerals. Indeed, fungal hyphae in the litter zone and A horizon of several different soils have been shown in scanning electron photomicrographs to carry calcium oxalate crystals attached to them, which is evidence of the extensive in situ production of oxalate by the fungus (Graustein et al., 1977). In one case, the basidiomycete fungus *Hysterangium crassum* was actually shown to weather clay in situ with the oxalic acid it produced (Cromack et al., 1979). Silica removal in nature is usually a slower process than in the laboratory because the conditions in natural environments are less favorable.

8.5 SILICA CYCLES

Silicon in nature may follow a series of cyclical biogeochemical transformations (Kuznetsov, 1975; Harriss, 1972). Silicates in rocks and soil are subject to the weathering action of biological, chemical, and physical agents. The extent of the contribution of each of these agents must depend on the particular environmental circumstances. Silica liberated in these processes may be leached away by surface or groundwater, and from these waters it may either be removed by chemical or biological precipitation at new sites, or it may be swept into bodies of fresh water or the sea. In either type of water body, silica will tend to be removed by biological agents. Upon their death, these biological agents will release their silica back into solution or their siliceous remains will be incorporated into the sediment, where some or all of the silica may later be returned to solution. The sediments of the ocean appear to be a sink for excess silica swept into the oceans since the silica concentration of seawater tends to remain relatively constant.

8.6 SUMMARY

The environmental distribution of silicon is significantly influenced by microbial activity. Certain microorganisms assimilate it and build it into cell-support structures. They include diatoms and some other chrysophytes, silicoflagellates, some xanthophytes, radiolarians, and actinopods. Silicon uptake rates by diatoms have been measured, but the mechanism by which silcon is assimilated is still only partially understood. Certain silicon-incorporating bacteria may provide a biochemical model for silicon assimilation in general. Silicon-assimilating microorganisms such as diatoms and radiolaria are important in the formation of siliceous oozes in oceans, and diatoms are important in forming such oozes in lakes.

Some bacteria and fungi are able to solubilize insoluble silicates. They accomplish this by forming chelators, acids, or bases, which react with silicates. These actions are important in the weathering of rock and in cycling silicon in nature.

9

Geomicrobial Transformations of Phosphorus Compounds

Phosphorus is an element fundamental to life, being a structural and functional component of all organisms. It is a component of such vital cell constituents as nucleic acids, nucleotides, phosphoproteins, phospholipids, teichoic acids (in gram-positive bacteria), and phytins (also known as inositol phosphates). In the pentavalent state, the element is capable of forming anhydrides in the form of organic or inorganic pyrophosphates (see Fig. 5.1). In this valence state it is also capable of forming anhydrides with organic acids and amines, and with sulfate (as in adenosine phosphosulfate). The phosphate anhydride bond serves to store biochemically useful energy. For example, a free-energy change $\Delta F°$ of -7.3 kcal per mole of anhydride bonds is associated with the hydrolysis of adenosine 5'-triphosphate to adenosine 5'-diphosphate and orthophosphate. Unlike many anhydrides, some of those of phosphate, such as adenosine 5'-triphosphate (ATP), are unusually resistant to hydrolysis in the aqueous environment. Chemical hydrolysis of these bonds requires 7 min of heating in dilute acid (e.g., 1 N HCl) at the temperature of boiling water (Lehninger, 1970, p. 290). At more neutral pH and physiological temperature, hydrolysis proceeds at an optimal rate only in the presence of appropriate enzymes (e.g., ATPase). The relative resistance of phosphate anhydride bonds is probably the reason ATP came to be selected in the evolution of life as a universal transfer agent of chemical energy in biological systems.

Phosphorus is found in all parts of the biosphere. Whether inorganic or organic, phosphorus compounds occur most commonly in the form of phosphate derivatives. Total phosphorus concentrations in mineral soil range from 0.05 to 0.25% P_2O_5 (Lawton, 1955). An average concentration in fresh water is 0.005 mg liter^{-1} (Bowen, 1966) and in seawater 0.09 mg liter $^{-1}$ (*Marine Chemistry*, 1971). The ratio of organic to inorganic phosphorus (P_{org}/P_i) varies widely in these environments. In mineral soil, P_{org}/P_i may range from 1:1 to 2:1 (Cosgrove, 1967, 1977).

9.1   ENZYMATIC BREAKDOWN AND SYNTHESIS
OF PHOSPHATE ESTERS AND ANHYDRIDES

An important source of free organic phosphorus compounds in the biosphere is
the breakdown of animal and vegetable matter, as well as its excretion by living
microbial cells (Shapiro, 1967) and by animals. Since organically bound phosphorus is for the most part not available to living organisms because it cannot be
taken into the cell in this form, it must first be freed from organic combination
through mineralization. This is accomplished through hydrolytic cleavage
catalyzed by phosphatases. In soil as much as 70-80% of the microbial population may be able to participate in this process (Dommergues and Mangenot,
1970, p. 266). Active organisms include bacteria and fungi, such as *Bacillus
megaterium, B. subtlilis, B. malabarensis, Serratia* spp., *Proteus* spp., *Arthrobacter* spp., *Streptomyces* spp., *Aspergillus* sp., *Penicillium* sp., *Rhizopus* sp.,
and *Cunninghamella* (Dommergues and Mangenot, 1970, p. 266). These organisms secrete or liberate upon their death, phosphatases with greater or lesser
substrate specificity (Skujins, 1967). Such activity has also been noted in the
marine environment (Ayyakkamin and Chandramotean, 1971).

Phosphate liberation from phytin generally requires the enzyme phytase

$$\text{Phytin} + 6H_2O \;\rightarrow\; \text{inositol} + 6P_i \tag{9.1}$$

Phosphate liberation from nucleic acid requires the action of nucleases, which
yield nucleotides, followed by the action of nucleotidases, which yield nucleo-
sides and inorganic phosphate

$$\text{Nucleic acid} \xrightarrow[\text{H}_2\text{O}]{\text{nucleases,}} \text{nucleotides} \xrightarrow[\text{H}_2\text{O}]{\text{nucleotidase,}} \text{nucleoside} + P_i \tag{9.2}$$

Phosphate liberation from phosphoproteins, phospholipids, ribitol and glycerol
phosphates requires phosphomono- and phosphodiesterases. The phosphodiesterases attack the diester linkages while the monoesterases attack the mono-
ester linkages (Mahler and Cordes, 1966, p. 513).

$$
\underset{\overset{|}{\text{OH}}}{\overset{\overset{\text{O}}{\|}}{R-O-P-O-R'}}
\xrightarrow[\text{diesterase}]{\text{H}_2\text{O,}}
\text{ROH} +
\underset{\overset{|}{\text{OH}}}{\overset{\overset{\text{O}}{\|}}{\text{HO}-P-O-R'}}
\xrightarrow[\text{monoesterase}]{\text{H}_2\text{O,}}
$$

$$
\underset{\overset{|}{\text{OH}}}{\overset{\overset{\text{O}}{\|}}{\text{HO}-P-\text{OH}}} + \text{HOR}' \tag{9.3}
$$

Synthesis of organic phosphates (monomers; phosphate esters) is an intra-cellular process and normally proceeds through a reaction between a carbinol group and ATP in the presence of an appropriate kinase. For example,

$$\text{Glucose} + \text{ATP} \xrightarrow{\text{glucokinase}} \text{glucose 6-phosphate} + \text{ADP} \tag{9.4}$$

Phosphate esters may also arise through phosphorolysis of certain polymers, such as starch or glycogen:

$$(\text{Glucose})_n + \text{H}_3\text{PO}_4 \xrightarrow{\text{phosphorylase}} (\text{glucose})_{n-1}$$
$$+ \text{glucose 1-phosphate} \tag{9.5}$$

$$\text{Glucose 1-phosphate} \xrightarrow{\text{phosphoglucomutase}} \text{glucose 6-phosphate} \tag{9.6}$$

Adenosine 5′-triphosphate (ATP) may be generated from ADP by adenylate kinase,

$$2\text{ADP} \rightarrow \text{AMP} + \text{ATP} \tag{9.7}$$

or by substrate-level phosphorylation, as in the reaction

$$\text{3-Phosphoglyceraldehyde} + \text{NAD}^+ + \text{P}_i \xrightarrow[\text{dehydrogenase}]{\text{triosephosphate}}$$
$$\text{1,3-diphosphoglycerate} + \text{NADH–H}^+ \tag{9.8}$$

$$\text{1,3-Diphosphoglycerate} + \text{ADP} \xrightarrow{\text{ADP kinase}} \text{3-phosphoglycerate}$$
$$+ \text{ATP} \tag{9.9}$$

It may also be generated by oxidative phosphorylation,

$$\text{ADP} + \text{P}_i \xrightarrow{\text{electron transport system}} \text{ATP} \tag{9.10}$$

or by photophosphorylation,

$$\text{ADP} + \text{P}_i \xrightarrow[\text{light}]{\text{photosynthetic system,}} \text{ATP} \tag{9.11}$$

Phosphate polymers are generally produced through reactions such as the following:

$$(\text{Polynucleotide})_{n-1} + \text{nucleotide triphosphate} \xrightarrow{\text{polymerase}}$$

$$(\text{polynucleotide})_n + \text{PP} \qquad (9.12)$$

$$\text{PP} \xrightarrow{\text{pyrophosphatase}} 2P_i \qquad\qquad\qquad\qquad (9.13)$$

9.2 NATURAL OCCURRENCE OF INORGANIC PHOSPHATES

Inorganic phosphorus may occur in soluble or insoluble form in nature. The most common inorganic form is orthophosphate (H_3PO_4). As an ionic species, the concentration of phosphate is controlled by its solubility in the presence of an alkaline earth cation such as Ca^{2+}, and of metal ions such as Mg^{2+}, Fe^{3+}, and Al^{3+} at appropriate pH values (see Table 9.1). In seawater, for instance, the soluble phosphate concentration (about 3×10^{-6} M, maximum) is controlled by Ca^{2+} ions (4.1×10^2 mg liter^{-1}), which form hydroxyapatite with phosphate in the prevailing pH range 7.4-8.1.

Insoluble phosphate occurs most commonly in the form of apatite [Ca_5-$(PO_4)_3(F,Cl,OH)$], in which the (F,Cl,OH) radical may be represented exclusively by F, Cl, or OH or any combination of these. In soil, phosphate may also occur as an aluminum salt (e.g., variscite, $AlPO_4 \cdot 2H_2O$) or the iron salts strengite ($FePO_4 \cdot 2H_2O$) or vivianite [$Fe_3(PO_4)_2 \cdot 8H_2O$].

Microbial Solubilization of Inorganic Phosphates Insoluble forms or inorganic phosphorus (calcium, aluminum, and iron phosphates) may be solubilized through microbial action. The mechanism by which the microbes accomplish this solubilization varies. It may be by (1) the production of inorganic and organic acids that attack the insoluble phosphates; (2) the production of chelators, such as 2-ketogluconate (Duff and Webley, 1959; see also Chapter 8),

Table 9.1 Solubility Products of Some Phosphate Compounds

Compound	K_2	Reference
$CaHPO_4 \cdot 2H_2O$	2.18×10^{-7}	Kardos (1955), p. 185
$Ca_{10}(PO_4)_6(OH)_2$	1.53×10^{-112}	Kardos (1955), p. 188
$Al(OH)_2HPO_4$	2.8×10^{-29}	Kardos (1955), p. 184
$FePO_4$	1.35×10^{-18}	From ΔF of formation

citrate, oxalate, and lactate, which can complex the cationic portion of the insoluble phosphate salts and thus force their dissociation; and (3) the production of hydrogen sulfide (H_2S), which can react with the iron phosphate and precipitate it as iron sulfide and in the process liberate phosphate, as in the reaction

$$2FePO_4 + 3H_2S \rightarrow 2FeS + 2H_3PO_4 + S^0 \qquad (9.14)$$

Table 9.2 lists some organisms active in phosphate solubilization.

Solubilization of inorganic phosphates has been noted directly in soil (Alexander, 1977; Dommergues and Mangenot, 1970; Chatterjee and Nandi, 1964). Besides orthophosphate, pyrophosphate (metaphosphate) may also be found in soil. It is readily hydrolyzed by pyrophosphatase, especially in flooded soil (Racz and Savant, 1972).

Microbial Phosphate Insolubilization Microorganisms can cause fixation or immobilization of phosphate, either by promoting the formation of inorganic precipitates or by assimilation into organic constituents of the cell [called transitory immobilization by Dommergues and Mangenot (1970) because of the ready resolubilization through mineralization upon death of the cell]. In soil and freshwater environments, the second of these two processes is often

Table 9.2 Some Organisms Active in Phosphate Solubilization

Organism	Mechanism of solubilization	Reference
B. megaterium	H_2S production; FeS precipitation	Sperber (1958b); Swaby and Sperber (1958)
Thiobacillus sp.	H_2SO_4 production from sulfur	Lipman et al. (1916)
Nitrifying bacteria	NH_3 oxidation to HNO_3	Dommergues and Mangenot (1970), p. 263
Pseudomonads, Arthrobacter, Erwinia-like bacterium	Chelate production; glucose converted to 2-ketogluconate	Duff et al. (1963); Sperber (1958a), Dommergues and Mangenot (1970), p. 262
Sclerotium	?	Dommergues and Mangenot (1970), p. 262
A. niger, A. flavus, Sclerotium rolfsii, Fusarium oxysporum, Cylindrosporium sp., Penicillium sp.	Organic acid production (e.g., citric acid)	Agnihotri (1970)

more important, although fixation of phosphate by Ca^{2+}, Al^{3+}, or Fe^{3+} is recognized. In a few marine environments (coastal waters or shallow seas) where phosphorite deposits occur, the precipitation mechanism may be more important.

Phosphorite Deposition   The exact role of microbes in inorganic phosphate precipitation is not always clear. In soil and sediment, their metabolic activity in first liberating reactive Ca, Al, and Fe ions from inorganic and organic combinations, and their influence on prevailing pH and E_h, must be important contributing factors in the formation of disseminate, insoluble calcium, aluminum, and iron phosphates (Patrick et al., 1973; Williams and Patrick, 1971). In marine environments in which localized concentrated phosphates occur as phosphorite (apatite-type mineral), the microbial role, although widely accepted (McConnell, 1965), is still poorly understoood. According to one model (Youssef, 1965), calcium phosphate (phosphorite) results from the mineralization of phytoplankton remains which have settled into a depression on the sea floor, leading to localized accumulation of phosphate, which then precipitates as a result of reaction with calcium of seawater. According to another model (Piper and Codispoti, 1975), carbonate fluorapatite $[Ca_{10}(PO_4, CO_3)_6 F_{2-3}]$ precipitation in the marine environment is dependent on denitrification by bacteria in the oxygen minimum layer of the ocean as it intersects with the ocean floor. The loss of nitrogen due to denitrification means lowered biological activity and can lead to excess accumulation of phosphate in this zone. The more acid conditions (pH 7.4-7.9) in the deeper waters keep phosphate dissolved and allow for upward movement in upwellings to regions where phosphate precipitation is favored (pH > 8) (Fig. 9.1). This model takes into account the conditions of marine apatite formation described by Gulbrandsen (1969) and helps to explain the occurrence of probable contemporary formation of phosphorite in regions of upwelling such as the continental margin of Peru (Veeh et al., 1973) and on the continental shelves of southwestern Africa (Baturin, 1972; Baturin et al., 1969). To explain the more extensive ancient phosphorite deposits, a periodic warming of the ocean can be invoked, which would reduce the oxygen solubility and favor more intense denitrification in deeper waters and therefore temporarily lessened biological activity, with a resultant increase in phosphate concentration leading to phosphate precipitaton (Piper and Codispoti, 1975). Sizable phosphorite deposits are associated with only six brief geological intervals: the Cambrian, Ordovician, Devonian-Mississipian, Permian, Cretaceous, and Cenozoic eras. Since in many instances these phosphorite deposits are associated with black shales and contain uranium in a reduced form (Altschuler et al., 1958), they are presumed to have accumulated under reducing conditions.

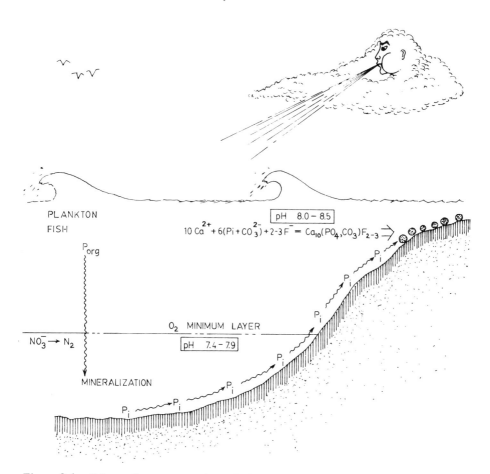

Figure 9.1 Schematic representation of phosphorite formation in the marine environment according to the model of Piper and Codispoti (1975). Note that the rising of P_i upslope is due to upwelling.

Some phosphate mineral accumulations of apatite and vivianite, such as those of the North Atlantic coastal plain, may have resulted to a significant extent from diagenetic processes in which apatite replaced calcite and vivianite replaced siderite (Adams and Burkhart, 1967). Microbes, particularly decomposers, may have been involved in supplying the orthophosphate needed for this transformation. Similar occurrences of phosphorite deposition through partial diagenesis have been found in deposits off Baja California and in a core from the eastern Pacific Ocean (d'Anglejan, 1967, 1968).

9.3 MICROBIAL OXIDATION AND REDUCTION
OF INORGANIC PHOSPHORUS COMPOUNDS

Phosphorus may also undergo redox reactions, some or all of which may be catalyzed by microbes. Of biologic interest are the $+5, +3, +1$, and -3 oxidation states, as in orthophosphate (H_3PO_4), orthophosphite (H_3PO_3), hypophosphite (H_3PO_2), and phosphine (PH_3), respectively. Phosphine formation through the reduction of phosphate by soil bacteria has been reported (Rudakov, 1972; Tsubota, 1959). Mannitol appeared to be a proper electron donor in this reaction, and phosphite and hypophosphite were claimed to be intermediates in the reduction (Rudakov, 1927; Tsubota, 1959). Recently, it was reported that iron phosphide (Fe_3P_2) was formed when a cell-free preparation of *Desulfovibrio* was incubated in the presence of steel in yeast extract broth under hydrogen gas (Iverson, 1968). Hydrogenase from *Desulfovibrio* may have been responsible for the formation of phosphine from phosphate contained in the yeast extract. The phosphine could then have reacted with ferrous iron to form the Fe_3P_2 (Iverson, 1968).

Questions have been raised about the ability of microbes to reduce phosphate. Liebert (1927) showed, on the basis of thermodynamic calculations using heats of formation, that the reduction of phosphate to phosphite by mannitol is an energy-consuming process and could therefore not serve a respiratory function. He calculated a ΔH of $+20$ kcal on the basis of the following equation:

$$C_6H_{14}O_6 + 13Na_2HPO_4 \longrightarrow 13Na_2HPO_3 + 6CO_2 + 7H_2O \qquad (9.15)$$

316 kcal   5,390 kcal   4,460 kcal   566 kcal   478 kcal

He also calculated a ΔH of $+483$ kcal for the reduction of phosphate to hypophosphite and a ΔH of $+1,147$ kcal for the reduction of phosphate to phosphine. These same conclusions can also be reached when free-energy changes (ΔF) are considered. In 1962, Woolfolk and Whiteley reported that they were unable to reduce phosphate with hydrogen in the presence of an extract of *Veillonella alcalescens* (formerly *Micrococcus lactilyticus*), even though this extract could catalyze the reduction of some other oxides with hydrogen. Skinner (1968) also questioned the ability of bacteria to reduce phosphate. He could not find such organisms in soils he tested. Interestingly, Barrenscheen and Beckh-Widmanstetter (1923) reported the production of hydrogen phosphide (phosphine; PH_3) from organically bound phosphate during putrefaction of beef blood. Thus, it may be that phosphate is reduced only when organically bound. This needs further investigation.

Reduced forms of phosphate may be oxidized by bacteria aerobically and anaerobically. Thus, *Bacillus caldolyticus,* a thermophile, can oxidize hypophosphite to phosphate aerobically (Heinen and Lauwers, 1974). The active enzyme system consists of an $(NH_4)_2SO_4$-precipitable protein fraction, NAD, and respiratory chain components. The enzyme system does not oxidize phosphite. Adams and Conrad (1953) first reported the aerobic oxidation of phosphite to phosphate by bacteria and fungi from soil. All phosphite that was oxidized by these strains was assimilated. None was released into the medium before death of the organisms. Phosphate added to the medium inhibited phosphite oxidation. Active organisms included *P. fluorescens, P. lachrymans, Aerobacter* (now known as *Enterobacter) aerogenes, Erwinia amylovora, Alternaria, Aspergillus niger, Chaetomium, Penicillium notatum,* and some actinomycetes. In later studies, Casida (1960) found that a culture of *P. fluorescens* strain 195 formed orthophosphate aerobically from orthophosphite in excess of its needs and released phosphate into the medium. The culture was heterotrophic, and its phosphite-oxidizing activity was inducible and stimulated by yeast extract. The enzyme system involved in phosphite oxidation was an orthophosphite-nicotinamide adenine dinucleotide oxidoreductase, which was inactive on arsenite, hypophosphite, nitrite, selenite, or tellurite and was inhibited by sulfite (Malacinski and Konetzka, 1966, 1967).

Recently, a soil bacillus was isolated that is capable of anaerobic oxidation of hypophosphite and phospite to phosphate (Foster et al. 1978). In a mixture of phosphite and hypophosphite, phosphite was oxidized first. Phosphate inhibited the oxidation of either phosphite or hypophosphite. The organism did not release phosphate into the medium.

Phosphite and hypophosphite have not been reported in detectable quantities in the natural environment. It has been suggested that microbial ability to utilize the compounds, especially anaerobically, may be a vestigial property developed at a time when the earth had a reducing atmosphere surrounding it which favored the occurrence of phosphite (Foster et al., 1978).

In many ecosystems phosphorus concentration is limiting to growth. The element follows cycles in which it finds itself alternately outside or inside living cells, in organic or inorganic form, free or fixed, dissolved or precipitated. Microbes play a central role in these changes of state, as outlined in Fig. 9.2 and as discussed in this chapter.

9.4 SUMMARY

Phosphorus is a very important element to all forms of life. It is used in cell structure as well as cell function. It plays a role in transducing biochemically useful energy. When free in the environment, it occurs primarily as organic phosphate esters and as inorganic phosphate. Some of the latter, such as calcium, aluminum, and iron phosphates, are very insoluble at neutral or alkaline pH. To be nutritionally utilizable, organic phosphates have to be enzymatically

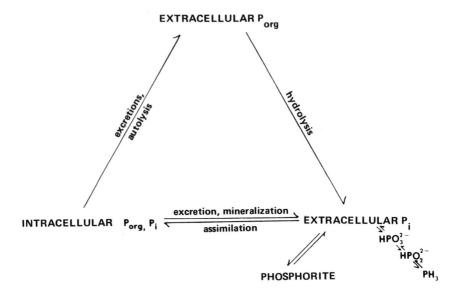

Figure 9.2   The phosphorus cycle. Phosphorite can be considered to be a natural reservoir of phosphorus.

hydrolyzed to liberate the orthophosphate. Microbes play a central role in this process. Microbes may also free orthophosphate from insoluble inorganic phosphates by production of organic or mineral acids or chelators, or, in the case of iron phosphates, the production of H_2S. Under some conditions, microbes may promote the formation of insoluble inorganic phosphates, such as those of calcium, aluminum, or iron. They have been implicated in phosphorite formation in the marine environment.

Microbes have been implicated in the reduction of pentavalent phosphorus to lower valence states. The experimental evidence for this is somewhat equivocal, however. Microbes have also been implicated in the oxidation of reduced forms of phosphorus to phosphate. The experimental evidence in this case is strong. It includes demonstration of enzymatic involvement. The significance of these redox reactions in nature is not presently clearly understood.

10

Geomicrobial Transformations of Arsenic and Antimony Compounds

10.1 ARSENIC

Arsenic is widely distributed in the upper crust of the earth, although mostly at very low concentration (Carapella, 1972). It rarely occurs in elemental form. More often it is found combined with sulfur, as in orpiment (As_2S_3) or realgar (AsS); with selenium, as in As_2Se_3; with tellurium, as in As_2Te_3; as sulfosalts, as in enargite (Cu_5AsS_4) or arsenopyrite (FeAsS); or as arsenides of heavy metals such as copper, iron, nickel, and cobalt, as in loellingite ($FeAs_2$). Sometimes the element is found in the form of arsenite minerals (arsenolite or claudetite, As_2O_3) or in the form of arsenate minerals [erythrite, $Co_3(AsO_4)_2 \cdot 8H_2O$; scorodite, $FeAsO_4 \cdot 4H_2O$; olivenite, $4CuOAsO_5 \cdot H_2O$]. Arsenopyrite is the most common and widespread mineral form of arsenic; orpiment and realgar are also fairly common. The ultimate source of arsenic on the earth's surface is igneous activity. On weathering of arsenic-containing igneous rocks, which may harbor 1.8 ppm of the element, the arsenic is dispersed through the upper lithosphere and the hydrosphere. The valence states in which arsenic usually is encountered in nature include $-3, 0, +2, +3$, and $+5$.

Arsenic concentration in soil may range from 0.1 to more than 1,000 ppm. The average concentration in seawater is given as 2.6 μg $liter^{-1}$ and in fresh water as 0.4 μg $liter^{-1}$ (Bowen, 1966). In atmospheric dust, arsenic may be found in the concentration range 50-400 ppm. Some living organisms may concentrate arsenic manyfold over its level in the environment. Thus, some algae have been found to accumulate arsenic 200-3,000 times in excess of its concentration in their growth medium (Lunde, 1973). Man may artificially raise the arsenic concentration in soil and water through the introduction of sodium arsenite (Na_2AsO_3) or cacodylic acid [$(CH_3)_2ASOOH$] as herbicide.

Biological Effects of Arsenic   Arsenic compounds are toxic to most living organisms. Arsenite (AsO_3^{3-} or AsO_2^-) has been shown to inhibit dehydrogenases such as pyruvate, α-ketoglutarate, and dihydrolipoate dehydrogenases

147

(Mahler and Cordes, 1966). Arsenate (AsO_4^{3-}) is known as an uncoupler of oxidative phosphorylation (Mahler and Cordes, 1966).

Both the uptake of arsenate and the inhibitory effect of arsenate on metabolism can be modified by phosphate (Button et al., 1973; Da Costa, 1971, 1972). A common transport mechanism for phosphate and arsenate seems to exist in the membranes of some organisms, although a separate transport mechanism for phosphate may also exist (Bennett and Malamy, 1970). In the latter case, phosphate does not affect arsenate uptake. The reversal by phosphate of metabolic inhibition by arsenate occurs at the site of oxidative phosphorylation (Da Costa, 1972). Thus, the effects of phosphate on arsenate uptake by cells and of arsenate inhibition of oxidative phosphorylation is best explained on the basis of a competitive effect. In one reported case with a fungus, *Cladosporium herbarium*, arsenite toxicity could also be modified by the presence of phosphate. In that instance, the investigator attributed the effect to the prior oxidation of arsenite to arsenate by the fungus (Da Costa, 1972). Phosphate can inhibit the formation by growing cultures of *Candida humicola* of trimethylarsine from arsenate, arsenite, and monomethyl arsinate, but not from dimethyl arsinate (Cox and Alexander, 1973). Phosphite can also suppress trimethylarsine production by the growing culture from monomethyl arsinate but not from arsenate or dimethyl arsinate, and hypophosphite can cause temporary inhibition of the conversion of arsenate, monomethyl arsonate, and dimethyl arsinate (Cox and Alexander, 1973). High antimonate concentrations lower the rate of conversion of arsenate to trimethylarsine by resting cells (Cox and Alexander, 1973).

Bacteria may develop genetically determined resistance to arsenic. This resistance may be determined by a locus on a plasmid, for example in *Staphylococcus aureus* (Dyke et al., 1970) and *Escherichia coli* (Hedges and Baumberg, 1973). Some of the resistant organisms have the capacity to metabolize arsenic.

Microbial Oxidation of Arsenite   Bacterial oxidation of arsenite to arsenate was first reported by Green in 1918. He discovered an organism, which he named *Bacillus arsenoxydans,* in arsenical cattle-dipping solution used for protection against insect bites. Quastel and Scholefield (1953) observed arsenite oxidation in soil perfusion experiments. They passed 2.5×10^{-3} M sodium arsenite solution through columns charged with Cardiff soil. They did not isolate the organism or organisms responsible but showed that 0.1% solution of NaN_3 inhibited the oxidation. The onset of arsenite oxidation occurred only after a lag. The length of this lag was increased when sulfanilamide was added with the arsenite. A control of pH was found important for sustained bacterial activity. They noted almost-stoichiometric O_2 consumption with arsenite oxidation.

Further investigations of arsenical cattle-dipping solutions led to the isolation of 15 arsenite-oxidizing strains of bacteria (Turner, 1949, 1954). These organisms were assigned to the genera *Pseudomonas, Xanothomonas*, and *Achromobacter*. *Achromobacter arsenoxydans-tres* has now been identified as being synonymous with *Alcaligenes faecalis* (Hendrie et al., 1974). This organism is described in the eighth edition of *Bergey's Manual of Determinative Bacteriology* (1974) as having frequently the capacity for arsenite oxidation.

Of the 15 isolates, *P. arsenoxydans-quinque* was studied in detail with respect to arsenite oxidation. Resting cells of this culture oxidized arsenite at an optimum pH of 6.4 and at an optimum temperature of 40°C (Turner and Legge, 1954). Cyanide, azide, fluoride, and pyrophosphate inhibited the activity. Under anaerobic conditions, 2,6-dichlorophenol indophenol, *m*-carboxyphenolindo-2,6-dibromophenol, and ferricyanide could act as electron acceptors. Pretreatment of the cells with toluene or acetone, or by desiccating them, rendered them incapable of oxidizing arsenite in air. The arsenite-oxidizing enzyme was "adaptable." Examination of arsenite oxidation by cell-free extracts of *P. arsenoxydans-quinque* suggested the presence of soluble "dehydrogenase" activity which under anaerobic conditions conveyed electrons from arsenite to 2,6-dichlorophenol indophenol (Legge and Turner, 1954). The activity was partly inhibited by 10^{-3} M *p*-chloromercuribenzoate. The entire arsenite-oxidizing system was believed to consist of "dehydrogenase" and an oxidase (Legge, 1954).

An arsenite-oxidizing soil strain of *Alcaligenes faecalis* was isolated in 1973 whose arsenite-oxidizing ability was found to be inducible by arsenite and arsenate (Osborne and Ehrlich, 1976). It oxidized arsenite stoichiometrically to arsenate (Table 10.1):

$$AsO_2^- + H_2O + \frac{1}{2}O_2 \rightarrow AsO_4^{3-} + 2H^+ \tag{10.1}$$

The oxidation involved an oxidoreductase with a bound flavin that passed electrons from arsenite to oxygen by way of cytochrome c and cytochrome oxidase

Table 10.1 Stoichiometry of Oxygen Uptake by *Alcaligenes faecalis* on Arsenite Based on the Reaction $AsO_2^- + H_2O + \frac{1}{2}O_2 \rightarrow AsO_4^{3-} + 2H^+$

NaAsO$_2$ added (μmol)	Oxygen uptake		
	Theoretical	Experimental	Percent of theoretical
19.25	9.63	8.79	91.3
38.50	19.25	18.48	96.0
57.75	28.88	27.05	93.7
77.00	38.50	37.05	96.2

Source: Osborne (1973); by permission.

(Osborne and Ehrlich, 1976). A similar strain was independently isolated and characterized by Phillips and Taylor (1976). Neither strain oxidizes arsenite strongly until the late exponential or stationary phase of growth is reached in batch culture (Phillips and Taylor, 1976; Ehrlich, 1978a). The strain of Osborne and Ehrlich may be able to derive maintenance energy from arsenite oxidation (Ehrlich, 1978a).

Microbial Oxidation of Arsenic Minerals Arsenic combinations with iron, copper, and sulfur are also bacterially oxidized. The simplest of these compounds, orpiment (As_2S_3), was found to be oxidized with the production of arsenite and arsenate by *Thiobacillus ferrooxidans* TM cultured in a mineral salts solution (9K medium without iron; see Silverman and Lundgren, 1959a) to which the mineral had been added in ground form (Ehrlich, 1963a) (Fig. 10.1).

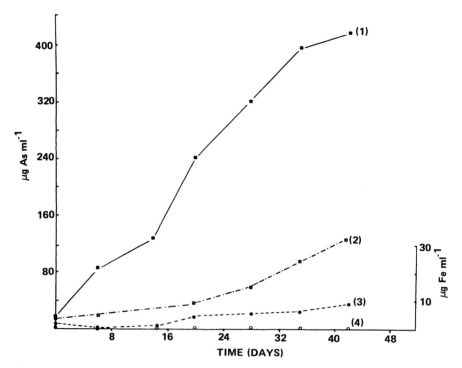

Figure 10.1 Bacterial solubilization of orpiment: (1) total arsenic released with bacteria (*T. ferrooxidans*); (2) total arsenic released without bacteria; (3) total iron released with bacteria (*T. ferrooxidans*); (4) total iron released without bacteria. (From Ehrlich, 1963a. Originally published in *Economic Geology*, Vol. 58, pp. 991-994, 1963.)

The initial pH was 3.5 and dropped to 2.0 in 35 days. In contrast, in an uninoculated control, in which orpiment oxidized spontaneously but slowly, the pH rose from 3.5 to 5. Realgar (AsS) was not attacked by *T. ferrooxidans* TM.

Arsenopyrite (FeAsS) and enargite (Cu_5AsS_4) were also oxidized by an iron-oxidizing *Thiobacillus* culture under the same test conditions as used with orpiment (Ehrlich, 1964a). With arsenopyrite, the arsenic was transformed to

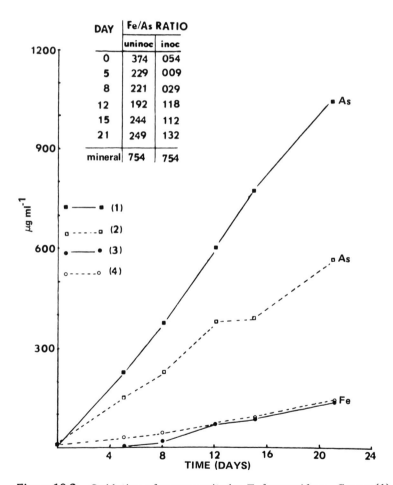

Figure 10.2 Oxidation of arsenopyrite by *T. ferrooxidans*. Curves (1) and (3) represent changes in inoculated flasks. Curves (2) and (4) represent changes in uninoculated flasks. (From Ehrlich, 1964a. Originally published in *Economic Geology*, Vol. 59, pp. 1306-1312, 1964.)

arsenite and arsenate. At least some of the iron was oxidized and extensively precipitated as iron arsenite and arsenate. The pH dropped from 3.5 to 2.5. Oxidation of arsenopyrite in the absence of bacteria was significantly slower (Fig. 10.2). With enargite the bacteria released cupric copper and arsenate into solution together with some iron which was present as an impurity in the mineral. With active bacteria, a pH drop from 3.5 to between 2 and 2.5 was observed, whereas in the absence of bacteria, a pH rise to 4.5 was usually observed. Some copper and arsenic remained out of solution in the experiments. The rate of enargite oxidation without bacteria was significantly slower, and may have followed a different course of reaction on the basis of the different rates of Cu and As solubilization.

Since these reports on the oxidation of arsenic compounds originally appeared, somewhat similar observations have been made elsewhere. The leaching by *T. ferrooxidans* of As from carbonaceous gold ores, containing 6% As, has been reported (Kamalov et al., 1973). As much as 90% of the As in the ore was removed in 10 days under some circumstances. Also, it was found possible to accelerate leaching of As from a copper-tin-arsenic concentrate 6-7 times, using an adapted strain of *T. ferrooxidans* (Kulibakin and Laptev, 1971). Finally, it was found that an iron-oxidizing *Thiobacillus* was able to accelerate As leaching from finely divided arsenopyrite by a factor of seven (Pol'kin and Tanzhnyanskaya, 1968). Pol'kin et al. (1973) reported additional observations on arsenopyrite leaching.

Microbial Reduction of Arsenic Compounds   Bacteria, fungi, and algae are able to reduce arsenic compounds. One of the first reports on arsenite reduction involved fungi (Gosio, 1897). Although originally the product of this reduction was thought to be diethylarsine (Bignelli, 1901), it was later shown to be trimethylarsine (Challenger et al. 1933). A bacterium from cattle-dipping tanks which reduced arsenate to arsenite was also described quite some time ago (Green, 1918). Recently, it has been reported that a strain of the green alga *Chlorella* can reduce a part of the arsenate it absorbs from the medium to arsenite (Blasco et al., 1971).

The fungus *Scopulariopsis brevicaulis* has been found to convert arsenite to trimethylarsine by a mechanism that includes the following steps (Challenger, 1951):

$$
\underset{\substack{\text{arsenious}\\\text{acid}\\(\text{As}+3)}}{\overset{\displaystyle O}{\overset{\|}{As-OH}}} \xrightarrow[\text{H}_2\text{O}]{\text{RCH}_3} \underset{\substack{\text{methylarsonic}\\\text{acid}\\(\text{As}+3)}}{\overset{\displaystyle O}{\overset{\|}{\underset{\underset{\text{OH}}{|}}{H_3C-As-OH}}}} \xrightarrow{\text{RCH}_3} \underset{\substack{\text{cacodylic}\\\text{acid}\\(\text{As}+1)}}{\overset{\displaystyle O}{\overset{\|}{\underset{\underset{\text{CH}_3}{|}}{H_3C-As-OH}}}} \xrightarrow{\text{RCH}_3} \underset{\substack{\text{trimethyl-}\\\text{arsine}\\(\text{As}-3)}}{As(CH_3)_3\uparrow}
$$

$$(10.2)$$

Methionine and other methyl compounds are the methyl donors (RCH_3).
Besides *S. brevicaulis,* other fungi, such as *Aspergillus, Mucor, Fusarium, Paecilomyces,* and *Candida humicola,* have also been found active in such reductions (Alexander, 1977; Cox and Alexander, 1973).

Bacterial reduction of arsenate to arsenite by H_2 has been studied using cell extracts of *Micrococcus (Veillonella) lactilyticus* and whole cells of *M. aerogenes* (Woolfolk and Whiteley, 1962). The active enzyme was hydrogenase. Arsine (AsH_3) was not formed in these reactions. *Pseudomonas* sp. and *Alcaligenes* sp., on the other hand, are able to reduce arsenate and arsenite anaerobically to arsine (Cheng and Focht, 1979). Dimethylarsine was produced from arsenate with H_2 by whole cells and cell extracts of the strict anaerobe *Methanobacterium* M.O.H. (McBride and Wolfe, 1971). The studies with extracts showed that the immediate methyl donor in this reaction was methylcobalamin (CH_3B_{12}). The reaction, moreover, required the consumption of ATP. 5-CH_3-Tetrahydrofolate or serine could not replace CH_3B_{12}, although CO_2 could when tested by isotopic tracer technique. The reaction sequence was found to be

$$AsO_4^{3-} \xrightarrow{2e-} AsO_2^- \xrightarrow{RCH_3} \underset{\underset{\text{methylarsonic}}{\underset{\text{acid}}{OH}}}{\overset{\overset{O}{\parallel}}{CH_3-\underset{|}{As}-OH}} \xrightarrow{2e-} \underset{\underset{\text{cacodylic}}{\underset{\text{(dimethyl-}}{\underset{\text{arsinic) acid}}{CH_3}}}}{\overset{\overset{O}{\parallel}}{CH_3-\underset{|}{As}-OH}} \xrightarrow{4e-}$$

$$\underset{CH_3}{\overset{|}{CH_3-As-H}} \qquad (10.3)$$

In an excess of arsenite, methylarsonic acid was not transformed further, the supply of CH_3B_{12} being limiting. In an excess of CH_3B_{12}, a second methylation step yielding cacodylic acid followed the first one. The last step was shown to occur in the absence of CH_3B_{12}. All steps were enzymatic. Cell extracts of *Desulfovibrio vulgaris* were also found to produce a volatile arsenic derivative, presumably an arsine (McBride and Wolfe, 1971).

Only a few direct observations of microbial reduction of arsenic compounds in nature have been reported. One concerned bacterial reduction of arsenate to arsenite in seawater (Johnson, 1972). In this study, bacteria from phytoplankton samples from Narragansett Bay and from Sargasso Sea water were able to reduce arsenate added to Sargasso Sea water. An arsenate reduction rate of 10^{-11} μmol cell^{-1} min^{-1} was measured after 12 hr of incubation at 20-22°C. Arsenic was not accumulated by the bacteria, and none was lost from the medium through volatilization. These observations may help to explain why the observed ratio of As^{5+}/As^{3+} was 10^{-1} to 10 instead of 20^{26}, as predicted under equilibrium conditions in oxygenated seawater at pH 8.1 (Johnson, 1972). In

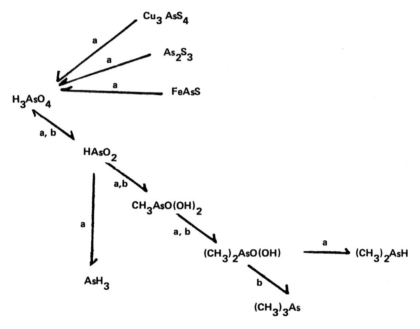

Figure 10.3   Transformations of arsenic compounds catalyzed by bacteria and fungi: a, bacteria; b, fungi.

another study, oxidation of arsenite to arsenate by aerated activated sludge has been reported (Myers et al., 1973). A stepwise reduction by activated sludge of arsenate to arsenite and, over a longer period, to even lower oxidation states has also been noted. *P. fluorescens* was an active reducer in this system under aerobic conditions (Myers et al.,1973). According to Cheng and Focht (1979), bacteria can reduce arsenate and arsenite mainly to arsine under anaerobiosis in soils. Unlike fungi, the bacteria produce mono- and dimethylarsine only when methyl arsonate or dimethyl arsinate are available.

Figure 10.3 summarizes those reactions involving arsenic compounds that are catalyzed by microorganisms. The oxidation of methylated arsine, although not indicated in the diagram, has been suggested by Cheng and Focht (1979).

10.2   ANTIMONY

Antimony is a rare element. Its average concentration in igneous rocks is 0.2 ppm, in shales 1.5 ppm, in sandstones 0.05 ppm, and in limestones 0.2 ppm

(Bowen, 1966). It may occur, among other forms, as stibnite (Sb_2S_3), hermesite (Sb_2S_2O), senarmontite (Sb_2O_3), jamesonite ($2PbS \cdot Sb_2S_3$), barlangevite, ($5PbS \cdot 2Sb_2S_3$), as sulfantimonides of copper, silver, and nickel, and sometimes in elemental form (Gornitz, 1972). Stibnite is the most common mineral. Antimony can exist in the oxidation states $-3, 0, +3,$ and $+5$.

Antimony concentrations in soil have been given as 2 to 10 ppm (?) (Bowen, 1966), in freshwater as 0.27 to 4.9×10^{-3} mg liter^{-1} (Kharkar et al. 1968), and in seawater at 0.3 μg liter^{-1} (*Marine Chemistry*, 1971).

Like arsenic compounds, antimony compounds tend to be toxic to most living organisms. The basis for this toxicity has not been clearly established.

Although microbial reduction of oxidized antimony compounds has not been reported so far, microbial oxidation of reduced antimony compounds has. Among the earliest observations is a report by Bryner et al. (1954) on the oxidation of the mineral tetrahedrite ($4Cu_2S \cdot Sb_2S_3$) by *T. ferrooxidans*. Lyalikova in 1961 observed oxidation of antimony trisulfide (Sb_2S_3) by *T. ferrooxidans*. In both cases oxidation proceeded under acid conditions (pH 2.45) (Kuznetsov et al.,1963). More recently, Silver and Torma (1974) reported on the oxidation of synthetic antimony sulfides by *T. ferrooxidans*, and Torma and Gabra (1977) on the oxidation of the mineral stibnite (Sb_2S_3) by the organism. The latter authors suspected that *T. ferrooxidans* oxidized trivalent antimony [Sb(III)] to pentavalent antimony [Sb(V)] but could offer no proof. Lyalikova et al. in 1972 reported on the bacterial oxidation of Sb-Pb sulfides, Sb-Pb-Te sulfides, and Sb-Pb-As sulfides with the formation of such minerals as anglesite and valentinite.

A new autotrophic organism, *Stibiobacter senarmontii*, capable of using Sb_2O_3 (senarmontite) or Sb_2O_4 as an energy source which it oxidizes to Sb_2O_5, has been discovered in Yugoslavia (Lyalikova, 1972; 1974; Lyalikova et al., 1976). The organism is a gram-positive, motile rod (0.5-1.8×0.5 μm) with a single polar flagellum and has the ability to form rudimentary mycelium in certain stages of development. It grows at neutral pH and generates acid when oxidizing Sb_2O_3 (pH range 7.5-5.5). When grown on reduced antimony oxide, the organism possesses the enzyme ribulose biphosphate carboxylase, indicating its chemolithotrophic propensity (Lyalikova et al., 1976). Antimony sulfide ores can thus be oxidized completely in two steps:

$$Sb_2S_3 \xrightarrow[\;(1)\;]{O_2} Sb_2O_3 \xrightarrow[\;(2)\;]{O_2} Sb_2O_5 \tag{10.4}$$

The first step is catalyzed by an organism such as *Thiobacillus* Y or *T. ferrooxidans* [see (1) above], and the second by *Stibiobacter senarmontii* [see (2) above] (Lyalikova et al.,1974).

10.3 SUMMARY

Although arsenic and antimony compounds are toxic to most forms of life, some microbes metabolize them. Arsenite has been found to be enzymatically oxidized by several different bacteria. The enzyme system is inducible. In laboratory experiments, *Alcaligenes faecalis* oxidizes arsenite most intensely only after having gone through active growth. The organism probably can derive maintenance energy from arsenite oxidation. Simple and compound arsenic sulfides are oxidized by *Thiobacillus ferrooxidans*. No evidence has been obtained, however, that trivalent arsenic is enzymatically oxidized to pentavalent arsenic in these cases.

Arsenite and arsenate have also been shown to be reduced by certain bacteria and fungi. When reducing the arsenic to the -3 oxidation state, the bacteria produce dimethylarsine, whereas the fungi produce trimethylarsine. Both of these arsines are volatile. Some bacteria may merely reduce arsenate to arsenite.

Antimony-containing compounds have also been shown to be microbially ozidized. *T. ferrooxidans* has been shown to attack a variety of antimony-containing sulfides. Although enzymatic oxidation of Sb(III) to Sb(V) has been claimed in the oxidation of Sb_2S_3 by *T. ferrooxidans,* clear proof is lacking. Generally, only the sulfide moiety and ferrous iron, if present in the mineral, are oxidized by this organism. Recently, *Stibiobacter senarmontii,* a chemolithotroph, was isolated from an ore deposit in Yugoslavia, which oxidizes the antimony in Sb_2O_3 or Sb_2O_4 to Sb_2O_5. Microbial reduction of oxidized antimony compounds has not been reported so far.

11

The Geomicrobiology of Mercury

The element mercury has long been known, from at least as far back as 1500 B.C. The physician Paracelsus (A.D. 1493-1541) attempted to cure syphilis by administering metallic mercury to sufferers of the disease, probably on the basis of intuitive or empirical knowledge that at an appropriate dosage, mercury was more toxic to the cause of the disease than to the patient. The true etiology of syphilis was, however, unknown to him. Only in recent times has the full lethal effect of mercury on human beings and other animals become very apparent. This came as a result of major physical impairments and death which were traced to accidental intake of mercury compounds in food and water, as in Japan (Minamata disease), Iraq, Pakistan, Guatemala, and the United States. In some cases, food made from seed grain had become tainted by mercury compounds used as fungistatic agents. The agents had been added to the seed grain to prevent damage by fungi before planting. The seed grain had not been intended for food use. In other cases, food such as meat had become tainted because the animals yielding the meat had taken water that had become polluted by mercury compounds or had eaten mercury-tainted feed. Tracing the fate of mercury introduced into the environment has revealed an intimate role of microbes in the interconversion of some mercury compounds.

The concentration of mercury in the earth's crust has been reported as 0.08 ppm (Jonasson, 1970). Its concentration in uncontaminated soils has been given as 0.07 ppm (Jonasson, 1970). Its concentration in fresh waters may range from 0.01 to 10 ppb, although concentrations as high as 1,600 ppb have been measured in waters in contact with copper deposits in the southern Urals (Jonasson, 1970). The maximum permissible level in potable waters in the United States has been set at 5 ppb. The average mercury concentration in seawater has been reported at $0.2 \, \mu g \, liter^{-1}$ (ppb) (*Marine Chemistry*, 1971).

Mercury can exist in nature as mercury metal or as mercury compounds. The metal is liquid at ambient temperature and has a significant vapor pressure (1.2×10^{-3} mm Hg at 20°C) and a heat of vaporization of $14.7 \, cal \, mol^{-1}$ at 25°C (Vostal, 1972). The most prevalent mineral of mercury is cinnabar (HgS), which is found in highest concentrations in volcanically active zones, such as the circum-Pacific volcanic belt, the East Pacific Rise, and the Mid-Atlantic Ridge. The occurrence of mercury metal is rarer. In water, inorganic mercury may exist as aquo, hydroxo, halido, and bicarbonate complexes of mercuric ion, but the mercuric ion may also be adsorbed to particulate or colloidal materials in suspension (Jonasson, 1970). In soil, inorganic mercury may exist in the form of elemental mercury vapor, at least in part adsorbed to soil matter. It can also exist as mercuric humate complexes at pH 3-6 or as $Hg(OH)^+$ and $Hg(OH)_2$ in the pH range of 7.5-8.0. The latter two species may be adsorbed to soil particles (Jonasson, 1970). Mercury in soil and water may also exist as methylmercury $[(CH_3)Hg^+]$, which may be adsorbed by negatively charged particles.

The local mercury level may be affected by human activity. Some industrial operations, such as the synthesis of certain chemicals like vinyl chloride and acetaldehyde, which employ inorganic mercury compounds as catalysts, or the electrolytic production of chlorine gas and caustic soda, which employs mercury electrodes, or the manufacture of paper pulp, which makes use of phenylmercuric acetate as a slimicide (Jonasson, 1970), may pollute the environment. In agriculture, organic compounds used as fungicides to prevent fungal attack of seed may pollute the soil. In mining, the exposure of mercury ore deposits and other deposits in which mercury is only a trace component leads to weathering and resultant solubilization and spread of some of the mercury into the environment.

As Jonasson (1970) has pointed out, in the past, inorganic mercury compounds were considered less toxic than organic mercury compounds, but since the discovery that inorganic mercury compounds can be converted into organic ones (e.g., methylmercury), this is no longer considered to be true. Living tissue has a high affinity for methylmercury. Fish have been found to concentrate it up to 3,000 times over the concentration found in water.

Microorganisms have in recent years been shown to be intimately involved in interconversions of inorganic and organic mercury compounds. The initial discoveries of the microbial activities were those of Jensen and Jerneloev (1969), who demonstrated the production of methylmercury from mercuric chloride $(HgCl_2)$ added to lake sediment samples and incubated for several days in the laboratory. They also noted the production of dimethylmercury $[(CH_3)_2Hg]$ from decomposing fish tissue, containing methylmercury or supplemented with Hg^{2+}, and incubated for several weeks. Later work established that methylation could be brought about by bacteria and fungi (see below).

11.1 SPECIFIC MICROBIAL INTERACTIONS WITH MERCURY

Microbial Methylation of Mercury   The mechanism of bacterial methylation of mercury ions was studied with extracts of a methanogenic culture in the presence of low concentrations of Hg^{2+}, which caused formation of $(CH_3)_2 Hg$ but little methane (CH_4), through preferential interaction between methylcobalamin and Hg^{2+} (Wood et al., 1968). Although the production of methylcobalamin depended on enzymatic catalysis, the production of $(CH_3)_2 Hg$ from the reaction of Hg^{2+} with methylcobalamin did not. The nonenzymatic nature of mercury methylation by methylcobalamin has been confirmed (Bertilsson and Neujahr, 1971; Imura et al., 1971; Schrauzer et al., 1971). The mechanism of mercury methylation by methylcobalamin can be summarized as follows (DeSimone et al., 1973):

$$Hg^{2+} \xrightarrow{\ CH_3 B_{12}\ } (CH_3)Hg^+ \xrightarrow{\ CH_3 B_{12}\ } (CH_3)_2 Hg \qquad (11.1)$$

The initial methylation of Hg^{2+} in this reaction sequence proceeds 6,000 times as fast as the second methylation (Wood, 1974).

Another organism capable of methylating inorganic mercuric compounds is *Clostridium cochlearium* (Yamada and Tonomura, 1972a,b). It was found to attack HgO, $HgCl_2$, $Hg(NO_3)_2$, $Hg(CN)_2$, $Hg(SCN)_2$, and $Hg(OOCCH_3)_2$.

Although in the laboratory, bacterial methylation of mercury appears to be favored by anaerobic conditions, partially aerobic conditions are needed in nature. This is because under anaerobic conditions in situ, biogenic H_2S is likely to prevail, and, as a result, mercuric mercury will exist most probably as HgS (Fagerstroem and Jerneloev, 1971; Vostal, 1972). HgS cannot be methylated without prior conversion to a soluble Hg^{2+} salt or HgO (Yamada and Tonomura, 1972c). Recently, it was discovered that H_2S, which in nature is frequently of biogenic origin, can transform preexisting methylmercury into dimethylmercury (Craig and Bartlett, 1978).

The fungus *Neurospora crassa* uses a different reaction in the methylation of mercury. This organism first complexes mercuric ion with homocysteine or cysteine and then, with the help of a methyl donor and the enzyme transmethylase, cleaves $(CH_3 Hg^+)$ from this complex (Landner, 1971).

$$Hg^{2+} + \begin{matrix} SH \\ | \\ CH_2 \\ | \\ CH_2 \\ | \\ CHNH_2 \\ | \\ COOH \\ \\ \text{homo-} \\ \text{cysteine} \end{matrix} \longrightarrow \begin{matrix} SHg^+ \\ | \\ CH_2 \\ | \\ CH_2 \\ | \\ CHNH_2 \\ | \\ COOH \\ \text{mercuryl} \\ \text{homo-} \\ \text{cysteine} \end{matrix} \xrightarrow[\text{transmethylase}]{\text{methyl donor,}} \begin{matrix} SH \\ | \\ CH_2 \\ | \\ CH_2 \\ | \\ CHNH_2 \\ | \\ COOH \\ \text{methyl} \\ \text{mercury} \end{matrix} + (CH_3)Hg^+ \qquad (11.2)$$

Suitable methyl donors may be betaine or choline but not $CH_3 B_{12}$.

Methylmercury is water-soluble as well as fat-soluble and is more readily taken up by living cells than mercuric ion. Owing to its lipid solubility, nervous tissue, especially the brain, has a high affinity for it. It is also bound by inert matter, especially negatively charged particles such as clays.

Dimethylmercury is volatile. It can thus enter the atmosphere from soil or water phases. The ultraviolet component of sunlight can, however, cause this compound to dissociate into elementary mercury, methane, and ethane.

Microbial Dephenylation and Demethylation of Mercury   Phenyl- and methylmercury can be microbiologically converted to volatile Hg^0 by bacteria in lake and estuarine sediments and in soil (Nelson et al. 1973; Spangler et al., 1973a,b; Tonamura et al., 1968). The bacteria most frequently involved are mercury-resistant strains of *Pseudomonas*. Although mercury-resistant strains of other genera may also be involved, they seem to be active on a much more limited scale (Nelson and Colwell, 1975). The microbial decomposition of phenyl- and methylmercury by these organisms can be written as follows:

$$\phi Hg^+ + H^+ + 2e^- \longrightarrow Hg^0 + \phi \tag{11.3}$$

$$CH_3Hg^+ + H^+ + 2e^- \longrightarrow Hg^0 + CH_4 \tag{11.4}$$

The enzymatic system for Hg^0 production from phenylmercuric acetate by a mercury-resistant pseudomonad was found to involve a reduced nicotinamide-adenine dinucleotide phosphate ($NADPH + H^+$)-generating system, glucose or arabinaose dehydrogenase, two carbon-mercury linkage splitting enzymes, and an inducible metallic mercury releasing enzyme that was shown to contain flavin-adenine dinucleotide (FAD) as a prosthetic group (Furukawa and Tonomura, 1971, 1972a,b; Tezuka and Tonomura, 1976, 1978). The inducibility of phenylmercury metabolism has also been demonstrated (Komura and Izaki, 1971; Nelson et al., 1973).

Microbial Diphenylmercury Formation   A case of microbial conversion of phenylmercuric acetate to diphenylmercury has been reported (Matsumura et al.,1971). The reaction can be summarized as follows,

$$2\phi Hg^+ \longrightarrow (\phi)_2 Hg + \text{unknown Hg compound and trace of } Hg^{2+} \tag{11.5}$$

Microbial Reduction of Mercuric Ion   Mercuric ion has been reported to be reduced by bacteria and fungi to volatile metallic mercury. Active organisms include strains of *Pseudomonas* spp. enteric bacteria, *Staphylococcus aureus*, and *Cryptococcus* (Brunker and Bott, 1974; Komura et al.,1970; Nelson et al., 1973; Summers and Lewis, 1973), and possibly *Bacillus, Vibrio*, coryneform bacteria, *Cytophaga, Flavobacterium, Achromobacter, Alcaligenes*, and

Acinetobacter (Nelson and Colwell, 1974). A significant number of *Pseudomonas* strains having this reducing ability were found to be unable to utilize glucose (Nelson and Colwell, 1974). The reduction mechanism involves the reaction of Hg^{2+} with reduced NAD (NADH + H^+) or reduced NADP (NADPH + H^+) in the presence of an FAD-containing enzyme and a sulfhydryl compound such as mercaptoenthanol, cysteine, dithiothreitol, or glutathione (Izaki et al., 1974).

$$Hg^{2+} + NADH + H^+ \xrightarrow[\substack{\text{sulfhydryl} \\ \text{compound}}]{\text{enzyme,}} Hg^0 + 2H^+ + NAD^+ \qquad (11.6)$$

Interestingly, this reduction of Hg^{2+} is generally associated with aerobic conditions (e.g., Nelson et al., 1973; Spangler et al.,1973a,b). Not all reduction of Hg^{2+} ion is biological, however. Chemical reduction of mercuric ion to metallic mercury may also occur in nature (Nelson and Colwell, 1975). Thus, humic acids have been shown to be effective reductants (Alberts et al. 1974).

Oxidation of Metallic Mercury   Native mercury (Hg^0) has been reported to be oxidizable to mercuric ion in the presence of certain bacteria (Holm and Cox, 1975). Whereas strains of *P. aeruginosa, P. fluorescens, E. coli,* and *Citrobacter* oxidized only small amounts of Hg^0, strains of *B. subtilis* and *B. megaterium* oxidized more significant amounts. In none of the cases was methylmercury formed. The observed oxidation was not enzymatic, but was due to reaction with metabolic products which acted as oxidants. Even yeast extract was found to be able to oxidize Hg^0.

11.2   GENETIC CONTROL OF MERCURY TRANSFORMATIONS

In general, the bacteria found capable of reducing inorganic mercury compounds to volatile mercury are relatively resistant to the toxicity of inorganic mercury compounds. This resistance in bacteria is usually plasmid-determined (i.e., R-factor- or sex-factor-linked) (Komura and Izaki, 1971; Loutit, 1970; Novice, 1967; Richmond and John, 1964; Smith, 1967; Schottel et al.,1974; Summers and Silver, 1972). Such plasmid-determined resistance can thus be transferred to susceptible cells through conjugation or phage transduction among gram-negative organisms, and through phage transduction among gram-positive organisms (Summers and Silver, 1972).

Neurospora crassa strains capable of methylating Hg^{2+} were obtained by selecting for mercury-resistant mutants (e.g., resistant to 225 ppm Hg^{2+}). Landner (1971) believes that his Hg^{2+}-resistant mutants may be constitutive mutants which have lost control over one of the last enzymes in methionine biosynthesis, so that methionine no longer interferes with methylation of Hg^{2+}, as it does in wild-type strains.

So far, no special trait of resistance to mercury has been reported for bacteria that can methylate mercuric ion. Since the methylation mechanisms in bacteria represent abnormal utilization of preexistent metabolic pathways (e.g., constitutive cobalamin formation in methanogenic bacteria), the acquisition of special traits for mercury resistance may not be necessary.

11.3   BIOLOGIC SIGNIFICANCE OF MERCURY TRANSFORMATIONS

The enzymatic attack of mercury compounds is not for the dissipation of excess reducing power (respiration) or the production of useful metabolites, but for the purpose of detoxification. The products of volatile mercury (Hg^0), and dimethyl- and diphenylmercury, owing to their high volatility and low water solubility, are thus readily lost to the atmosphere from the normal habitat of microbes and other creatures. The metabolic transformation to volatile forms of mercury protects not only the organisms actively involved in the conversion but also coinhabitants which lack this ability and are more susceptible to mercury poisoning.

11.4   A MERCURY CYCLE

On the basis of the foregoing discussion on the physiological interaction of microbes with mercury compounds, it is apparent that microbes play an important role in the movement of mercury in nature, especially in the soil, sedimentary, and aqueous environments. The main result of microbial action on mercury seems to be its volatilization, whether it involves reduction of mercuric ion or methyl- or phenylmercury compounds to volatile Hg^0, or whether it involves conversion of mercuric ion to dimethylmercury or of phenylmercuric ion to diphenylmercury. However, in the case of dimethylmercury synthesis, an intermediate monomethylmercury ion, is produced which is more toxic to susceptible forms of life than is Hg^{2+}, probably because of its greater lipid solubility combined with its positive charge. Mercuric and methyl- or phenylmercury ions, because of their positive charge, may be readily bound to negatively charged particles, in which case they become apparently less accessible or inaccessible to microbial attack and are not available for absorption by living organisms in general. They may also be complexed by, for example, humic substances. This probably explains why mercury concentrations in soil are often higher in the topsoil than in the subsoil (Anderson, 1967). Volatile forms of mercury metal may pass into the atmosphere but may also be adsorbed to solid surfaces of inert particles. Dimethylmercury is readily dissociated into volatile Hg^0, methane,

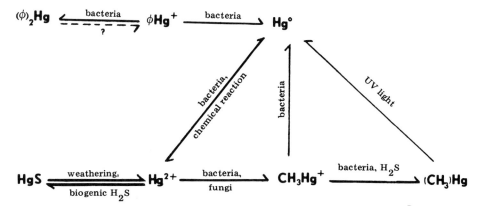

Figure 11.1   Mercury transformations by microbes and by chemical and physical agents.

and enthane by ultraviolet light from the sun, as previously mentioned. A mercury cycle is outlined in Figure 11.1.

Mercuric sulfide of volcanic origin on exposure to air and moisture slowly oxidizes to mercuric sulfate and may become disseminated in soil and water through groundwater movement. Humic substances may reduce the Hg^{2+} to Hg^0, as may bacteria and fungi. The volatile mercury (Hg^0) may be adsorbed to soil or sediment components or lost to the atmosphere. Some Hg^{2+} may also become methylated through the action of bacteria and fungi. Some positively charged methylmercury is readily fixed by negatively charged soil and sediment particles and thereby becomes unreactive, but that which is not fixed can be further disseminated by water movement. Some methylmercury may be further methylated to volatile dimethylmercury, which readily escapes into the atmosphere from soil and water. Methylmercury may also be bacterially reduced to volatile mercury. Phenylmercury, which is usually of man-made origin, may be similarly reduced to volatile mercury by bacteria in soil, but it may also be bacterially converted to diphenylmercury. Mercuric ion may be converted to mercuric sulfide by bacterially produced H_2S. Only this last reaction seems to be strictly dependent on anaerobiosis.

11.5   SUMMARY

Mercuric ion (Hg^{2+}) may be methylated by bacteria and fungi to methylmercury [$(CH_3)Hg^+$]. At least some fungi use a methylation mechanism different from bacteria. Methylmercury is water-soluble and more toxic than mercuric ion.

Methylmercury may be fixed by sediment. Some bacteria can methylate methylmercury, forming volatile dimethylmercury. This may constitute a detoxification when it occurs in soil or sediment, because this compound can escape into the atmosphere. Methylmercury as well as phenylmercury can be enzymaticaly reduced to volatile metallic mercury (Hg^0) by some bacteria. Phenylmercury can be microbially converted to diphenylmercury. It is very important that mercuric ion can be enzymatically reduced to metallic mercury by bacteria and fungi. This may also constitute a detoxification, because the product is volatile. It may be a more important reaction in nature than the methylation of methylmercury. Metallic mercury has also been found to be oxidized to mercuric mercury by bacteria. However, this reaction is not enzymatic but rather is the result of interaction with metabolic products. Microbes metabolizing mercury are generally resistant to its toxic effects. The mercury cycle in nature is thus under the influence of microorganisms.

12

The Geomicrobiology of Iron

Iron is the fourth most abundant element in the earth's crust, and is the most abundant element in the earth as a whole (see Chapter 2). Its average concentration in the crust is 5.0% (Rankama and Sahama, 1950). It is found in a number of minerals in rocks, soils, and sediments. Table 12.1 lists mineral types in which Fe is a major or minor structural component.

The primary source of iron accumulations on the earth's surface is volcanic activity. However, weathering of iron-containing rocks is often an important phase in the formation of local iron accumulations (sedimentary ore deposits).

Iron is a very reactive element. It can exist in oxidation states of 0, +2, and +3. At pH values greater than 5, its ferrous form (+2) is readily oxidized in air to the ferric form (+3). Under reducing conditions, ferric iron is readily reduced to the ferrous state. In acid solution, metallic iron readily oxidizes to ferrous iron with the production of hydrogen.

$$Fe^0 + 2H^+ \rightarrow Fe^{2+} + H_2 \tag{12.1}$$

Ferric iron precipitates in alkaline solution and dissolves in acid solution.

Iron is important biologically. Cells use it catalytically in the enzymatic transfer of electrons, as in respiration in which the heme iron of cytochromes and other nonheme iron proteins are involved in the transfer of electrons to a terminal acceptor, or in photosynthesis in which ferredoxin, a nonheme iron protein and some cytochromes are involved. Cells also employ iron in the heme groups of the enzymes catalase and peroxidase, which catalyze reactions involving H_2O_2. Microorganisms capable of nitrogen fixation employ ferredoxin and several other nonheme iron proteins in N_2 reduction. All these catalytic requirements make iron nutritionally important. Since in environments of neutral pH, ferric iron precipitates readily from solution, and since in this insoluble form iron cannot be readily taken up by cells, a number of microorganisms have acquired the ability to synthesize chelators which help to keep ferric iron in solution or which may return it to solution in sufficient quantities

165

Table 12.1   Iron-Containing Minerals

Igneous (primary) minerals	Secondary minerals	Sedimentary minerals
Pyroxenes	Montmorillonite[a] $(OH)_4 Si_8 Al_4 O_{20} \cdot nH_2O$	Siderite $(FeCO_3)$
Amphiboles		Goethite $(Fe_2O_3 \cdot H_2O)$
Olivines	Illite $(OH)_4 K_y (Al_4 Fe_4 Mg_4 Mg_6)$- $(Si_{8-y} Al_y)O_{20}$	Limonite $(Fe_2O_3 \cdot nH_2O)$
Micas		Hematite (Fe_2O_3)
		Magnetite (Fe_3O_4)
		Pyrite, marcasite (FeS_2)
		Pyrrhotite $(Fe_n S_{n+1})$; n = 5-6
		Ilmenite $(FeO \cdot TiO_2)$

[a] Montmorillonite contains iron by lattice substitution for aluminum.

to be nutritionally effective. Examples of such chelators, known collectively as **siderophores**, are enterobactin or enterochelin, a catechol derivative from *Salmonella typhimurium* (Pollack and Neilands, 1970); aerobactin, a hydroxamate derivative produced by *Aerobacter aerogenes* (Gibson and Magrath, 1969); and rhodotorulic acid, a hydroxamate derivative produced by fungi (Neilands, 1974) (Fig. 12.1). The iron can be taken up by these organisms in the chelated form.

As will be discussed below, ferrous iron may also serve as a major energy source to certain bacteria, whereas ferric iron may serve as a terminal electron acceptor for some other bacteria. Indeed, it has been postulated that in one stage in the evolution of life on earth, ferrous iron was an important reductant in primordial, oxygen-producing photosynthesis by reacting with the toxic oxygen produced in the process. This reaction gave rise to the banded iron formations in the sedimentary record 3.3 to 2 billion years ago (Cloud, 1973). They consist of cherty magnetite (Fe_3O_4) and hematite (Fe_2O_3) (see the discussion later in this chapter). From a biogeochemical viewpoint, large-scale microbial iron oxidation and reduction are important because they lead to extensive precipitation and solubilization of iron in the biosphere. They are discussed in detail below.

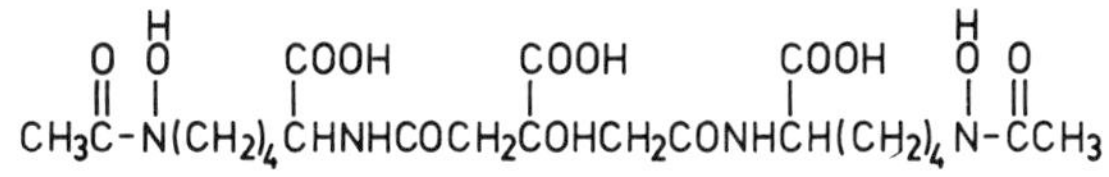

$$CH_3\overset{O}{\underset{\parallel}{C}}-\overset{H}{\underset{\mid}{N}}(CH_2)_4\overset{COOH}{\underset{\mid}{C}}HNHCOCH_2\overset{COOH}{\underset{\mid}{C}}OHCH_2CONH\overset{COOH}{\underset{\mid}{C}}H(CH_2)_4\overset{H}{\underset{\mid}{N}}-\overset{O}{\underset{\parallel}{C}}CH_3$$

AEROBACTIN   (BACTERIAL)

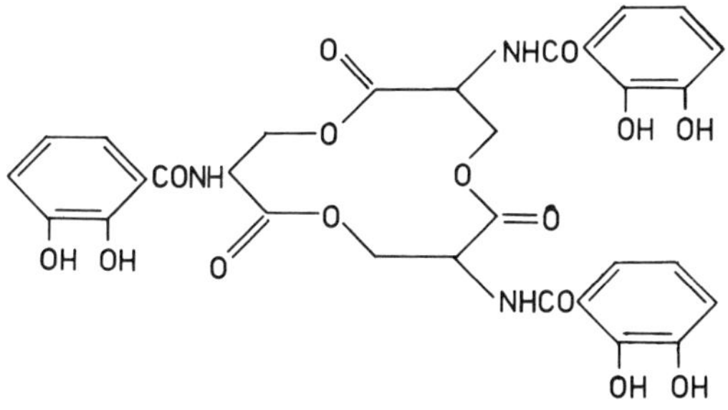

ENTEROCHELIN   (BACTERIAL)

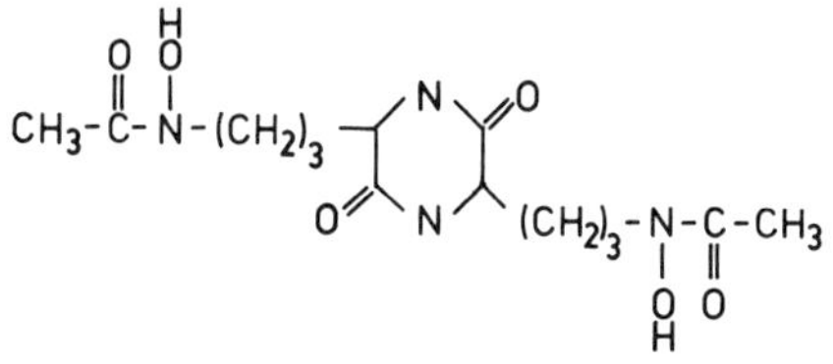

RHODOTORULIC ACID (FUNGAL)

Figure 12.1   Examples of siderophores.

12.1   FERROUS IRON OXIDATION

Thiobacillus ferrooxidans   Microbes may promote iron oxidation, but this does not mean that the oxidation is always enzymatic. Because of the tendency of ferrous iron to autoxidize in aerated solution at pH values above 5, it is difficult to demonstrate enzyme-catalyzed iron oxidation in near-neutral solutions. At this time, incontrovertible evidence for enzyme-catalyzed iron oxidation involving bacteria exists only at pH values below 5. *Thiobacillus ferrooxidans* (Fig. 12.2), a gram-negative, motile rod (0.5×1.0 μm), first isolated by Colmer et al.

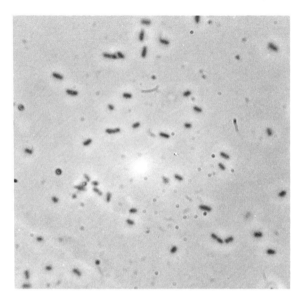

Figure 12.2 *Thiobacillus ferrooxidans* (5,170X). Cell suspension viewed by phase contrast.

(1950) and named and characterized as a chemolithotroph by Temple and Colmer (1951), is an example of a microorganism that is able to derive energy and reducing power from iron, and carbon from CO_2, at acid pH. When using inorganic nitrogen as its nitrogen source, it grows better on NH_3-N than on NO_3-N (Temple and Colmer, 1951; Lundgren et al., 1964). However, it can also use the amino acids alanine, glutamic acid, and lysine to satisfy its complete nitrogen requirements. The amino acids arginine or histidine are able to replace its requirement for NH_3-N partially, whereas the amino acids proline, tryptophane, and methionine are unable to satisfy any of its need for NH_3-N (Lundgren et al., 1964). A recent report indicates that at least one strain is capable of nitrogen fixation (Mackintosh, 1978).

Morphologically, the cells of *T. ferrooxidans* exhibit the multilayered cell wall typical of gram-negative bacteria (Avakyan and Karavaiko, 1970; Remsen and Lundgren, 1966) (Fig. 12.3). They do not contain special internal membranes like those found in nitrifiers and methylotrophs. Cell division is mostly by constriction, but occasionally also by partitioning (Karavaiko and Avakyan, 1970).

T. ferrooxidans is acidophilic. Iron oxidation by resting cells is optimal at pH 3 to 3.6 (full range, 2.5-4.2) (Silverman and Lundgren, 1959b). The pH

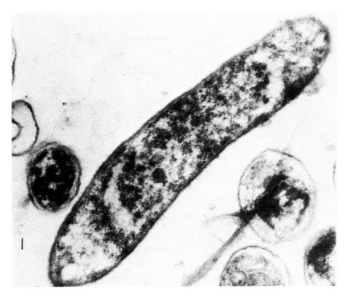

Figure 12.3   *Thiobacillus ferrooxidans* (30,000X). Electron photomicrograph of a thin section (courtesy of D. G. Lundgren).

optimum for growth of the organism is 2.5-5.8 (full range, pH 1.4-6.0) (*Bergey's Manual*, 1974). The organism is mesophilic (i.e., its optimum temperature range for growth is 15-20°C) (*Bergey's Manual*, 1974), although at least one strain has exhibited an optimum around 28°C (Silverman and Lundgren, 1959a). Interestingly, this same strain oxidizes iron optimally without growth at 37°C. Another strain has been described that oxidizes iron optimally at 40°C (Landesman et al., 1966).

Iron is not the only possible inorganic energy source for *T. ferrooxidans*. It can also use reduced forms of inorganic sulfur and metal sulfides as energy sources (see Chapters 14 and 16). Some strains of the organism can be adapted to grow heterotrophically with glucose as energy and carbon sources.

Two related acidophilic, iron-oxidizing bacteria, *Ferrobacillus ferrooxidans* (Leathen et al., 1956) and *F. sulfooxidans* (Kinsel, 1960), have been described in the literature. These organisms are presently considered to be synonymous with *T. ferrooxidans* (Unz and Lundgren, 1961; Ivanov and Lyalikova, 1962; Hutchinson et al., 1966; Kelly and Tuovinen, 1972; *Bergey's Manual*, 1974).

Laboratory study of *T. ferrooxidans* depends on ease of culturing. Table 12.2 lists the ingredients of four liquid media, of which the 9K and T&K media are the most suitable. Cultivation on solid media, solidified with agar or silica gel,

Table 12.2 Media for Cultivating *T. ferrooxidans*

Ingredients	Quantity of ingredients (g liter^{-1})			
	9K[a]	T&K[b]	L[c]	T&C[d]
$(NH_4)SO_4$	3.0	0.4	0.15	0.5
KCl	0.1	–	–	–
K_2HPO_4	0.5	0.4	0.05	–
$MgSO_4 \cdot 7H_2O$	0.5	0.4	0.5	1.0
$Ca(NO_3)_2$	0.01	–	0.01	–
$FeSO_4 \cdot 7H_2O$	44.22	33.3	1.0	129.1

[a]9K medium of Silverman and Lundgren (1959a). $FeSO_4$ is dissolved in 300 ml of distilled water and filter-sterilized. The remaining salts are dissolved in 700 ml of distilled water and autoclaved after adjustment of the pH with 1 ml of 10 N H_2SO_4.

[b]Medium of Tuovinen and Kelly (1973). The salts are dissolved in 0.11 N H_2SO_4. The medium is sterilized by filtration.

[c]Medium of Leathen et al. (1956).

[d]Medium of Temple and Colmer (1951). The pH of this medium is adjusted to 2.0-2.5 with H_2SO_4. The medium is sterilized by filtration.

has been at best only partially successful. A more uniformly successful method employs certain membrane filters, especially the Sartorius or Millipore types, as a support for medium for the organism (Tuovinen and Kelly, 1973). Filter membranes carrying *T. ferrooxidans* cells on one surface are incubated on T&K medium solidified with 0.4% of Japanese agar at an initial pH of 1.55 and 20°C. Rust-colored colonies of 1.0-1.5 mm diameter develop after 2 weeks.

Heterotrophic Growth of T. ferrooxidans   Although *T. ferrooxidans* is generally considered to be a chemolithotroph (autotroph), a few strains have been adapted to grow heterotrophically on glucose as the exclusive energy and carbon source (Lundgren et al., 1964; Shafia and Wilkinson, 1969; Shafia et al., 1972; Tabita and Lundren, 1971a). Although the strain of Lundgren et al. (1964) could be reverted to an autotrophic mode of nutrition by reculturing in mineral salts medium with ferrous iron as the energy source, the strain of Shafia and co-workers could not.

Consortia with T. ferrooxidans   Recently, Zavarzin (1972), Guay and Silver (1975), and Mackintosh (1978) reported the presence of acidophilic satellite organisms unable to oxidize iron in a few cultures of *T. ferrooxidans*. Zavarzin obtained his satellite organism in Leathen's medium without $(NH_4)_2SO_4$ but with added yeast extract (0.01-0.02%) and adjusted to pH 3-4. This organism was morphologically distinct from *T. ferrooxidans*, being a single rod. It also differed in that it could not oxidize iron even though it required its presence at high concentration in the medium. It was acidophilic (optimum pH range 2-3). It required yeast extract but could not grow in an excess of it. It appeared to be able to use glucose as an energy source, and also citric acid, sucrose, fructose, ribose, maltose, xylose, succinic and fumaric acids, mannitol, and ethanol. The organism was originally found in peat from an acid bog. It seemed to resemble *Acetobacter acidophilum* but grew at lower pH and exhibited only weak acetic acid- and ethanol-forming ability.

The satellite culture of Guay and Silver (1975) was derived from *T. ferrooxidans* TM by subculturing in 9K medium with increasing amounts of glucose concentrations from 0.01 to 1.0% and concomitant decreases in Fe^{2+} from 9,000 ppm to 10 ppm in four steps. The satellite was isolated on 9K agar at pH 4.5. The culture thus obtained was named *Thiobacillus acidophilus*. It was a gram-negative, motile rod (0.5-0.8 μm $\times$ 1.0-1.5 μm) and thus morphologically not very distinct from *T. ferrooxidans*. Physiologically, however, it appeared to be a facultative autotroph incapable of oxidizing Fe^{2+}, thiosulfate, sulfite, sulfide, or metal sulfides, but able to oxidize elemental sulfur (S^0). It could also use D-ribose, D-xylose, L-arabinose, D-glucose, D-fructose, D-galactose, D-mannitol, sucrose, citrate, malate, dl-aspartate, and dl-glutamate as carbon and energy sources in 9K medium without Fe^{2+}. It could not use D-mannose, L-sorbose, L-rhamnose, ascorbic acid, lactose, D-maltose, cellobiose, trehalose, D-melibiose, raffinose, acetate, lactate, pyruvate, glyoxalate, fumarate, succinate, mandelate, cinnamate, phenylacetate, salicylate, phenol, benzoate, phenylalanine, tryptophane, tyrosine, or proline. The satellite organism of Mackintosh was merely described as unable to produce brown precipitate of iron, thus resembling *T. acidophilus*.

Like *T. ferrooxidans*, *T. acidophilus* is acidophilic (pH range 1.5-6.0; optimum pH 3.0). Its DNA has a GC ratio (62.9-63.2%) distinctly different from that of *T. ferrooxidans*, whose GC ratio if 56.1%. Differences in the concentration of some key enzymes, such as glucose 6-phosphate dehydrogenase, 6-phosphogluconate dehydrogenase, fructose 1,6-disphosphae aldolase, isocitric dehydrogenase, α-ketoglutarate dehydrogenase, NADH:acceptor oxidoreductase, thiosulfate oxidizing enzyme, and rhodanese in *T. ferrooxidans* and *T. acidophilus* were also noted. What is puzzling about *T. acidophilus* is that it was carried for years as a satellite of *T. ferrooxidans* in 9K medium without an organic supplement. Earlier studies (Schnaitman and Lundgren, 1965) had shown that *T. ferrooxidans* TM growing in 9K medium liberates some of the CO_2 that it fixes as pyruvate and as some other unknown substance(s). However, the maximum pyruvate concentration detected was only 12 μg per 100 ml. That of the other substances was somewhat greater. Since *T. acidophilus* cannot use pyruvate as a carbon and energy source, the only other possible carbon and energy source must be one or more of the unidentified substances released into 9K medium. It is a real question, however, whether the concentration of this material, less than 1 ppm, would be enough to satisfy the demands of *T. acidophilus*, even as a small population. This problem needs to be investigated further.

The Energetics of Ferrous Iron Oxidation   Oxidation of Fe^{2+} does not furnish much energy on a molar basis, for instance compared to glucose. In the past, estimates of the ΔF for iron oxidation have ranged around 10 kcal mol^{-1}, as calculated for example by Baas Becking and Parks (1927) from the equation

$$4FeCO_3 + O_2 + 6H_2O \rightarrow 4Fe(OH)_3 + 4CO_2 \quad (\Delta F_{298} = -40 \text{ kcal}) \quad (12.2)$$

A more recent examination of the question of the free-energy yield of iron oxidation by Lees et al. (1969), who assumed that the reaction proceeded as follows:

$$Fe^{2+} + H^+ + \frac{1}{4}O_2 \rightarrow Fe^{3+} + \frac{1}{2}H_2O \tag{12.3}$$

and who took into account effects of pH and ferric iron solubility upon the reaction, led to the following equation for calculating the molar free energy ΔF between pH 1.5 and 3:

$$\Delta F = -1.3(7.7 - pH - 0.17) \tag{12.4}$$

From this equation it may be calculated that the ΔF at pH 2.5 is -6.5 kcal mol^{-1}, barely enough for the synthesis of 1 mol of ATP (which requires about 7 kcal mol^{-1}). If we assume that it takes 120 kcal of energy to assimilate 1 mol of carbon at 100% efficiency (Silverman and Lundgren, 1959b), approximately 18.5 mol of Fe would have to be oxidized to incorporate this much carbon. However, *T. ferrooxidans* is not 100% efficient in using the energy available from iron oxidation. An early experimental estimate of the true efficiency of carbon assimilation at the expense of Fe^{2+} oxidation was 3.2% (Table 12.2) (Temple and Colmer, 1951). At that efficiency level, it would take 577 mol of Fe. Taking the most recent determination of efficiency (Silverman and Lundgren, 1959b) in Table 12.3, a consumption of 90.1 mol of Fe^{2+} to assimilate 1 mol of carbon would be predicted. This is greater than the consumption of 50 mol observed by Silverman and Lundgren (1959b) and raises the question as to whether the assumption that it takes 120 kcal in these cells to assimilate 1 mol of carbon is correct. However, this estimate of 90.1 mol of Fe^{2+} per mole of carbon approaches the results of another experiment (Beck, 1960), in which about 100 mol of Fe^{2+} had to be oxidized to fix 1 mol of CO_2 by a strain of

Table 12.3   Estimates of Free-Energy Efficiency of Carbon Assimilation
by Iron-Oxidizing Thiobacilli

Efficiency (%)	Age of cells	Reference
3.2	17 days	Temple and Colmer (1951)
30	?	Lyalikova (1958)
4.8-10.6	?	Beck and Elsden (1958)
20.5 ± 4.3	Late log phase	Silverman and Lundgren (1959b)

T. ferrooxidans. No matter what the actual efficiency of energy utilization, the observed molar ratios of Fe^{2+} oxidized to CO_2 assimilated illustrate that large amounts of iron have to be oxidized to satisfy the energy requirements for growth of these organisms.

Iron-Oxidizing Enzyme Systems   The enzymatic mechanism of iron oxidation has been examined. Results of a kinetic study with whole cells revealed an apparent K_m for iron oxidation of 5.4×10^{-3} M in an unbuffered system and 2.2×10^{-3} M in a system buffered by β-alanine-sulfate buffer.* Changes in pH as well as changes in SO_4^{2-} concentration affected the value for V_{max} but not for K_m. A requirement for SO_4^{2-} ion in iron oxidation had previously been demonstrated (Lazaroff, 1963). Even when the cells were chloride-adapted, chloride ion could not fully replace SO_4^{2-} ion (Lazaroff, 1963; see also Kamalov, 1967). In at least one instance, SO_4^{2-} could be partially replaced by HPO_4^{2-} or $HAsO_4^{2-}$ but not by BO_3^-, MoO_4^{2-}, NO_3^-, or Cl^- ions. Formate and MoO_4^{2-} ions inhibited iron oxidation (Schnaitman et al., 1969).

A search for the enzymatic components involved in Fe^{2+} oxidation has been undertaken by a number of investigators. The iron-oxidizing system appears to be constitutive and independent of the sulfur-oxidizing system in *T. ferrooxidans* (Margalith et al., 1966; Beck and Brown, 1968; Duncan et al., 1967). An early study of the Fe^{2+}-oxidizing enzyme system in cell-free extracts of a strain of *T. ferrooxidans* revealed the presence of cytochrome c and cytochrome a_1, and either traces or no cytochrome b (Vernon et al., 1960). These cytochromes were suspected to have been present originally in the cell membrane in the intact cells. Both cytochromes c and a_1 were reduced in intact cells in the

*K_m is defined by the Michaelis-Menton equation

$$v = \frac{V_{max} [S]}{[S] + K_m}$$

where v is the reaction velocity, V_{max} the maximal velocity, [S] the initial substrate concentration, and K_m is a constant (Segel, 1975).

presence of Fe^{2+}. Spectrophotometric study of another strain of *T. ferro-oxidans* revealed the involvement of cytochromes c and c_1 and cytochrome oxidase in Fe^{2+} oxidation (Tikhonova et al., 1967). In still another study (Blaylock and Nason, 1963), a strain of *T. ferrooxidans* was subjected to cell fractionation for examination of the components of its iron-oxidizing enzyme system. This study revealed the presence of an iron-cytochrome c reductase which was involved in the transfer of electrons from Fe^{2+} to cytochrome c. The electrons from cytochrome c were shown to be conveyed to oxygen via a cytochrome oxidase system involving cytochrome a. Although cytochrome b was present in this strain, it was not involved in Fe^{2+} oxidation. An iron-cytochrome c reductase from this organism was later purified and characterized (Din et al., 1967a). It had a pH optimum at pH 7.0. Its apparent K_m for cyto-chrome c was 9.5×10^{-5} M and for $FeSO_4 \cdot 7H_2O$, 1×10^{-3} M. The enzyme contained ribonucleic acid (RNA) and nonheme iron. Its activity was inhibited nonspecifically by p-chloromercuribenzoate, N-ethylmaleimide, and atabrine.

It has been suspected for some time that Fe^{2+} may have to be chelated to be oxidized because the E_h for the Fe^{2+}/Fe^{3+} couple is too high (+700 mV) for oxidation via the cytochrome respiratory chain. Chelation can have the effect of lowering the E_h of the Fe^{2+}/Fe^{3+} couple. According to Dugan and Lundgren (1965), sulfate may help to bind Fe^{2+} to the cell envelope as an iron-oxygen complex in preparation for the action of iron-cytochrome c reductase upon it. The binding site in the cell envelope may be phosphatidylserine (Agate and Vishniac, 1970). Coenzyme Q may be involved as a carrier in the enzymatic transfer of the electron from the iron to cytochrome c (Dugan and Lundgren, 1964). The reaction sequence for this mechanism of iron oxidation by *T. ferrooxidans* may be summarized in the following equation:

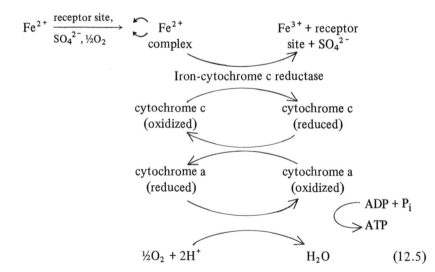

$$Fe^{2+} \xrightarrow[SO_4^{2-},\ \frac{1}{2}O_2]{\text{receptor site,}} \begin{array}{c} Fe^{2+} \\ \text{complex} \end{array} \qquad \begin{array}{c} Fe^{3+} + \text{receptor} \\ \text{site} + SO_4^{2-} \end{array}$$

Iron-cytochrome c reductase

cytochrome c   cytochrome c
(oxidized)   (reduced)

cytochrome a   cytochrome a
(reduced)   (oxidized)

$ADP + P_i$
ATP

$\frac{1}{2}O_2 + 2H^+$ H_2O (12.5)

The free energy from the oxidation of iron which is made available to the cell is trapped in high-energy phosphate bonds in oxidative phosphorylation of ADP with P_i to form ATP. One such phosphorylation site exists in the electron transport system for iron oxidation, as shown in the reaction sequence 12.5.

A further modifiction of this mechanism for iron oxidation by *T. ferrooxidans* has now been proposed (Ingledew et al., 1977). It incorporates the most up-to-date ideas of how ATP may be formed by oxidative phosphorylation (the chemiosmotic hypothesis of P. Mitchell; see Hinkle and McCarty, 1978). According to this scheme, ferrous iron is oxidized in the periplasmic space of the cell envelope by a copper protein, rusticyanin, which transfers its electrons to cytochrome c. This cytochrome shuttles between the periplasmic space and the cell membrane, where it passes its electrons to cytochrome a_1 on the inner surface of the plasma membrane. At the same time that electrons are passed inward through the membrane, protons and oxygen pass in the same direction to be combined into water by cytochrome a_1. The passage of the protons inward through the membrane results from a pH gradient (pH 2 outside, pH 5-6 inside) and occurs through channels associated specifically with adenosine triphosphatase (ATPase) on the inside of the membrane, enabling this enzyme to synthesize ATP from ADP and P_i (Fig. 12.4). Rusticyanin has been purified and

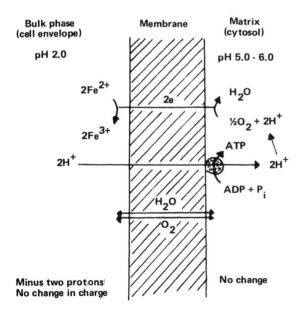

Figure 12.4 Scheme for oxidative phosphorylation, at the expense of the acid environment, in *Thiobacillus ferrooxidans*. The acid pH in the bulk phase during ion oxidation is maintained, at least in part, through the hydrolysis of the ferric iron produced. (Redrawn from Ingledew et al., 1977; by permission.)

shown to be capable of reacting with Fe^{2+} directly (Cox and Boxer, 1978). This scheme of ATP synthesis coupled to iron oxidation does not explain the requirement of sulfate by the organism, nor does it explain the significance of the iron-cytochrome c reductase of Din et al. (1967a). It does, however, present an acceptable explanation of the energy-coupling mechanism.

Iron as a Source of Reducing Power for CO_2 Assimilation   Not all electrons removed from Fe^{2+} in its oxidation are transferred to O_2 by the cell. A significant portion has to be used in the reduction of fixed CO_2. Such electrons are shunted from the Fe^{2+} via cytochrome c to nicotinamide adenine nucleotides (NAD and NADP) (Aleem et al., 1963). Since the cytochrome c couple has a much higher E_h (+254 mV) than does the NAD/NADH couple (E_h −320 mV) or the NADP/NADPH couple (E_h' −324 mV), energy in the form of ATP is required to transfer electrons from cytochrome c to the NAD or NADP. This process is called **reverse electron transport**. In one strain of *T. ferrooxidans* cytochrome c, cytochrome c_1, cytochrome b, and a flavin have been identified as participants in the reverse electron transport system (Tikhonova et al., 1967). Arsenate and amytal were inhibitory to this system.

The major mechanism of CO_2 assimilation in *T. ferrooxidans* involves the Calvin cycle (Din et al., 1967b; Gale and Beck, 1967; Maciag and Lundgren, 1964). A minor CO_2 fixation mechanism involving phosphoenolypyruvate carboxylase also exists in the organism (Din et al., 1967b). The latter enzyme system is needed for the formation of certain amino acids. In the Calvin cycle, CO_2 is fixed by ribulose 1,5-diphosphate obtained from ribulose 5-phosphate as follows (see also Chapter 5).

$$\text{Ribulose 5-phosphate + ATP} \xrightarrow{\text{phosphoribulokinase}}$$
$$\text{ribulose 1,5-diphosphate + ADP} \tag{12.6}$$

$$\text{Ribulose 1,5-diphosphate} + CO_2 \xrightarrow[\text{carboxylase}]{\text{ribulose biphosphate}}$$
$$2(\text{3-phosphoglycerate}) \tag{12.7}$$

Each 3-phosphoglycerate is then reduced to 3-phosphoglyceraldehyde:

$$\text{3-Phosphoglycerate + NADPH + H}^+ \text{ + ATP} \xrightarrow[\text{dehydrogenase}]{\text{3-phosphoglyceraldehyde}}$$
$$\text{3-phosphoglyceraldehyde + NADP}^+ \text{ + ADP + P}_i \tag{12.8}$$

The 3-phosphoglyceraldehyde is then converted by a series of steps to various cell constituents as well as to catalytic amounts of ribulose 5-phosphate.

Phosphoenolpyruvate carboxylase catalyzes the fixation of CO_2 by phosphoenolpyruvate, which is formed from 3-phosphoglycerate as follows:

$$3\text{-Phosphoglycerate} \xrightarrow{\text{phosphoglyceromutase}} 2\text{-phosphoglycerate} \qquad (12.9)$$

$$2\text{-Phosphoglycerate} \xrightarrow{\text{enolase}} \text{phosphoenolpyruvate} \qquad (12.10)$$

The phosphoenolpyruvate is then combined with CO_2:

$$\text{Phosphoenolpyruvate} + CO_2 \xrightarrow{\text{PEP carboxylase}} \text{oxalacetate} + P_i \qquad (12.11)$$

Whether a functional tricarboxylic acid (TCA) cycle exists in *T. ferrooxidans* when growing autotrophically on iron is presently uncertain. Tabita and Lundgren (1971b) found it only in glucose-grown cells.

Other Acidophilic Iron-Oxidizing Bacteria   Other acidophilic bacteria capable of oxidizing ferrous iron enzymatically have been discovered in recent years. They have not yet been examined in the detail that *T. ferrooxidans* has. Most or all of them are chemolithotrophic. One of these organisms is a nonspore-forming, thermophilic, acidophilic, nonmotile, pleomorphic organism (1-1.5 μm in diameter) (Brierley and Brierley, 1973; Brierley and Murr, 1973) (Fig. 12.5). It resembles *Sulfolobus* (Brock et al., 1972). The GC content of its DNA is 57 ± 3 mol%, compared to *Sulfolobus* with a GC content of 60-68 mol%. The cells of this organism lack a rigid cell wall. Instead, the cells are surrounded by a membrane that is covered by an amorphous layer. In laboratory media, the organism grows best at pH 3.0. Its temperature range for growth is between 45 and 70°C. It can survive exposure to 80°C for at least 2 hours. Nutritionally, it can satisfy its energy requirements by the oxidation of iron or sulfur. Cells grown on iron medium for 9 days cannot oxidize S^0, but cells grown on S^0 can oxidize iron, although not at the maximum rate. The addition of yeast extract to a concentration of 0.02% enhances its oxidation of iron and sulfur slightly. The organism can grow in the absence of added yeast extract, but its growth is much slower.

Sulfolobus acidocaldarius itself has now been shown to be able to oxidize Fe^{2+} at temperatures as high as 80-85°C but not at 90°C (Brock et al., 1976). This organism has been shown to oxidize Fe^{2+} in acid hot springs having a total iron concentration of 1-200 ppm, of which about 50% is ferrous.

Other thermophilic, acidophilic, gram-negative, iron-oxidizing bacteria have recently been reported (Brierley and LeRoux, 1977; Brierley and Lockwood, 1977; J. A. Brierley, 1978; Brierley et al., 1978). They are still incompletely characterized. At least one type oxidizes iron at 50 and 55°C. It needs small

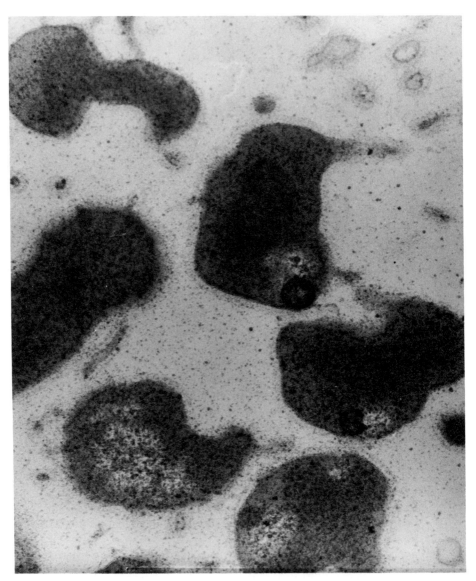

Figure 12.5 Thin sections of *Sulfolobus* sp. (28,760X). (Courtesy of J. A. Brierley and C. L. Brierley.)

amounts of yeast extract, cysteine, or glutathione for growth. It resembles other thiobacilli (J. A. Brierley, 1978). A new genus of sporeforming, thermophilic, acidophilic, iron- and sulfur-oxidizing bacteria, *Sulfobacillus thermosulfido-oxidans*, was recently described (Golovacheva and Karavaiko, 1978).

Another recently reported acidophilic, iron-oxidizing bacterium is *Leptospirillum ferrooxidans* (Markosyan, 1972; Balashova et al., 1974). This organism was isolated from a copper deposit in Armenia. It consists of a vibrioid cell with a polar flagellum, and has a diameter of about 25 nm. Involution cells may exhibit a spiral shape. Tori form at pH values less than 2. In the laboratory the organism grows best in Leten's (Leathen's) medium at pH 2-3 (Kuznetsov and Romanenko, 1963). The organism oxidizes iron for energy but cannot oxidize elemental sulfur and is incapable of growth in organic media. Its cells have been shown to contain an active ribulose biphosphate carboxylase, typical of chemolithotrophs (autotrophs), which use the Calvin cycle for CO_2 assimilation.

An acid-tolerant *Metallogenium* has been reported from mesoacidic, iron-bearing groundwaters (Walsh and Mitchell, 1972a). It is a filamentous organism consisting of branching filaments (0.1-0.4 μm in diameter and > 1 μm long), usually encrusted with iron. The organism tolerates a pH range from 3.5 to 6.8, with an optimum at pH 4.1. In the laboratory, the addition of 0.024 M phthalate at pH 4.1, but not of acetate, citrate, or phosphate, was important for observing iron oxidation and growth at initial Fe^{2+} concentrations greater than 100 mg liter^{-1}. The growth medium was constituted as follows: $(NH_4)_2SO_4$, 0.1%; KH_2PO_4, 0.001%; $FeSO_4 \cdot 7H_2O$, 9%; distilled water; pH adjustment to 4.1. It is not yet clear whether this organism is an autotroph or a heterotroph. Care has to be taken in identifying this organism. Some inorganic ferric iron precipitates may resemble *Metallogenium* morphologically (Ivarson and Sojak, 1978).

Bacterial Iron Oxidation at Neutral pH   Although convincing evidence has been developed for enzymatic iron oxidation by iron bacteria growing at acid pH, unequivocal evidence for such oxidation by iron bacteria which grow preferentially near netural pH is mostly lacking. The strongest evidence to date for enzymatic oxidation of iron by bacteria at neutral pH has been presented for the genus *Gallionella*, especially the species *Gallionella ferruginea* (Fig. 12.6). This organism was first described by Ehrenberg in 1836. As presented in *Bergey's Manual* (1974), it consists, when in mature form, of a bean-shaped cell with a lateral stalk consisting of twisted bundles of fibrils. The bean-shaped cell on the lateral stalk was first recognized as an integral part of the organism by Cholodny (1924). The stalks may branch and carry a bean-shaped cell at the tip of each branch. The stalk is usually anchored to a solid surface. It may be heavily encrusted with ferric hydroxide. The cells, which may form one or two polar flagella, may detach from their stalk, swim away as swarmers, and seek a new site for attachment and a stalked growth habit. *Gallionella* sp. is a so-called gradient organism (i.e., it grows best under low oxygen tension) [1 mg of O_2 per milliliter according to *Bergey's Manual* (1974); 0.1-0.2 mg of O_2 per milliliter according to Hanert (1968)] and over a pH range of 6-7. Its low oxygen requirement probably explains why this organism can catalyze Fe^{2+}

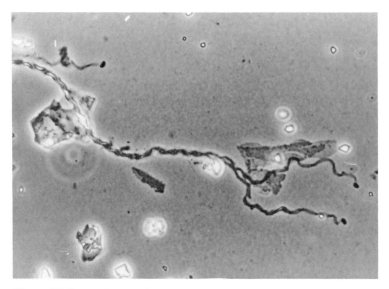

Figure 12.6 *Gallionella* (1280X). Note the small cell attached laterally at the tip of each of the twisted stalks. (Courtesy of P. Hirsch.)

oxidation effectively. The iron under these partially reduced conditions aut-oxidizes only slowly (Wolfe, 1964, p. 83).

A widely accepted method for the isolation and propagation of *Gallionella* is that of Kucera and Wolfe (1957). Their medium consists of NH_4Cl, 0.1%; K_2HPO_4, 0.05%; $MgSO_4$, 0.02%; and freshly prepared ferrous sulfide suspension making up 10% of the total volume of the medium. A small amount of CO_2 is bubbled through the salts solution prior to the addition of the ferrous sulfide. The medium is placed in test tubes which are stoppered to prevent loss of CO_2. A redox gradient is established in the culture medium as oxygen from the air diffuses into it. *Gallionella*, when growing on this medium, occupies the lower two-thirds of the volume above the ferrous sulfide. The organism will not grow anaerobically, whether nitrate is added or not. The temperature range for growth observed by Kucera and Wolfe (1957) was 12-30°C. Tap water or natural water had to be used in preparing media for subculture because distilled water apparently did not supply a required component.

Owing to the complex growth habit of *Gallionella*, determination of its growth rate becomes a problem. This has been solved for individual development in microculture by microscopically following stalk elongation and twisting (Hanert, 1974a). In other words, growth is followed by measuring increase in mass of the organism. An elongation rate in the first generation of 40-50 $\mu m\,hr^{-1}$

has been obtained. This is 2-4 times faster than in natural environments. Stalks do not elongate further after three or four divisions of the apical cell. Stalk lengthening occurs at the tip where the apical cell is attached. The stalk twists as it lengthens, owing to the rotation of the apical cell. The rotation occurs at a constant rate. In the natural environment, the problem of measuring the extent of development of *Gallionella* has also been solved by showing a dependence of growth rate on the rate of attachment to a solid surface such as a submerged microscope slide, and on the rate of stalk elongation, as shown in the relationship (Hanert, 1973)

$$V_t = \frac{b_{\bar{v}} \times l_{\bar{v}}}{2} \; t^2 \tag{12.12}$$

Here $b_{\bar{v}}$ is the average rate of attachment, $l_{\bar{v}}$ the average rate of stalk elongation, and t the length of the growth period, which should not be longer than 10 hr if this relationship is to hold. V_t is a measure of the amount of growth at time t.

The rate of iron oxidation by *Gallionella* in the natural environment may be measured by submerging a microscope slide at the site of *Gallionella* development for a desired length of time, then removing the slide and measuring the amount of iron deposited on it (Hanert, 1974b). Iron deposition can be expressed in terms of the amount of iron per unit surface area of the slide which was submerged and on which iron was laid down.

Gallionella is probably a chemolithotroph (autotroph). The strongest evidence in support of this notion is (1) the fact that the organism will grow in a mineral salts medium in the absence of a significant amount of organic carbon (Kucera and Wolfe, 1957), (2) that it will not grow without oxidizable iron in the medium (Hanert, 1968), and (3) that it assimilates significant quantities of $^{14}CO_2$ from $NaH^{14}CO_3$ added to the iron sulfide medium (Hanert, 1968). A quantitative demonstration of CO_2 uptake coupled to ferrous iron oxidation remains to be made, as does an examination for the presence of ribulose biphosphate carboxylase.

Another group of iron bacteria to which enzymatic iron oxidation above pH 6 has been attributed are certain of the sheathed bacteria from iron-containing freshwater environments (Fig. 12.7). The evidence for such enzymatic oxidation of iron is, however, up to now equivocal (Van Veen et al., 1978). The sheathed bacteria that deposit iron according to *Bergey's Manual* (1974) are listed in Table 12.4.

Of the organisms listed in Table 12.4, *Leptothrix* spp. sometimes classed with *Sphaerotilus* (Pringsheim, 1949; Stokes, 1954; Hoehnl, 1955), have been most directly examined for enzymatic iron oxidation. Winogradsky (1888) first reported that *L. ochracea* (probably *L. discophora*, according to Cholodny, 1926)

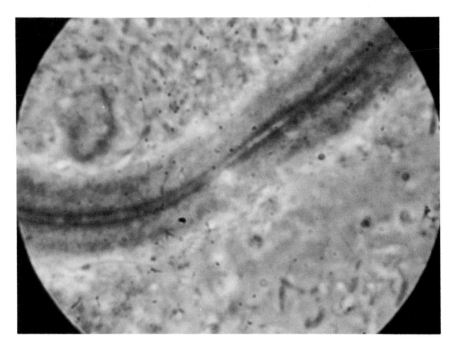

Figure 12.7   Portion of a sheathed iron bacterium (*Leptothrix*) (5,240X). (Courtesy of E. J. Arcuri.)

could oxidize iron. He found that he could grow the organism in hay infusion only if he added ferrous carbonate. The iron was oxidized and deposited in the sheath of the organism. He inferred from this observation that the organism was an autotroph. Molisch (1910) and Pringhseim (1949) disagreed with Winogradsky's conclusion about iron oxidation by *L. ochracea,* believing that the organism merely deposited autoxidized iron in its sheath. However, Lieske (1919) confirmed Winogradsky's earlier observtions of growth on ferrous carbonate in very dilute organic solution and suggested that the organism might be mixotrophic. Cholodny (1926), Sartory and Meyer (1947), and Praeve (1957) also made observations similar to those of Winogradsky and Lieske. So far, any claim for enzymatic iron oxidation by *Leptothrix* rests mainly on the requirement for growth of ferrous iron in dilute medium and on the oxidation of this ferrous iron during growth. However, Praeve (1957) also showed a stimulation of oxygen uptake by the organism when in the presence of Fe^{2+} as the only exogenous, oxidizable substrate in Warburg experiments. Significantly, he found empty sheaths unable to take up oxygen on Fe^{2+}. Dubinina (1978a,b) recently reported that *L. pseudoochracea*, as well as *Metallogenium* and *Arthrobacter siderocapsulatus*, oxidized Fe^{2+} with metabolically produced H_2O_2 through catalysis by catalase of the organism. It may be that different mechanisms of enzymatic iron oxidation exist.

Table 12.4   Iron-Depositing Sheathed Bacteria

Organism	Distinguishing features
Sphaerotilus natans	Rarely deposits iron oxides and never manganese oxides on sheath. Exhibits rod-shaped cells in a sheath attached to a surface by a holdfast. Rods may become subpolarly flagellated swarmers and leave the sheath. Sheath formation is more pronounced in dilute than in rich media. Van Veen et al. (1978) do not regard this organism as an iron bacterium.
Leptothrix spp.	Deposit iron and/or manganese oxides on their sheath. Sheaths are shorter and more brittle than those of *S. natans*. Cells may become flagellated swarmers and leave sheath, an occurrence that is more frequent than with *S. natans*.
Crenothrix polyspora	This organism forms filaments of up to 1 cm long, which are attached to surfaces. The filaments may be swollen at the free end. Sheaths are thin and may become encrusted with iron or manganese oxides at their base. Cells vary from cylindrical to disk-shaped. They divide by forming cross septa. At the tip cells may also divide by longitudinal septation. Cells may leave the sheath, but no motile swarmers are formed.
Clonothrix sp.	This organism forms filaments of up to 1.5 cm long. In one species they may taper toward the tip. Filaments may be attached to a surface or be free. They have a distinct sheath that may be encrusted with iron or manganese oxides. Cells in a sheath are cylindrical. On reproduction, they separate and leave from a broken sheath. The organism has not been cultivated in artificial media.
Lieskeella bifida	Filaments of rod-shaped cells with rounded ends. Usually, two filaments are wound around each other surrounded by a common slime capsule (sheath?) which may bear granulated deposits of iron oxide. The organism exhibits creeping motility. It has not been cultivated.

Source: Based on descriptions in *Bergey's Manual* (1974) and Van Veen et al. (1978).

For all other sheathed bacteria listed in Table 12.4, enzymatic iron oxidation is at most presumptive, based on gross morphological similarities with *Leptothrix* and the observation of oxidized iron deposits on their sheaths (capsules). It is quite possible that the organisms merely deposit oxidized iron on their sheaths (see below).

Recently, a wall-less bacterium, *Mycoplasma laidlawii* (now known as *Acholeplasma laidlawii*), has been reported to oxidize iron (Balashova and Zavarzin, 1972). The organism was cultured in a salt-free meat-peptone medium containing iron wire or powder. Ferric iron was formed during active growth and, in part, precipitated on the cells of the organism. Catalase was found to depress ferric oxide production, suggesting that H_2O_2 played a role in the oxidation process. It is interesting that in this instance, catalase did not accelerate iron oxidation as found by Dubinina (1978a). It is not clear from the report whether other enzymes played a direct role in the oxidation of iron by this organism.

Nonenzymatic Iron Oxidation Many different kinds of microorganisms can promote iron oxidation indirectly (i.e., nonenzymatically). They accomplish this generally by affecting the oxidation-reduction potential in the environment and/or the pH, thus favoring oxidation by oxygen in the air (autoxidation) or some other chemical. Among the first to recognize this phenomenon were Harder (1919), Winogradsky (1922), and Cholodny (1926). Starkey and Halvorson (1927) demonstrated the indirect oxidation of iron in laboratory experiments with bacteria and explained their findings in terms of reaction kinetics involving the oxidation of ferrous iron and the solubilization of ferric iron by acid. From their work it can be inferred that any organism which raises the pH of a medium by forming ammonia from protein or protein-derived material (equation 12.3) or by consuming salts of organic acids (equation 12.14) can promote ferrous iron oxidation in an aerated medium:

$$RCHCOOH + H_2O + \frac{1}{2}O_2 \longrightarrow RCCOOH + NH_4^+ + OH^- \qquad (12.13)$$
$$\underset{NH_2}{|} \qquad\qquad\qquad \underset{O}{\|}$$

amino acid   ketoacid

$$C_3O_3H_5^- + 3O_2 \longrightarrow 3CO_2 + 2H_2O + OH^- \qquad (12.14)$$

lactate

A more specialized case of indirect microbial iron oxidation is that associated with algal photosynthesis. This photosynthetic process may promote ferrous iron oxidation in two ways: (1) by generating oxygen and (2) by raising the pH of the waters in which the algae grow. The rise in pH is explained by equations 12.15 and 12.16:

$$2HCO_3^- \longrightarrow CO_3^{2-} + CO_2 + H_2O \qquad (12.15)$$

$$CO_3^{2-} + H_2O \longrightarrow HCO_3^- + OH^- \qquad (12.16)$$

Reaction 12.15 is promoted by CO_2 assimilation in photosynthesis, which may be summarized as follows:

$$CO_2 + H_2O \rightarrow CH_2O + O_2 \qquad (12.17)$$

Reaction 12.17 also explains the source of the oxygen. Its genesis causes a rise in E_h because of increased saturation or even supersaturation of the water. This will initiate or accelerate the oxidation of ferrous iron by oxygen in water.

Ferrous iron may be protected from chemical oxidation at elevated pH and E_h by chelation with oxalate, citrate, humic acids, and tannins. In that instance, bacterial breakdown of the ligand will free the ferrous iron, which then oxidizes spontaneously to ferric iron. This has been demonstrated in the laboratory with a *Pseudomonas* and a *Bacillus* strain (Kullmann and Schweisfurth, 1978). These cultures do not derive any energy from iron oxidation but rather from the oxidation of the ligand.

The production of ferric iron from the oxidation of ferrous iron at pH values above 5 usually leads to precipitation of the iron. But the presence of chelating agents such as humic substances, citrate, and the like can prevent the precipitation. Unchelated ferric iron tends to hydrolyze at higher pH values and may form ferric hydroxide.

$$Fe^{3+} + 3H_2O \rightarrow Fe(OH)_3 + 3H^+ \qquad (12.18)$$

Ferric hydroxide is relatively insoluble and will settle out of suspension. It may crystallize and dehydrate, forming limonite ($Fe_2O_3 \cdot nH_2O$), goethite ($Fe_2O_3 \cdot H_2O$), or hematite (Fe_2O_3).

Ferric Iron Precipitation by Microbes   Ferric iron may also be removed from solution by adsorption to surfaces of cells or to inanimate matter. The incrustation by ferric hydroxide or ferric oxides of many bacteria such as *Gallionella* spp., sheathed bacteria, *Siderocapsa* (Fig. 12.8) (now identified as *Arthrobacter sidercapsulatus*) (Dubinina and Zhadnov, 1975), *Naumanniella, Ochrobium, Siderococcus, Pedomicrobium* (Fig. 13.8), *Herpetosyphon, Seliberia* (Fig. 12.9), *Toxothrix* (Krul et al., 1970), *Archangium*, and protozoans such as *Anthophysa, Siderodendron, Bikosoeca*, and *Siphonomonas* can be attributed to this phenomenon. Recently, iron incrustation of *Acinetobacter* has been observed (MacRae and Celo, 1975). This organism caused a precipitation of colloidal iron at pH 6 and 7.6. The precipitated iron seemed to block respiration by the cells.

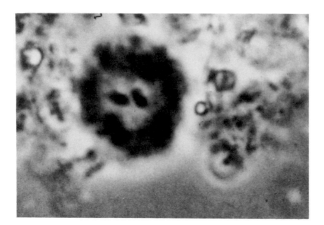

Figure 12.8 *Siderocapsa geminata* Skuja (1956) (7,000X). Specimen is from filtered water from Pluss See, Schleswig-Holstein, West Germany. Note the capsule surrounding the pair of bacterial cells. (Courtesy of W. C. Ghiorse and W.-D. Schmidt.)

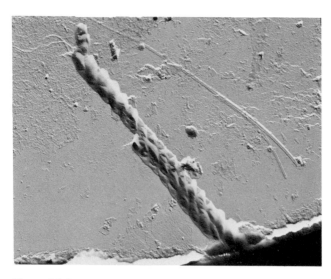

Figure 12.9 *Seliberia* sp. from forest pond neuston (8,200X). Courtesy of W. C. Ghiorse.)

Ferric ion precipitation on microbes is not dependent on direct or indirect oxidation of ferrous iron, although it may be coupled to it as in the case of *Gallionella* and *Leptothrix* and some budding bacteria (Hirsch, 1968). Precipitation of ferric iron may in some cases be associated with the biological destruction of the ligand of a ferric iron complex, especially if organic. The intact ligand keeps ferric iron in solution. *Pedomicrobium, Naumanniella, Siderocapsa,* and *Seliberia* have been especially implicated in ligand destruction of iron chelates. *Pedomicrobium* may also be able to oxidize ferrous iron itself after destroying its ligand (Aristovskaya and Zavarzin, 1971). Except possibly for *Pedomicrobium*, these organisms cannot be claimed to be iron-oxidizing bacteria.

12.2 FERRIC IRON REDUCTION

Occurrence of Ferric Iron Reduction Among Bacteria   Ferric iron in nature may be microbiologically reduced to ferrous iron. As in the case of iron oxidation, this reduction may be enzymatic or nonenzymatic. Ferric iron has been found to influence fermentative metabolism of bacteria as a result of its ability to act as terminal electron acceptor. Thus, Roberts (1947) showed a change in fermentation balance when comparing the action of *Bacillus polymyxa* on glucose anaerobically in the presence and absence of iron (Table 12.5). The ferric iron in these experiments was supplied as freshly precipitated, dialyzed ferric hydroxide suspension obtained in a reaction of ferric chloride and an excess of potassium hydroxide. The suspension had a pH of 7.8. It seemed to act as an alternative electron acceptor in the fermentation and in this way changed the quantities of certain products formed from glucose. Thus, in the presence of iron, less H_2 and CO_2, more ethanol, and less 2,3-butylene glycol were formed in either organic or synthetic medium than in the absence of iron. Also, more glucose was fermented in the presence of iron than in its absence in either medium.

Bromfield (1954a) showed that besides *B. polymyxa*, growing cultures of *B. circulans* can also reduce ferric iron. Depending on the medium, even *Escherichia freundii, Aerobacter* sp., and *Paracolobactrum* could do so. But he inferred from his results that the reduction of iron was not directly involved in the oxidation of the substrate, which is, however, at variance with his results with resting cells (Bromfield, 1954b). He found that completely anaerobic conditions were not required to obtain ferric iron reduction by the bacteria. But as the degree of aeration of the cultures was increased, ferrous iron became reoxidized due to autoxidation.

Bromfield (1954b) also showed that washed cells of *B. circulans, B. megaterium,* and *A. aerogenes* reduced the ferric iron of several ferric compounds

Table 12.5   Fermentation Balances for *Bacillus polymyxa* Growing in Two
Different Media in the Presence and Absence of Ferric Hydroxide[a]

Products	Synthetic medium[b] (mol/100 mol glucose)		Organic medium[c] (mol/100 mol glucose)	
	$- Fe(OH)_3$	$+ Fe(OH)_3$	$- Fe(OH)_3$	$+ Fe(OH)_3$
CO_2	199	170	186	178
H_2	51	31	53	33
HCOOH	11	12	9	12
Lactic acid	17	19	14	7
Ethanol	72	82	78	94
Acetoin	0.5	1	1	2
2,3-Butylene glycol	64	51	49	44
Acetic acid	0	0	0	0
Iron reduced	–	42	–	1
Glucose fermented (mg/100 ml)	1,029	2,333	1,334	2,380
C recovery (%)	112.1	101.8	98.8	97.2
O/R index	1.06	1.0	1.06	1.03

[a] Incubation was for 7 days at $35°C$.

[b] Glucose, 2.4%; asparagine, 0.5%; K_2HPO_4, 0.08%; KH_2PO_4, 0.02%; KCl, 0.02%; $MgSO_4 \cdot 7H_2O$, 0.5%.

[c] Glucose, 2.4%; peptone, 1%; K_2HPO_4, 0.08%; KH_2PO_4, 0.02%; KCl, 0.02%; $MgSO_4 \cdot 7H_2O$, 0.5%.

Source: Roberts (1947); by permission.

[$FeCl_3$, $Fe(OH)_3$, $Fe(lactate)_3$] in the presence of such suitable hydrogen donors as glucose, succinate, and malate. He was able to inhibit the reduction by boiling the cells, or by adding chloroform or toluene to the reaction mixture, but he did not observe inhibition with either azide or cyanide. He interpreted the ferric iron reduction as being associated with dehydrogenase activity. He felt, however, that the reduction of insoluble ferric iron (e.g., ferric hydroxide) could only have occurred in the presence of a complexing agent. From more recent studies, it is clear that although some complexing agents may speed up the rate of iron reduction, as α, α-dipyridyl did in his experiment, it is not essential (see later discussions in this chapter).

Cultural Indication of Enzymatic Ferric Iron Reduction   Bromfield's experimental observations have been confirmed and extended in a study of bacterial isolates from sediment from several lakes in the Karelian peninsula in the U.S.S.R. (Troshanov, 1968, 1969). *B. circulans* from these lake sediments was the most active among the isolates, but iron-reducing activity was also associated with *Pseudomonas liquefaciens, Bacillus mesentericus, B. cereus, B. centrosporus,* and *Micrococcus* spp. *B. mycoides* and *B. polymyxa* were only slightly active. All the reducers of iron in this study also reduced oxides of manganese, but the reverse was not true. The effect of oxygen on iron reduction depended on the culture. Some organisms, such as *B. circulans,* reduced iron more readily anaerobically; others, including *B. polymyxa,* did not. Insoluble iron [Fe(OH)$_3$] of bog ore was reduced more slowly than was soluble iron (FeCl$_3$), but an active organism such as *B. circulans* still reduced insoluble iron to a significant degree. *B. polymyxa* caused extensive reduction of FeCl$_3$ in laboratory experiments, but the same organism was inactive on Fe(OH)$_3$ or bog ore. Nitrate added to the medium inhibited ferric iron reduction. A physiological explanation for this inhibitory effect of nitrate is offered by the work of Ottow (1968, 1969a, 1970a) and Hammann and Ottow (1974). These investigators showed that ferric iron reduction by bacteria such as Enterobacteriaceae, Bacillaceae, *Pseudomonas,* and *Micrococcus,* which possess nitrate reductase A, is inhibited in the presence of nitrate and chlorate. This inhibitory effect is competitive. Since chlorate as well as nitrate had previously been shown to be reducible in the presence of nitrate reductase A (Pichinoty, 1963), it is inferred that nitrate reductase A can also catalyze ferric iron reduction. Appropriate amounts of nitrate in the medium thus prevent ferric iron reduction by these organisms by displacing the iron from the enzyme.

Ottow also found some bacteria that reduce ferric iron but lack nitrate reductase. These include *B. pumilus, B. sphaericus, Clostridium saccharobutyricum,* and *C. butyricum.* Iron reduction by them is not inhibited by nitrate (Fig. 12.10) or chlorate. Similarly, ferric iron reduction by mutants lacking nitrate reductase (Nit$^-$), isolated from wild-type strains possessing this enzyme (Nit$^+$), is insensitive to inhibition by nitrate or chlorate (Ottow, 1970a). Most of these Nit$^-$ mutants reduced iron less rapidly than did their wild-type parent, but a Nit$^-$ mutant of *B. polymyxa* reduced iron more intensely than its wild-type parent. These findings lead to the conclusion that two ferric-iron-reducing enzyme systems exist, one that seems to involve nitrate reductase A, and the other, as yet unidentified, that is incapable of reducing nitrate (Ottow, 1970a). This conclusion is supported by the observation that the size of the population of the soil microflora capable of reducing iron is usually greater than the size of

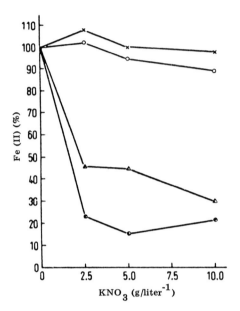

Figure 12.10   Effect of KNO_3 concentration on dissolved Fe(II) production from Fe_2O_3 reduced by two Nit⁻ clostridia and two Nit⁺ strains of *Bacillus polymyxa*. Fe(II) is reported in percent of nitrate-free controls. Incubation at each nitrate concentration was for 6 days at 37°C under anaerobic conditions. X——X, *C. saccharobutyricum* S_{10}; ○——○, *C. butyricum*; ▲——▲, *B. polymyxa* S_{29} (aerogenic); ●——●, *B. polymyxa* S_{471} (anaerogenic). (From Hammann and Ottow, 1974; with permission from Verlag Chemie International Inc.)

the population capable of reducing nitrate (Ottow and Ottow, 1970). Interestingly, it was recently reported that a branched pathway exists in *Staphylococcus aureus* through which nitrate is reduced by a cytochrome b-requiring branch and ferric iron is reduced by a branch that precedes cytochrome b. Nitrate in this strain of *S. aureus* inhibits ferric *iron* reduction because it presumably draws on available electrons more easily than ferric iron (Lascelles and Burke, 1978).*

*Recent studies of microbial iron uptake by means of siderophores suggest that it will be necessary in the future to distinguish between assimilatory and dissimilatory ferric iron reduction, as is done now in the cases of nitrate and sulfate reduction. In assimilatory ferric iron reduction, ferric iron is reduced by the cell for the purpose of intracellular release from siderophores (Brown and Ratledge, 1975; Cox, 1980; Ernst and Winkelmann, 1977; Tait, 1975) and involves only such quantities of iron as are needed for the synthesis of iron prophyrins, Fe-S proteins, and so on. In dissimilatory ferric iron reduction, iron is reduced at the membrane level as a form of anaerobic respiration and involves relatively large quantities of iron.

Among autotrophs, *Thiobacillus thiooxidans, T. ferrooxidans,* and *Sulfolobus acidocaldarius* can reduce ferric iron with S^0 as electron donor (Brock and Gustafson, 1976). *T. thiooxidans* can bring about this reduction aerobically. *T. ferrooxidans* forms Fe^{2+} only anaerobically because aerobically it can reoxidize it to ferric iron. *S. acidocaldarius* can accumulate Fe^{2+} microaerophilically (at 70°C) because under this limited oxygen availability it does not reoxidize Fe^{2+}.

Fungal Reduction of Ferric Iron   Some fungi also seem to have a capacity for reducing ferric iron. Lieske reported such a phenomenon (see Starkey and Halvorson, 1927). Ottow and von Klopotek (1969) observed hematite (Fe_2O_3) reduction by *Alternaria tenuis, Fusarium oxysporum,* and *F. solani,* all of which possessed an inducible nitrate reductase. Fungi incapable of reducing nitrate were also incapable of reducing iron. Nitrate was found to inhibit ferric iron reduction by the active fungi. It was, therefore, concluded that nitrate reductase functions also in the fungi in the reduction of ferric iron, a mechanism thus similar to that in many bacteria. It raises a question, however, as to where in the fungi the nitrate reductase is located to be able to act on hematite.

Types of Ferric Compounds Reduced Bacterially   The ease with which ferric iron is reduced by microbes depends in part on the form in which it occurs. In one study (Ottow, 1969a), the following order of decreasing reducibility was found: $FePO_4 \cdot 4H_2O > Fe(OH)_3 >$ lepidocrocite (γ-FeOOH) $>$ goethite (α-FeOOH) $>$ hematite (α-Fe_2O_3). In another study (De Castro and Ehrlich, 1970), marine *Bacillus* 29A, using glucose as the electron donor, was found to solubilize larger amounts of Fe from limonite and goethite than from hematite. In initial stages of the reduction, the order of decreasing activity of the three forms of ferric iron was goethite $>$ limonite $>$ hematite. The organism did not reduce ferric iron significantly when it occurred in ferromanganese nodules from the deep sea. Iron-reducing activity was also demonstrated with a cell extract of *Bacillus* 29A.

Nonenzymatic Reduction of Ferric Iron by Microbes   Starkey and Halvorson (1927) tried to explain ferric iron reduction in nature as an indirect effect of microbes in their environment. They argued that by causing a drop in pH and a lowering of oxygen tension, ferric iron would be changed to ferrous iron according to the relationship

$$4Fe^{2+} + O_2 + 10H_2O \rightleftarrows 4Fe(OH)_3 + 8H^+ \qquad (12.19)$$

in which Fe^{3+} was considered to be an insoluble phase [$Fe(OH)_3$]. From this equation they derived the relationship

$$Fe^{2+} = K \frac{[H^+]^2}{[O_2]^{\frac{1}{4}}} \qquad (12.20)$$

Others (Ottow, 1969a, 1970a; Hammann and Ottow, 1974; Munch and Ottow, 1977) have argued convincingly that this mode of ferric iron reduction by microbes is not significant in nature. Unlike Starkey and Halvorson (1927), they could not establish a strict correlation between iron solubilization and a drop in pH. It must be said, however, that although enzymatic reduction of ferric iron in nature seems to be a prominent phenomenon, indirect mechanisms of ferric iron reduction cannot be ignored. Chemical reductants may be produced by bacteria that can reduce ferric iron. For instance, H_2S produced by sulfate-reducing bacteria may reduce ferric to ferrous iron before precipitating ferrous sulfide (Berner, 1962).

$$\underset{\text{goethite}}{2HFeO_2} + 3H_2S \xrightarrow{\text{(at pH 7-9)}} 2FeS + S^0 + 4H_2O \qquad (12.21)$$

$$2HFeO_2 + 3H_2S \xrightarrow{\text{(at pH 4)}} FeS + FeS_2 + 4H_2O \qquad (12.22)$$

Another example is the reduction of ferric iron by formate, a metabolic product of a number of bacteria (e.g., *Escherichia coli*):

$$2Fe^{3+} + HCOOH \rightarrow 2Fe^{2+} + 2H^+ + CO_2 \qquad (12.23)$$

12.3   MICROBIAL IRON TRANSFORMATIONS IN NATURE

The Iron Cycle   Microbial transformations of iron play an important role in cycling of iron in nature (Fig. 12.11). Iron is introduced into the cycle through the weathering of iron-containing minerals in rocks, soils, and sediments. This weathering action is partly promoted by bacterial action and partly by chemical activity (Bloomfield, 1953a,b). The microbial action involves in many cases the interaction of the minerals with metabolic end products (Berthelin and Kogblevi, 1972, 1974; Berthelin and Dommergues, 1972; Berthelin et al., 1974). The liberated iron, if ferrous, may be biologically or nonbiologically oxidized to ferric iron under partially or fully aerobic conditions. The oxidation may be immediately followed by precipitation of the iron as a hydroxide, oxide, phosphate, or sulfate. If complexing agents such as humic substances abound, the ferric iron may be converted to soluble complexes and be dispersed from its site of formation. In podzolic soils (spodosols) of temperate climates, this complexed iron may be transported from the upper A horizons to the B horizons. In hot, humid climates, ferric iron is likely to precipitate at the site of its release from iron-containing soil mineral, owing to intense microbial action which rapidly and fairly completely oxidizes available organic matter, thus preventing

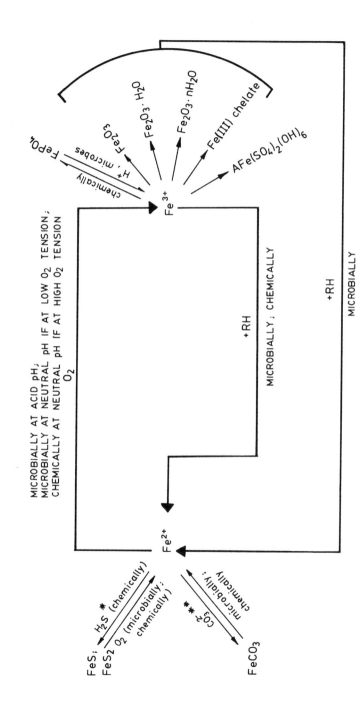

Figure 12.11 The iron cycle. *, often of biogenic origin; **, may be of biogenic origin; RH, electron donor.

in many cases extensive formation of soluble organic ferric complexes. The iron precipitates tend to cement soil particles together in a process known as **lateriza-tion**. Aluminum hydroxide liberated in the weathering process may also be precipitated and contribute to laterization (Brooks and Kaplan, 1972, p. 75; Merkle, 1955, p. 204).

Under anaerobic conditions, a process known as **gleying** may occur in water-logged soils. In this process, the affected soil becomes sticky and takes on a gray or light greenish-blue coloration (Alexander, 1977, p. 377). Although once associated with H_2S production, this is no longer thought to be the primary cause of gleying (Bloomfield, 1950). Waterlogged soil has been observed to bleach before sulfate reduction becomes active (Takai and Kamura, 1966). Bloomfield (1951) suggested that gleying was at least in part due to plant degradation products, although he had earlier demonstrated gleying under artificial conditions employing a sugar medium that was allowed to ferment. Ottow (1969b, 1970b, 1971) favors a microbial reduction of iron in which ferric iron is reduced by acting as an electron acceptor. He summarized the chemical transformation as follows (Ottow, 1971):

$$\text{Energy source} \xrightarrow[\text{dehydrogenase}]{\text{substrate}} e^- + H^+ + ATP + \text{end products} \qquad (12.24)$$

$$Fe(OH)_3 + 3H^+ + e^- \longrightarrow Fe^{2+} + 3H_2O \qquad (12.25)$$

$$2Fe(OH)_3 + Fe^{2+} + 2OH^- \longrightarrow Fe_3(OH)_8 \qquad (12.26)$$
$$\text{gray-green mixed}$$
$$\text{oxide of gley}$$

Formation of Ochre and Other Types of Sedimentary Iron Ores Intense forma-tion of ochre, an iron oxide, in drainage pipes and channels from hydromorphic (waterlogged) soils (e.g., marshes) has been observed in various parts of the world (Hanert, 1974b; Spencer et al., 1963; Ivarson and Sojak, 1978). The participation of bacteria in the precipitation of the iron (ochre formation) has been demonstrated by the submerged slide technique pioneered by Cholodny (Hanert, 1974b). Iron-depositing bacteria attach to the slides and can be seen by phase-contrast microscopy. According to one viewpoint, the primary ferric iron precipitation appears to be caused by *Gallionella, Leptothrix, Toxothrix,* and unicellular, iron-precipitating organisms and is the most important aspect of ochre formation (Hanert, 1974b). Secondary iron precipitation in this case may result from iron precipitation on bacterially secreted slimes, which seem to have a special affinity for iron. Some chemical iron precipitation may also occur but is usually of minor importance in ochre formation. *Gallionella fer-ruginea* has been found to be especially active in cases of very intense ochre formation (Hanert, 1974b). The source of the iron that is precipitated in the drainage pipes and channels is the waterlogged soil where microbial activity mobilizes iron through reduction of ferric iron to ferrous iron with appropriate organic compounds derived mostly from decaying vegetation on and in the soil. According to another viewpoint (Ivarson and Sojak, 1978) both biotic and abiotic processes contribute to ochre formation, but the degree to which each process contributes is difficult to assess at this time.

The above type of ochre formation may well be a model for many cases of sedimentary iron ore formation, including bog iron ore and lake iron ore. Bog iron ore consists chiefly of an earthy mixture of ferric hydroxides (limonite) having a yellow to red or dark brown color and may be associated with peat (Pettijohn, 1949). Lake iron ores often appear in the form of flat disks or concretions of ferric hydroxide or as a layer of bedded limonite with a soft or hard porous texture and yellowish color (Pettijohn, 1949). Iron ore may also consist of iron oxides cementing together the particles of sandy sediment. Harder (1919) attributed many of these deposits to the action of iron-oxidizing, iron-precipitating, and iron-complex-destroying bacteria.

Crerar et al. (1979) attribute bog iron deposits in the southern New Jersey coastal plain, the Pine Barrens, to the biogeochemical action of *Thiobacillus ferrooxidans, Leptothrix ochracea, Crenothrix polyspora, Siderocapsa geminata,* and an iron-oxidizing *Metallogenium* sp. (see Walsh and Mitchell, 1972a). The iron is oxidized and precipitated from acid surface waters, which exhibit a pH range of 4.3-4.5 in summer, by the iron-oxidizing bacteria and accumulates as limonite impregnating sands and silts. The process is going on at the present time. The source of the iron is glauconite and to a much lesser extent pyrite in underlying sedimentary formations, from which it is released into groundwater— whether with microbial help has not been ascertained as yet. The iron is brought to the surface by groundwater base flow which feeds local streams. Biocatalysis of iron oxidation is considered to be essential to account for the rapid rate of iron oxide formation in the acid waters. However, *T. ferrooxidans* is probably least important because it is only infrequently encountered, perhaps because the environmental pH is above its optimum.

Another case of contemporary biogenic iron ore deposition is that occurring in a bay of the Greek volcanic island Palaea Kameni in the Aegean Sea (Puchelt et al., 1973). Here in situ observations demonstrated that much of all of the iron deposition is associated with the growth of *Gallionella.* The source of the iron attacked by the organism appears to be exhalative ferrous carbonate.

In considering the origin of ancient iron ore deposits, it has been proposed that a number were biogenically formed. Gruner (1922) expressed the opinion that the iron ores of the Biwabik Formation of the Mesabi Range in Minnesota were of biogenic origin. According to him, weathering processes mobilized iron from igneous and sedimentary rocks. The iron then was picked up by waters rich in organic matter and carried to sites of deposition. At these sites, "iron bacteria" and algae caused precipitation of the iron by mechanisms previously discussed in this chapter. Gruner actually found remnants of algae and bacteria in chert from the Biwabik Formation. If correctly interpreted, the localized precipitation of the iron by microorganisms must be attributed to specially favorable environmental conditions which encouraged "blooms" of the responsible organisms.

Origin of the Banded Iron Formations   The Banded Iron Formations (BIF) from the geologic past, such as are found today, for instance, in the Animiki Formation of the Lake Superior region, which contains iron-rich and iron-poor layers of silica (chert), may have resulted from the buildup of ferrous iron in the surrounding environments and subsequent oxidation and precipitation by iron bacteria and cyanobacteria (Fig. 12.12) (Cloud, 1973, 1974). The geological age of these formations generally falls in the range 2×10^9 to 3.3×10^9 years. Some of the iron in the iron-rich bands of the BIF occurs as magnetite (Fe_3O_4). Cloud has postulated that the source of this iron may have been volcanic in the oldest of these deposits (Archean), and that some may have been brought to the site of deposition by upwelling of iron-bearing, deep, stagnant waters, or by leaching of adjacent low-lying deposits in the case of the younger deposits (Proterophytic). The solubilized iron remained in the ferrous state because the environment was still a reducing one (i.e., the atmosphere did not yet contain free oxygen). Atmospheric oxygen would not appear until the end of the BIF period. Although Cloud did not implicate bacteria in the mobilization of the iron, such action is not inconceivable in the light of what we know about the capabilities of modern bacteria. Cloud (1974) did suggest that silica may have been mobilized by bacteria or cyanobacteria through their solubilizing action on siliceous rocks (see Chapter 8). Owing to an absence of silica-depositing organisms (diatoms, silicoflagellates, and radiolarians—all eukaryotes—had not yet evolved), supersaturated silica solutions may have resulted which favored silica precipitation. The banding of the BIF is attributed by Cloud to episodic microbial growth and activity, or to a fluctuating ferrous iron supply. The iron-rich bands are pictured by him to be the result of reaction of ferrous iron with oxygen evolved by the first primitive oxygen-producing photosynthetic organisms (cyanobacteria?) growing in these ferrous iron-containing waters. Since oxygen is extremely toxic to unprotected cells (see Chapter 2), the ferrous iron thus served to protect the oxygen producers as well as other forms of life, none of which either produced or consumed oxygen. Not until the evolution of aerobic respiratory systems and of peroxydismutases could life exist in an oxygen-containing atmosphere. Oxygen-respiring organisms probably first began to evolve with the introduction of oxygen into the atmosphere later during this time.

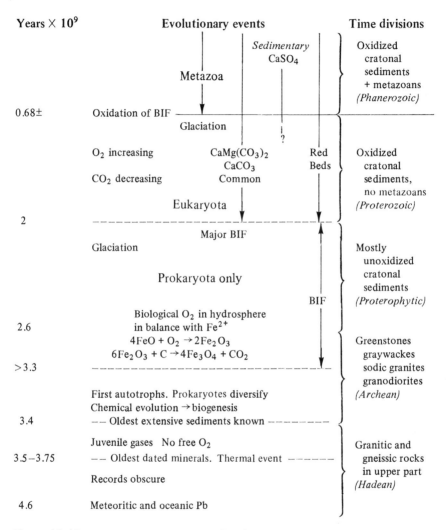

Figure 12.12   Banded Iron Formation (BIF) episode in relation to other geochemical events during the evolution of the earth. (From Cloud, 1974; by permission.)

Chemically, the iron-rich bands of the BIF can be explained on the basis that the ferrous oxide was first transformed to ferric oxide (Cloud, 1974),

$$4FeO + O_2 \rightarrow 2Fe_2O_3 \tag{12.27}$$

But in the presence of an excess of organic carbon in a reducing atmosphere, some iron oxide was partially reduced (Cloud, 1974),

$$6Fe_2O_3 + C_{org} \rightarrow 4Fe_3O_4 + CO_2 \tag{12.28}$$

and accumulated, therefore, as magnetite in the BIF deposits. Bacteria may have evolved during this time which acquired special capacities for precipitating iron and even for partially reducing ferric iron (reaction 12.28). Observations of nannofossils, including structures resembling photosynthesizing cyanobacteria and iron-precipitating bacteria in BIF, lend support to Cloud's hypothesis about the origin of the BIF (Fig. 12.13) (e.g., Cloud, 1965; Licari and Cloud, 1968; LaBerge, 1967).

Origin of the Red Beds   If the explanation of the origin of the BIF is accepted, the subsequent origin of the Red Beds [accumulations of hematite (Fe_2O_3)], which generally appear for the first time in the geologic record at the end of the BIF-forming episode or just thereafter (about 2×10^9 years ago), can be interpreted on the basis of the emergence of the more intensive oxygen-producing photosynthesizers in a developing oxidizing environment which kept iron fully oxidized. Based on fossil evidence, microbes may well have been involved in iron precipitation in Red-Bed formation.

Origin of Some Siderite Deposits   Some iron deposits occur as ferrous carbonate (siderite). They occur under reducing conditions in neutral to slightly alkaline environments, conditions that can be generated through active microbial intervention. The neutral-to-alkaline conditions can be generated through the decay of nitrogenous organic matter forming ammonia among the end products. Any CO_2 formed in the decay process would be trapped at least in part as carbonate, which could then react with available ferrous iron and precipitate. Limited oxygen availability in the presence of excess organic carbon will lead to anoxic conditions and the emergence of a reducing environment. Sulfate may have to be specially excluded to prevent the formation by sulfate-reducing bacteria of H_2S that would precipitate the iron as sulfide and prevent formation of iron carbonate (Sellwood, 1971). In siderite beds in the Yorkshire Lias in England, compacted clay sediment is the apparent barrier to sulfate ions (Sellwood, 1971).

Microbial Pyrite Oxidation   Iron sulfide ores (e.g., pyrite and marcasite, i.e., FeS_2) which may be associated with bituminous coal seams or with metal sulfide deposits, may be subject to bacterial attack when in contact with aerated waters. The reaction may be summarized as follows:

$$2FeS_2 + 2H_2O + 7O_2 \rightarrow 2FeSO_4 + 2H_2SO_4 \tag{12.29}$$

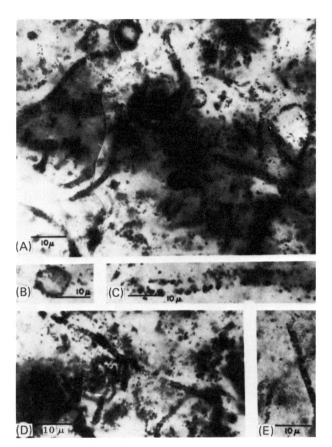

Figure 12.13   Gunflint microbiota in the stromatolites of the Biwabik Iron Formation, Corsica Mine, Minnesota. (A) *Gunflintia* filaments and *Huroniospora spheroides* replaced by hematite are abundant in some of the dark laminations in stromatolitic rocks at the Corsica Mine. (B) *Huroniospora,* Corsica Mine. (C) Wide filament, Corsica Mine. (D, E) *Gunflintia* filaments, Corsica Mine. (From Cloud and Licari, 1968; by permission.)

This reaction can also proceed in the absence of bacteria, but at a slower rate. *Thiobacillus ferrooxidans,* and perhaps other acidophilic iron-oxidizing bacteria, are the agents that can attack the crystal lattice of the FeS_2 directly (Silverman and Ehrlich, 1964). FeS_2 is also oxidized by Fe^{3+}:

$$FeS_2 + 14Fe^{3+} + 8H_2O \rightarrow 15Fe^{2+} + 2SO_4^{2-} + 16H^+ \tag{12.30}$$

The origin of the ferric iron consumed in this reaction is the biological oxidation of ferrous iron, originally from reaction 12.29, later also from reaction 12.30,

$$2Fe^{2+} + \frac{1}{2}O_2 + 5H_2O \rightarrow 2Fe(OH)_3 + 4H^+ \qquad (12.31)$$

It can be seen that acid is generated in this reaction. It accounts for the acidity of the water deriving from areas of active pyrite oxidation. At high sulfate concentration in the solution, some of the ferric iron may be transformed into basic ferric sulfate or jarosite (Duncan and Walden, 1972), as summarized in the equation

$$A^+ + 3Fe(OH)_3 + 2SO_4^{2-} + 3H^+ \rightarrow AFe(SO_4)_2(OH)_6 + 3H_2O \qquad (12.32)$$

A^+ may represent either K^+, NH_4^+, or H_3O^+. The formation of jarosite consumes half the acid produced from FeS_2 oxidation. Potassium jarosite has been formed by *T. ferrooxidans* in some soils, shales, and metamorphic rocks containing yellow-brown deposits (Ivarson, 1973). Such deposits have also been seen in streams receiving acid drainage from bituminous coal mines and from some metal sulfide mines. The pH of such streams may range from pH 2 to 4.5, and sulfate-ion concentration may range from 1,000 to 20,000 mg liter^{-1}, while ferrous iron content is normally 0 mg liter^{-1} (Lundgren et al., 1972).

It has been proposed that a *Metallogenium*-like organism which has been isolated from acid mine drainage, may be the dominant iron-oxidizing organism in ferrous iron-containing environments until the pH drops below 3.5, after which the more acid-tolerant *T. ferrooxidans* and, possibly, *Sulfolobus* and *Leptospirillum ferrooxidans* take over (Walsh and Mitchell, 1972b). *L. ferrooxidans*, however, will be able to oxidize only Fe^{2+}. It cannot apparently attack FeS_2 directly because it cannot oxidize reduced sulfur. Whether *Sulfolobus* can oxidize FeS_2 remains to be experimentally established. It does oxidize chalcopyrite ($CuFeS_2$) and molybdenite (MoS_2) (Brierley and Murr, 1973). The *Metallogenium*-like organism does not appear essential to lowering the pH in a pyritic environment to make it favorable for the acidophilic iron oxidizers. *T. ferrooxidans* itself is capable of it, at least in pyrite-containing coal or overburden (Kleinmann and Crerar, 1979). It may accomplish this by initial direct attack or pyrite, creating an acid microenvironment from which the organism then spreads.

12.4 SUMMARY

Iron may be enzymatically oxidized by microorganisms, serving them as an energy source. Best evidence for this comes from the study of the acidophilic iron oxidizers, like *Thiobacillus ferrooxidans, Sulfolobus* spp., and *Leptospirillum ferrooxidans*. The first has been most extensively studied. The reason why the most convincing evidence of microbial iron oxidation has come from the study of acidophiles is that ferrous iron is least susceptible to autoxidation below pH 5. Some bacteria growing near neutral pH may, however, also oxidize iron. The stalked bacterium *Gallionella ferruginea* may use it chemolithotrophically, whereas the sheathed bacteria *Leptothrix* spp. may use it mixotrophically. Both of these probably require a partially reduced environment for this activity. Both *Gallionella* and *Leptothrix* may also precipitate iron oxide passively on their stalk or sheath, respectively.

Iron may also be oxidized nonenzymatically by microorganisms by raising the redox potential and/or pH of their environment. This may be the consequence of ammonia production; the consumption or organic salts, some of which may chelate ferrous iron; or photosynthesis.

Ferric iron precipitation may not involve iron oxidation. Ferric iron can be stabilized in solution by chelation. Naturally produced chelators that may solubilize extensive amounts of ferric iron include oxalate, citrate, humic acids, and tannins. Ferric iron precipitation may result from microbial destruction (mineralization) of the chelators. A number of bacteria have been found active in this manner. Ferric iron may be locally concentrated by adsorption to the cell surface. A variety of microorganisms, including bacteria and protozoans, have been found capable of this.

Ferric iron may be enzymatically reduced to ferrous iron with a suitable electron donor, the ferric iron serving as a terminal electron acceptor. Nitrate reductase A is one enzyme that may be involved. However, at least one other enzyme may also be independently active. Both bacteria and fungi have been implicated in this reaction. Not all microbial ferric iron reduction is enzymatic. Some may be the result of reaction with metabolic end products such as H_2S or formate.

Both oxidative and reductive reactions of iron by microbes play important roles in the iron cycle in nature. They affect the mobility of iron as well as local accumulation of iron. The formation of some sedimentary iron deposits have been directly attributed to microbes acting on iron. Ochre formation has been associated with bacterial iron oxidation, whereas gleying has been associated with bacterial iron reduction. Acid mine drainage has been traced to the action of acidophilic iron bacteria attacking the pyrite in the coal seams of bituminous coal deposits on exposure to air and moisture and oxidizing both ferrous iron and sulfide-sulfur to ferric iron and sulfuric acid.

13

The Geomicrobiology of Manganese

Manganese is one of the elements of the first transition series, which includes, in order of increasing atomic number from 21 to 29, the elements Sc, Ti, V, Gr, Mn, Fe, Co, Ni, and Cu. Electronically, these elements differ mostly in the degree to which their d orbitals are filled (Latimer and Hildebrand, 1940). Their increasing oxidation states are attributed to removal of 4s and 3d electrons (Sienko and Plane, 1966).

The abundance of manganese in the earth's crust is 0.1% (Alexandrov, 1972, p. 670). The element is, therefore, less than $\frac{1}{50}$ as abundant as iron. Its distribution in the crust is by no means uniform. In soils, for instance, its concentration can range from 0.002 to 10% (Goldschmidt, 1954). The element can exist in the oxidation states of 0, +2, +3, +4, +6, and +7. However, in nature only the +2 and the +4 oxidation states are commonly found. The +3 state also occurs to some extent. Of these, only the +2 state can occur as a free ion in solution, whereas the +3 and +4 states occur in the form of insoluble oxides. The +4 oxides tend to be amphoteric (Latimer and Hindebrand, 1940), a property that accounts for their affinity for various cations, especially for the heavy metals such as Co, Ni, and Cu. Mn(IV) oxides have long been known as scavengers (Geloso, 1972; Goldberg, 1954). They are frequently associated with ferric iron in nature.

Manganous ion is more stable than ferrous ion under similar conditions of pH and E_h. Based on equilibrium computations, manganese should exist predominantly as Mn^{2+} below pH 5.5 and as Mn(IV) above pH 5.5 if the E_h is approximately 0.8 V and the Mn^{2+} ion concentration is 100 ppm in the presence of 100 ppm HCO_3^- ions. At an E_h below 0.5 V, Mn^{2+} may dominate up to pH 7.8-8.0 (Hem, 1963). Although in theory, 0.1 ppm of Mn^{2+} ions in aqueous solution should readily autoxidize when exposed to air at pH values above 4, they usually do not do so until the pH exceeds 8. Apart from Mn^{2+} concentration and E_h effects, one explanation for this resistance of Mn^{2+} ions to oxidation is the high energy of activation of the reaction (Crerar and Barnes, 1974). Another explanation is that the Mn^{2+} may be extensively complexed and thereby stabilized by such inorganic ions as Cl^-, SO_4^{2-}, and HCO_3^- if present (Hem, 1963; Goldberg and Arrhenius, 1958), or by organic compounds such as amino acids, humic acids, and others (Graham, 1959; Hood, 1963; Hood and Slowey, 1964).

In nature, manganese is found as a major or minor component of a number of minerals. It is an essential element in more than 100 minerals (*Mineral Facts and Problems,* 1965, p. 556). Maor accumulations occur in the form of oxides, carbonates, and silicates. Among the oxides, psilomelane $[(Ba,Mn(II)_2Mn(IV)_8-O_{16}(OH)_4]$, birnessite (MnO_2), pyrolusite (MnO_2), manganite $[MnO(OH)]$, todorokite $[(Mn^{2+}, Mg^{2+}, Ba^{2+}, Ca^{2+}, K^+, Na^+)_2 Mn_5^{4+}O_{12} \cdot 3H_2O]$, hausmannite (Mn_3O_4), and braunite $[(MnSi)_2O_3]$ are important examples. Of the carbonates, rhodochrosite $(MnCO_3)$ is important. Of the silicates, rhodonite $(MnSiO_3)$ is important. Among minerals that contain manganese as a minor constituent are found ferromagnesian minerals such as pyroxenes and amphiboles (Trost, 1958), or micas such as biotite (Lawton, 1955, p. 59), which are all of igneous origin. The oxides, carbonates, and silicates of manganese are either secondary minerals derived from the weathering of the primary, igneous minerals, or authigenic minerals resulting from the precipitation of dissolved manganese.

Manganese is an important trace element in biological systems. It is essential in microbial, plant, and animal nutrition. It is required as an activator by a number of enzymes, such as isocitric dehydrogenase or malic enzyme, and may replace Mg^{2+} ion as an activator, for example in enolase (Mahler and Cordes, 1966, pp. 245, 532). It is also required in photosynthesis, where it functions in the production of oxygen from water by photosystem II. As will be discussed below, Mn(II) may serve as an energy source to some autotrophic and mixotrophic bacteria, and Mn(IV) may serve as a terminal electron acceptor in the respiration of some autotrophic and heterotrophic bacteria. As in the case of iron, the most important geomicrobial interactions with manganese are manganese precipitating and solubilizing processes, including oxidations and reduction. These interactions will be considered in this chapter.

13.1 MICROBIAL OXIDATION OF MANGANESE

Enzymatic Oxidation   Like iron oxidation, manganese oxidation by microbes may be enzymatic or nonenzymatic. Beijerinck (1913; see Bromfield, 1956), who was first to report on microbial manganese oxidation, suggested that it may be associated with autotrophic growth. Lieske (1919) and Sartory and Meyer (1947) suggested that it may also be associated with mixotrophic growth. Either way, an enzymatic oxidation was implied. A conclusion to be drawn from all studies to date is that enzymatic manganese oxidation is restricted to bacteria.

Enzymatic manganese oxidation by bacteria proceeds by at least three different mechanisms. Two involve the oxidation of free Mn^{2+} ions and one involves the oxidation of Mn^{2+} prebound to $Mn(IV)$ oxide. The oxidation of free Mn^{2+} ions may be catalyzed by a manganese oxidase that conveys electrons to oxygen by a cytochrome pathway:

$$Mn^{2+} + \tfrac{1}{2}O_2 + H_2O = MnO_2 + 2H^+ \tag{13.1}$$

This oxidase may be constitutive or inducible. The oxidation of free Mn^{2+} ions may also be catalyzed by catalase in a reaction with metabolically produced H_2O_2:

$$Mn^{2+} + H_2O_2 = MnO_2 + 2H^+ \tag{13.2}$$

The oxidation of Mn^{2+} prebound to $Mn(IV)$ is catalyzed by a manganese oxidase that shunts electrons to oxygen by a cytochrome pathway:

$$MnMnO_3 + \tfrac{1}{2}O_2 + 2H_2O = 2H_2MnO_3 \tag{13.3}$$

This manganese oxidase is not capable of oxidizing free Mn^{2+} ions and has so far always been found to be constitutive. Oxidation catalyzed by either manganese oxidase may generate metabolically useful energy (ATP) for the cell (Ali and Stokes, 1971; Ehrlich, 1976, 1978a). Indeed, at least one manganese oxidizing bacterium, *Sphaerotilus discophorus* (now *Leptothrix discophora*) appears to be a facultative autotroph (Ali and Stokes, 1971). The standard free-energy change at pH 7.0 ($\Delta F'$) for reaction 13.1 is -16.3 kcal mol^{-1} (Ehrlich, 1978a), enough to generate at least one ATP, and possibly two. In Table 13.1, bacteria are classified which are capable of enzymatic manganese oxidation according to the mechanism they employ.

The oxidation mechanisms involving free Mn^{2+} ions are best suited to soil and freshwater environments, where Mn^{2+} ion concentrations may exceed 0.001 g liter^{-1}, whereas those involving prebound Mn^{2+} are best suited to marine environments where Mn^{2+} ion concentrations are only around 2 μg liter^{-1}. The oxidation mechanism involving H_2O_2 and catalase is probably used for detoxification by the organisms that employ it (Dubinina, 1978b).

Manganese Oxidation by Arthrobacter (Formerly Corynebacterium) Strain B

Arthrobacter strain B isolated from soil by Bromfield (1956) is able to oxidize Mn^{2+} in batch culture at 27°C, but only after the onset of the stationary phase of growth, which occurs after 15 hr of incubation. The culture medium for this process contains distilled water (100 ml); K_2HPO_4 (0.005 g); $MgSO_4 \cdot 7H_2O$ (0.002 g); $(NH_4)_2SO_4$ (0.01 g); Ca_3PO_4 (0.01 g); Difco yeast extract (0.005 g),

Table 13.1 Some Bacteria That Oxidize Manganese Enzymatically

Free Mn^{2+} attacked			Mn^{2+} prebound to Mn(IV) oxide attacked
Energy derived	No energy derived	Not known if energy derived	Energy derived
H. manganoxidans[a]	*A. siderocapsulatus*	*Arthrobacter* B	*Arthrobacter* 37
L. discophora	*L. pseudoochracea*	*A. citreus*	*Oceanospirillum*
	Metallogenium	*A. globiformis*	*Vibrio*
		A. simplex	Unidentified cultures
		Citrobacter freundii E$_4$	
		Hyphomicrobium T37	
		Pedomicrobium	
		Pseudomonas E$_1$	
		Pseudomonas spp.	

[a] A newly discovered organism that must oxidize Mn^{2+} in order to grow (Eleftheriadis, 1976).

and $MnSO_4 \cdot 4H_2O$ (0.005 g). Because omission of the yeast extract prevents growth, the organism does not grow autotrophically on manganese. It is able to oxidize Mn^{2+} in a resting cell suspension, prepared from a culture in the early stationary phase with the cells suspended in distilled water containing 0.005% $MnSO_4 \cdot 4H_2O$ and incubated at 40°C. The highest temperature at which resting cells oxidize Mn^{2+} is 45°C, suggesting that a heat-labile enzyme is involved. The addition of various organic substances does not enhance the rate of Mn^{2+} oxidation. The oxidation by resting cells is prevented by boiling the cells or by addition of one of a number of chemicals to the reaction mixture (Table 13.2). Bromfield (1956) has concluded that the inhibition by the copper and mercury salts indicates that the oxidation is enzymatic, and, since azide and cyanide are inhibitory, that heavy-metal enzymes are involved. He has discounted cytochrome oxidase involvement, however, because the oxidation by the cells is not inhibited by carbon monoxide in the dark. He has stated (Bromfield, 1956) that the enzyme oxidizing Mn^{2+} is intracellular, but has apparently not considered the possibility of its location in the plasma membrane. A membrane location would eliminate the necessity for postulating outward diffusion of oxidized manganese, an unlikely process because of the insolubility of Mn(IV) oxides. Although not explicitly stated, the enzyme system appears to be constitutive in *Arthrobacter* strain B. Quantitatively, Mn^{2+} oxidation by this organism proceeds as shown in Figure 13.1, where the oxidation is measured in terms of Mn^{2+} removed from solution as insoluble oxide.

Manganese Oxidation by Leptothrix discophora   The sheathed bacterium *Sphaerotilus discophorus*, presently known as *Leptothrix discophora*, has been shown to oxidize Mn^{2+} in resting-cell suspension prepared with cells obtained from an organic medium containing 0.05% $MnSO_4 \cdot H_2O$ (Fig. 13.2A) (Johnson and Stokes, 1966). The suspension consumed 43 μg of oxygen in 3 hr. Resting cells obtained from the same growth medium lacking $MnSO_4 \cdot H_2O$ could not oxidize Mn^{2+} (Fig. 13.2B), an observation that has been confirmed by Hogan (1973) with a different strain. This leads to the conclusion that Mn^{2+} oxidation by *L. discophora* is inducible. However, this inducibility has been questioned by Van Veen (1972) because he was able to obtain Mn^{2+} oxidation with cells grown in the absence of $MnSO_4 \cdot H_2O$. He suggested that phosphate ions, which he found to inhibit Mn^{2+} oxidation, may have interfered in the experiments of Johnson and Stokes (1966). However, both Johnson and Stokes and Hogan used phosphate whether induced or uninduced cells were tested. Therefore, Van Veen's objections seem unjustified. Since Van Veen used three strains of *L. discophora* different from the strains of Johnson and Stokes and Hogan, it may have been that his strains were constitutive in their Mn^{2+}-oxidizing capacity.

The Mn^{2+}-oxidizing capacity of *L. discophora* is destroyed by preheating the cells for 5 min at 93°C (Johnson and Stokes, 1966). Mercuric chloride at a concentration of 10^{-4} M fails to inhibit manganese oxidation. The product of

Table 13.2   Effect of Various Poisons and Treatments on the Oxidation of
Manganese by Cell Suspensions of *Corynebacterium* B

Treatment	Final concentration	Incubation time (min)	Oxidation[a]
Control	—	10	++
Boiled cells	—	24 hr	−
$HgCl_2$	M/2,700	24 hr	−
$CuCl_2$	M/300	24 hr	−
$CuCl_2$	M/2,000	1 hr	+
$MnSO_4$	M/4,500	10	++
$MnSO_4$	M/450	10	++
$MnSO_4$	M/225	10	+
$MnSO_4$	M/45	10	±
CO in dark	—	10	++
KCN	M/13,000	10	+
KCN	M/1,300	10	+
KCN	M/650	10	±
KCN	M/65	10	−
NaN_3	M/13,000	10	+
NaN_3	M/1,300	10	−
H_2O_2	M/400	10	++
$KCN + H_2O_2$	M/13,000 + M/400	10	−
$NaN_3 + H_2O_2$	M/13,000 + M/400	10	−
Ethanol	33% (wt/vol)	20	+
Methanol	33% (wt/vol)	20	±
Benzene	Saturated solution	20	−
Toluene	Saturated solution	20	+
CCl_4	Saturated solution	20	++
Chloroform	Saturated solution	20	++
n-Octyl alcohol	Saturated solution	20	++

[a]Color relative to control produced on addition of 1 drop of benzidine reagent to cell
system.
Source: Bromfield (1956); by permission.

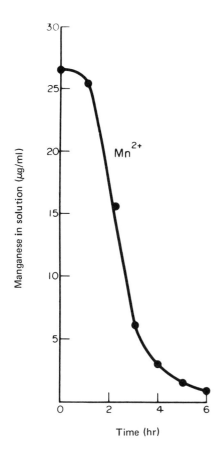

Figure 13.1   Oxidation of 0.5 mM Mn^{2+} by bacterial cell suspensions of *Arthrobacter* strain B (formerly *Corynebacterium* strain B) at pH 7.0. The mixture (80 ml) contained 4 mg of cells per milliliter and 27 μg of Mn per milliliter. (Reprinted with permission from *Soil Biol. Biochem.*, Vol. 8, S. M. Bromfield and D. J. David, Sorption and oxidation of manganous ions and reduction of manganese oxide by cell suspensions of a manganese oxidizing bacterium, Copyright 1976, Pergamon Press, Ltd.)

oxidation is a dark brown substance, soluble in HCl with the evolution of chlorine:

$$MnO_2 + 4HCl \longrightarrow MnCl_4 + 2H_2O \tag{13.4}$$

$$MnCl_4 \xrightarrow{\quad H_2O \quad} Mn^{2+} + 2Cl^- + Cl_2 \tag{13.5}$$

These findings lead to the inference that *L. discophora* oxidizes Mn^{2+} enzymatically.

Further evidence that *L. discophora* can oxidize Mn^{2+} enzymatically has been provided by the observation that the organism can grow autotrophically in a mineral salts medium containing $MnSO_4$ as energy source and traces of thiamin, biotin, and cyanocobalamin as growth factors (Ali and Stokes, 1971). The composition of the medium in which autotrophic growth was noted was

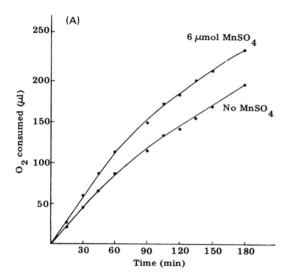

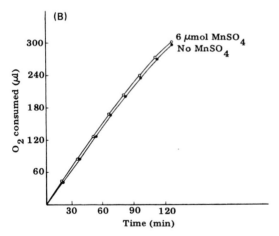

Figure 13.2 Manganese oxidation by *Sphaerotilus discophorus* strain 35 (presently known as *Leptothrix discophora*). (A) Adapted cells. (B) Unadapted cells. Note lack of stimulation of O_2 consumption by $MnSO_4$ with unadapted cells. (From Johnson and Stokes, 1966; by permission.)

0.1% $(NH_4)_2 SO_4$, 0.05% ferric ammonium citrate, 0.05% $K_2 HPO_4$, 0.02% $MgSO_4 \cdot 7H_2O$, 0.01 or 0.02% $MnSO_4 \cdot H_2O$, 0.005% $CaCl_2$, 0.001% $FeCl_3 \cdot 6H_2O$, 0.2 μg of thiamin, 0.02 μg of biotin, and 0.01 μg of cyanocabalamin per milliliter of medium. The pH of the medium was adjusted to 7.4 and the medium was autoclaved for 12 min at 121°C. Growth was measured by three criteria: increase in cell nitrogen, increase in cell protein, and increase in cell DNA. The growth curves for the organism by all three critiera were more or less parallel. Van Veen (1972) has criticized this study, claiming that in the presence of Mn^{2+} the cells release a protein that is precipitated by Mn^{2+} during their growth. This protein is not included in the final harvest of cells grown in the absence of added Mn^{2+} and thus could account for a greater apparent cell yield for cultures grown with than without Mn^{2+}. This criticism cannot be accepted as valid in view of the observed difference in total DNA yield with time between cells incubated with and without added manganese, as observed by Ali and Stokes (1971).

Some evidence suggesting that *L. discophora* can grow mixotrophically has also been presented (Ali and Stokes, 1971). When 0.2% vitamin-free casamino acids are substituted for $(NH_4)SO_4$ and the cyanocobalamin is omitted, growth of the organism appears to be progressively stimulated by the addition of up to 0.05% $MnSO_4 \cdot H_2O$. Greater concentrations of $MnSO_4 \cdot H_2O$ than 0.05% are inhibitory. Hajj and Makemson (1976) have, however, expressed doubt that these observations indicate mixotrophic growth because they found that the analytic techniques for protein and DNA used in the earlier work are affected by incompletely removed MnO_2. Yet even correcting for any errors in analysis in the earlier work, the finding of increases in total DNA cannot be explained except in terms of growth stimulation by Mn^{2+}. Hajj and Makemson in their own growth experiment found manganese to be oxidized mainly in the later logarithmic and stationary growth phases. They also found manganese to be somewhat toxic to growth, as had been previously noted by Hogan (1973).

The enzymatic nature of Mn^{2+} oxidation by *L. discophora* has been shown in cell-free extracts (Hogan, 1973). The Mn^{2+} oxidizing activity is associated with a particulate cell fraction that sediments at 48,000 *g*. Although the involvement of a cytochrome oxidase has been shown spectrophotometrically and by means of complete inhibition of oxygen uptake on Mn^{2+} by 10^{-5} M cyanide (KCN) and 10^{-4} M azide (NaN_3), the expected involvement of cytochrome c has so far not been demonstrated, despite the fact that cytochromes b and c are present in the particles. The full identification of all electrons transport components involved in Mn^{2+} oxidation by *L. discophora* has thus not been made.

Oxidation of Free Mn^{2+} Ions by Other Bacteria   Douka (1977; 1980) has demonstrated enzymatic Mn^{2+} oxidation by soil isolates of *Pseudomonas* sp. and *Citrobacter freundii*. She was able to demonstrate this oxidation in cell-free extracts. The enzyme system involved appears to be an oxidase. Dubinina (1978a,b) has reported that *Leptothrix pseudoochracea, Arthrobacter siderocapsulatus*, and *Metallogenium* oxidize Mn^{2+} with metabolically produced H_2O_2 catalyzed by catalase in the cells.

Oxidation of Prebound Mn^{2+}   A distinctly different kind of constitutive Mn(II)-oxidizing enzyme system has been detected in a number of marine bacteria isolated from ferromanganese nodules and deep-sea sediments. Growing cells, resting cells, and cell extracts of *Arthrobacter* 37, for instance, have been found to oxidize Mn^{2+} (Ehrlich, 1963b, 1966, 1968). The oxidation by this and some other marine isolates differs, however, from that carried on by the previously described soil and freshwater bacteria in that Mn^{2+} must first be bound to Mn(IV) oxide in the form of MnO$_2$, a synthetic Mn-Fe oxide, or ferromanganese nodule substance before it is oxidized. No oxidation of free Mn^{2+} ions has been detected. The oxidation of Mn(II) by these organisms, therefore, includes the following two steps:

$$Mn^{2+} + H_2MnO_3 \longrightarrow MnMnO_3 + 2H^+ \tag{13.6}$$

$$MnMnO_3 + \tfrac{1}{2}O_2 + 2H_2O \rightarrow 2H_2MnO_3 \tag{13.7}$$

Only the second of the two reactions is directly catalyzed by the Mn(II)-oxidizing system of the bacteria, but the progress of the first reaction is controlled by the second, and thus indirectly affected by the catalytic effect of the bacteria. Since the rate of Mn^{2+} adsorption is much greater than the rate of oxidation of the adsorbed manganese, the oxidation reaction 13.7 is the rate-controlling step in the overall conversion of Mn^{2+} to H$_2$MnO$_3$ if the supply of Mn^{2+} is not limited. A probable explanation for why only adsorbed Mn^{2+} is microbiologically oxidized is that the energy of activation for oxidizing the adsorbed ion is less than that for the free ion (Crerar and Barnes, 1974).

Whether manganese oxidases that act on free Mn^{2+} ions are stimulated by Mn(IV) oxide has not been widely examined so far. The one of a gram-positive bacillus isolated in this author's laboratory is not stimulated by the addition of MnO$_2$ (Ehrlich, unpublished data), whereas that of *Bacillus* sp. isolated by LaRock (1969) was stimulated by the addition of synthetic Mn-Fe oxide.

Although Mn(IV) oxide is probably the preferred binding agent for Mn^{2+} oxidation by marine bacteria, Ehrlich (1978b) reported that some marine sediments contain components which, if pretreated with ferric iron as FeCl$_3$ · 6H$_2$O, may substitute for Mn(IV) oxide as binding agent for Mn^{2+} upon which *Oceanospirillum* BIII 45 may act.

The optimum pH for Mn(II) oxidation by *Arthrobacter* 37 is 7.5 with the full range of activity being 6.5-8 (Ehrlich, 1970). The optimum temperature for cell-free enzyme preparations of this organism is 17.5°C (Ehrlich, 1968). At about 24°C, its enzyme functions well at a hydrostatic pressure of 476 atm but not at 578 atm. The oxidation of Mn(II) by *Arthrobacter* 37 has been shown to be oxygen-dependent (Ehrlich, 1968). The activity appears to be membrane-associated, since after Sephadex filtration the activity is found associated with large particles. Furthermore, with extracts from another bacterial isolate, *Oceanospirillum* strain BIII 45, similar activity can be sedimented at 29,000 g. Electron photomicrographs of this latter preparation have shown it to consist of membrane vesicles (Fig. 13.3). Demonstrating the location of Mn(II) oxidase to be in the plasma membrane is important because it explains why it is unnecessary for Mn^{2+} ions to enter the cell to be oxidized. Indeed, no evidence has been found that manganese-oxidizing bacteria concentrate manganese intracellularly (Yang, 1974). Mn(II) oxidiation by *Arthrobacter* 37 is inhibited by 10^{-6} M $HgCl_2$ and by 10^{-5} M *p*-chloromercuribenzoate (Ehrlich, 1968). It is also inhibited by 0.01 M KCN (tested on whole cells only), 10^{-4} M NaN_3, 3.5×10^{-5} M 2-n-nonyl-4-hydroxyquinoline-N-oxide, 7.3×10^{-6} M antimycin A, and 10^{-4} M dicumarol, but not by 10^{-3} M atabrine (Ehrlich, 1968; unpublished results). The inhibition pattern suggests that the Mn(II) oxidases may include coenzyme Q, cytochrome b, cytochrome c, and cytochrome oxidase. Spectrophotometric evidence for the involvement of at least cytochrome c in Mn(II) oxidation in two different strains of *Oceanospirillum* has been obtained (Arcuri, 1978) (Fig. 13.4). A soluble component is required to bring about the cytochrome reduction in membrane preparations

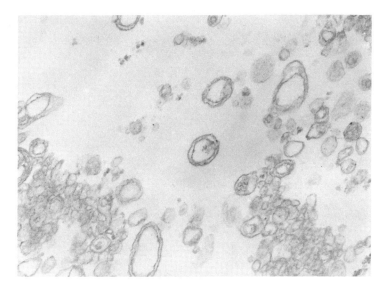

Figure 13.3 Sections of membrane vesicles of *Oceanospirillum* BIII 45 prepared by sonication (39,900X). The vesicles are capable of coupling ATP synthesis to Mn(II) oxidation. (Electron photomicrograph prepared by W. C. Ghiorse.)

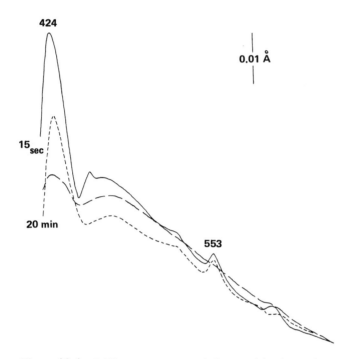

Figure 13.4   Difference spectra of the combined membrane and intracellular fractions of *Oceanospirillum* BIII 82. ————, air-oxidized minus air-oxidized spectrum, base line; - - - - -, difference spectrum 20 min after NaN₃ was added to the sample cuvette, showing endogenous activity; ————, difference spectrum 15 sec after MnSO₄ was added to both cuvettes, showing cytochrome c-type reduction by Mn(II). Total protein in each cuvette was 18.28 mg. (From Arcuri and Ehrlich, 1979; by permission.)

from these cultures (Arcuri, 1978; Arcuri and Ehrlich, 1979). The participation of an electron transport chain in the oxidation of Mn(II) by the marine bacteria suggests that they may derive useful energy from the oxidation. Energy coupling to Mn(II) oxidation has actually been observed with *Arthrobacter* 37 as well as six other marine bacteria—of which at least two have also been shown to be sensitive to KCN, NaN₃, antimycin A, and 2-n-nonyl-4-hydroxyquinoline-N-oxide—by demonstrating that the size of the ATP pool in starved cells increases upon oxidation of Mn(II) by these cells (Ehrlich, 1978a). ATP synthesis has also been shown to occur with membrane vesicles of all these cultures when incubated with Mn(II), MnO₂, and ADP in synthetic salt mixture with added inorganic phosphate (Ehrlich, 1976, 1978a; unpublished results). On the basis of the inhibition studies with antimycin A, 2-n-nonyl-4-hydroxyquinoline-N-oxide, azide, and cyanide, it may be inferred that they should be able to derive as many as two molecules of ATP for every Mn(II) oxidized (Ehrlich, 1976). The organisms are not autotrophs and must be using Mn(II) oxidation as part of mixotrophic nutrition.

Cultural Manifestation of Microbial Manganese Oxidation   Organisms that can oxidize free Mn^{2+} ions usually cause browning of liquid culture medium (e.g., *L. discophora*). Colonies of such organisms on agar media containing a suitable manganese compound frequently turn brown. The browning of colonies of *Arthrobacter* sp. (Bromfield, 1956) on $MnSO_4$-containing agar medium is affected by varying the concentrations of sucrose and yeast extract in it (Fig. 13.5) (Bromfield, 1974). Increasing both yeast extract and sucrose concentrations from 0 to 1% in the presence of 0.005% $MnSO_4 \cdot 4H_2O$ in the agar medium increases both growth and manganese oxide deposition in the colonies. Increasing the sucrose concentration only causes deposition of manganese oxide as specks within, below, and around colonies of the *Arthrobacter*. Increasing the yeast extract concentration only, on the other hand, confines any manganese oxide formed by the organism to its colonies as numerous small specks. The manganese oxide outside the colonies is thought to result from diffusion away from the colonies of manganese complexes and subsequent precipitation as oxide due to more alkaline conditions.

Organisms of the type that oxidize Mn(II) only when bound to Mn(IV) oxides do not cause browning of their medium.* It must be stressed that deposition of brown Mn(IV) oxides in a liquid medium or in colonies is no proof of enzymatic Mn(II) oxidation, because organisms that cause such oxidation nonenzymatically may also accumulate brown manganese oxides in their colonies.

Nonenzymatic Manganese Oxidation   Nonenzymatic manganese oxidation is promoted by a number of organisms. In many others that cause manganese oxidation, no attempt has so far been made to look for a manganese oxidizing enzyme system. Table 13.3 lists bacteria and fungi that have been reported to oxidize manganese, but for which the mechanism of oxidation is either unknown or presumed to be nonenzymatic. The table suggests that manganese-oxidizing bacteria are more common in fresh water, but no concerted study to show such differences has been made.

---

*Nealson and Ford (1980) reported recently that manganese precipitation by a marine bacillus (strain SG-1) was enhanced in liquid medium by solid surfaces such as those of calcite crystals, sand grains, frosted glass, and glass beads. On agar medium this organism precipitates manganese without addition of these solid surfaces. It remains to be clarified whether the surfaces play a role in bacterial attachment, in adsorbing Mn(IV) oxide formed by the bacteria, in Mn^{2+} oxidation, or in a combination of these factors.

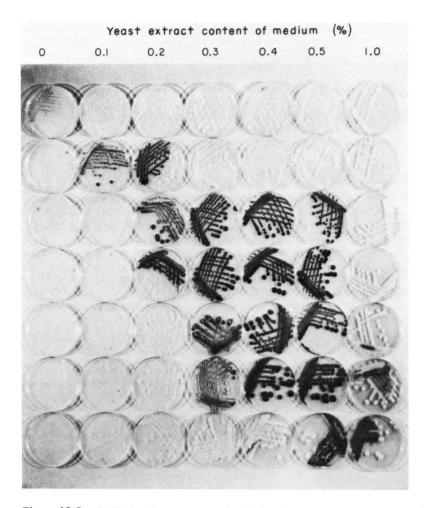

Figure 13.5 Oxidation by manganese by *Arthrobacter* sp. on media containing various mixtures and concentrations of sucrose and yeast extract. The concentrations of sucrose were 0, 0.1, 0.2, 0.3, 0.4, 0.5, and 1.0% in ascending order from top to bottom. (Reprinted with permission from *Soil Biol. Biochem.*, Vol. 6, S. M. Bromfield, Bacterial oxidation of manganous ions as affected by organic substrate concentrations and composition, Copyright 1974, Pergamon Press, Ltd.)

As in the case of nonenzymatic iron oxidation, nonenzymatic manganese oxidation may be promoted through the genesis of environmental pH and E_h conditions that favor autoxidation of Mn^{2+}, such as pH values greater than 8 and E_h values greater than 0.5V at Mn^{2+} ion concentrations of 0.01 ppm.

The pH-E_h conditions that are required for the autoxidation of Mn^{2+} are much more restricted than those for iron. Nonenzymatic manganese oxidation may also be promoted through production of one or more metabolic end products which cause chemical oxidation of Mn^{2+}. According to Soehngen (1914), a large number of microorganisms can cause nonenzymatic manganese oxidation when in the presence of hydroxycarboxylic acids such as citrate, lactate, malate, gluconate, and tartrate. He indicated that metabolic utilization of a part of such acid causes a rise in pH of the culture medium, and that at alkaline pH, residual hydroxycarboxylic acid catalyzes the oxidation of Mn^{2+}. In apparent agreement, Van Veen (1973) found that with *Arthrobacter* 216 hydroxycarboxylic acids are required for Mn^{2+} oxidation. However, Van Veen (1973) felt that microorganisms with specific manganese-oxidizing capacity are more important in soil Mn^{2+} oxidation.

Another example of nonenzymatic Mn^{2+} oxidation is furnished by an actinomycete, *Streptomyces* sp., from Australian soil in its transformation of Mn^{2+} to manganese oxide in soil agar in a pH range 5-6.5 (Bromfield, 1978, 1979). Apparently, the actinomycete produces a water-soluble, extracellular compound which is responsible for the oxidation. The oxidation in this case is thought to protect the organism from inhibition by Mn^{2+} ions (Bromfield, 1978). *Pseudomonas* MnBl has been reported to produce a Mn^{2+}-oxidizing protein intracellularly which is consumed in its reaction with Mn^{2+} and is thus not an enzyme. The oxidation does not require O_2. It proceeds optimally at pH 7.0. The formation of the protein depends on cessation of growth by the culture at the end of its exponential phase but does not require added Mn^{2+} in the medium (Jung and Schweisfurth, 1979).

As shown in Table 13.3, several types of fungi can promote manganese oxidation. The manganese oxide is frequently precipitated on the fungal hyphae (Fig. 13.6). The oxidation is suspected to be nonenzymatic. Recently, the zygospores of an alga, *Chlamydomonas*, from soil were reported to become encrusted with Mn(IV) oxide, presumably through oxidation of Mn^{2+} (Schulz-Baldes and Lewin, 1975). Fungi and algae (excluding cyanobacteria), being eukaryotic, do not carry their respiratory system in the plasma membrane as prokaryotes do, but in special organelles called mitochondria. It seems unlikely that mitochondria in intact cells of these organisms catalyze oxidation of Mn^{2+} because Mn(IV) oxide would be expected to accumulate in the cells, which has not been reported. Since fungi produce one or more hydroxycarboxylic acids as end products of their metabolism, especially from carbohydrate, it is quite possible that they oxidize Mn^{2+} through these acids. Whether *Chlamydomonas* zygospores carry out this mechanism needs to be investigated.

Microbial Precipitation of Preformed Manganese Oxides   Some microorganisms cause precipitation of preformed oxidized manganese, often through adsorption

Table 13.3   Bacteria and Fungi that Oxidize Mn(II) by Indirect Means or
by an As-Yet-Unknown Mechanism

Organism	Habitat	References
Bacteria		
Metallogenium sp.	Soil, fresh water	Perfil'ev and Cabe (1961); Zavarzin (1961)
Bacillus sp.	Fresh water	LaRock (1969); Van Veen (1973)
Pseudomonas spp.	Soil, fresh water	Schweisfurth (1968, 1973); Zavarzin (1962); Douka (1977)
Hyphomicrobium sp.	Fresh water	Tyler and Marshall 1967a);
Pedomicrobium spp.	Soil	Aristovskaya (1961)
Kuznezovia	Fresh water	Perfil'ev and Gabe (1964)
Caulococcus	Fresh water	Perfil'ev and Gabe (1964)
Leptothrix spp.	Fresh water	De Toni and Trevisan (1889)
Crenothrix	Fresh water	Cohn (1870)
Clonothrix	Fresh water	Roze (1896)
Siderocapsa spp.	Fresh water	Molisch (1910)
Naumanniella	Fresh water	Dorff (1934)
Aerobacter sp.	Soil (may require symbiosis with *Serratia* and *Pseudomonas*)	Domergues and Mangenot (1970); Skerman and Bromfield (1949); Alexander (1977)
Nocardia	Soil	Schweisfurth (1968); Timonin (1950a,b); Bromfield and Skerman (1950); Tyler (1970)
Streptomyces	Soil	Timonin (1950a,b); Timonin et al. (1972); Bromfield (1978)
Citrobacter freundii	Soil	Douka (1977)

Table 13.3 *(Continued)*

Organism	Habitat	References
Symbiotic combinations		
Chromobacterium and *Corynebacterium*	Soil	Bromfield and Skerman (1950)
Flavobacterium and *Corynebacterium*	Soil	Bromfield and Skerman (1950)
Pseudomonas and *Pseudomonas*	Soil	Zavarzin (1962)
Fungi		
Coniothyrium carpaticum	Fresh water	Schweisfurth (1971)
Cladosporium spp.	Soil	Schweisfurth (1971)
Curvularia spp.	Soil	Schweisfurth (1971)
Helminthosporium	Soil	Schweisfurth (1971)
Oidium manganifera	Soil, fresh water	Thiel (1925)
Pleospora sp.	Soil	Schweisfurth (1971)
Aspergillus spp.	Soil?	Schweisfurth (1971)
Phoma sp.	Soil	Schweisfurth (1971)
Botrytis sp.	Soil	Schweisfurth (1971)
Graphium sp.	Soil	Schweisfurth (1971)
Mycogone sp.	Soil	Schweisfurth (1971)
Cryptococcus sp.	Soil	Schweisfurth (1971)
Papulospora manganica	Soil	Schweisfurth (1971)
Periconis sp.	Soil	Schweisfurth (1971)
Philophora sp.	Fresh water	Schweisfurth (1971)
Sporocybe chartoikoon	Soil	Schweisfurth (1971)
Trichocladium sp.	Soil	Schweisfurth (1971)
Verticillium	Soil	Schweisfurth (1971)
Tringsheimia	Soil	Schweisfurth (1971)

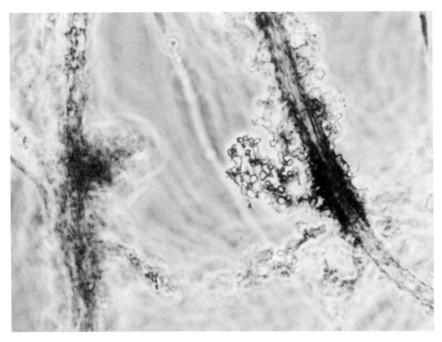

Figure 13.6   Manganese oxide deposition on fungal hyphae (990X). (Courtesy of R. Schweisfurth.)

to a surface structure or on extracellular material such as slime. The precipitation may or may not follow prior oxidation of Mn^{2+} by the organism. Among prokaryotes, such precipitation has been especially observed among sheathed bacteria and *Metallogenium* (Fig. 13.7), *Caulococcus, Kusnezovia, Pedomicrobium* (Fig. 13.8), *Hyphomicrobium* (Fig. 13.9), *Siderocapsa* (Fig. 12.8) (now believed to be an *Arthrobacter*), *Naumanniella,* and the Mn^{2+}-oxidizing fungi.

Figure 13.7   *Metallogenium.* (A) Microaccretions of *Metallogenium* from Lake Ukshezero in a split peloscope after 50 days; growth in central part of microzone (900X). From Perfil'ev and Gabe, 1965; with permission from Plenum Publishing Corporation.) (B) *Metallogenium* (arrows) from filtered water from Pluss See, Schleswig-Holstein, West Germany (2,450X). (Courtesy of W. C. Ghiorse and W.-D. Schmidt.)

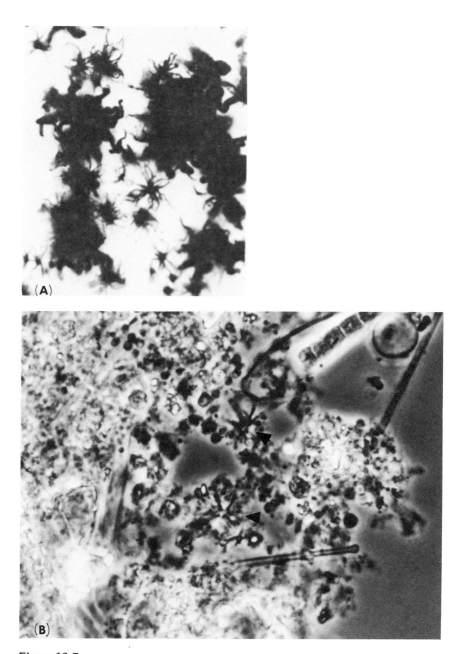

Figure 13.7

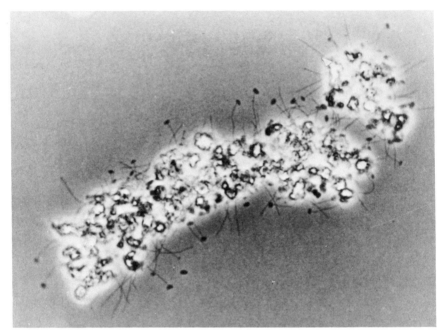

Figure 13.8 *Pedomicrobium* in association with manganese oxide particles (2,800X). (Courtesy of W. C. Ghiorse.)

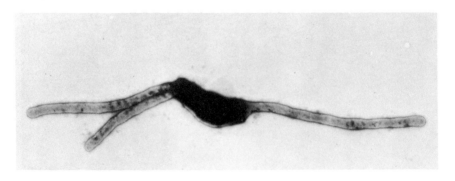

Figure 13.9 Manganese-oxidizing *Hyphomicrobium* sp. isolated from a Baltic Sea iron-manganese crust (15,600X). (Courtesy of W. C. Ghiorse.)

13.2 MICROBIAL REDUCTION OF Mn(IV)

Enzymatic Reduction by Soil and Freshwater Bacteria   Adeny in 1894 showed that manganese dioxide, formed in sewage by addition of potassium permanganate, was reduced to manganous carbonate. He attributed this reduction to the action of bacteria and thought it analogous to bacterial nitrate reduction. Troshanov in 1968 isolated a number of Mn(IV)-reducing bacteria from reduced horizons of several Karelian lakes (U.S.S.R.). The isolates included *Bacillus circulans, B. polymyxa, Pseudomonas liquefaciens, B. mesentericus, B. mycoides, Micrococcus* sp., *B. cereus, B. centrosporus,* and *B. filaris,* among others. Some of these strains could reduce both manganese and iron oxides, but strains that could reduce only manganese oxides were encountered most frequently. Nitrate did not inhibit Mn(IV) reduction the way it did Fe(III) reduction by *Bacillus circulans, B. polymyxa, P. liquifaciens,* and *B. mesentericus* (Troshanov, 1969). Anaerobic conditions stimulated Mn(IV) reduction by *B. centrosporus, B. mycoides, B. filaris,* and *B. polymyxa* but not by *B. circulans, B. mesentericus,* or *Micrococcus albus.* All carbohydrates tested were not equally good sources of reducing power for a given organism (Troshanov, 1969). Although a significant part of the Mn(IV) seemed to be reduced enzymatically, it was believed that some of it was reduced chemically since some reduction of oxidized manganese in lake ores by glucose and xylose and by some organic acids (e.g., acetic and butyric acids) was observed at a pH of 4.3-4.4 under sterile conditions (Troshanov, 1968). Such reaction of glucose with Mn(IV) oxides is, however, dependent on the form of the oxide; hydrous Mn(IV) oxides react readily, whereas crystalline MnO_2 does not. Recently, Bromfield and David (1976) reported that *Arthrobacter* strain B can reduce manganese oxides at lowered oxygen tension.

Enzymatic Reduction by Marine Bacteria   *Bacillus* 29 has been shown to reduce MnO_2 using glucose as the source of reducing power (Trimble and Ehrlich, 1968). The glucose does not react significantly with reagent-grade MnO_2 in the experiments. Cultures of *Bacillus* 29 previously grown in the absence of added $MnSO_4$ do not start to reduce MnO_2 in growth experiments until 1 day after inoculation. Anaerobic conditions (incubation in an oxygen-free atmosphere of nitrogen) before MnO_2 reduction is initiated prevents reduction during the whole period of the experiment. Anaerobic conditions after MnO_2 reduction is initiated permit continuation of the reduction at a slightly reduced rate. MnO_2 reduction once initiated by a culture growing in the presence of MnO_2 is quickly inhibited by the addition to the medium of $HgCl_2$ to a final concentration of 10^{-3} M. During the MnO_2 reduction, the organism forms more lactate and pyruvate from glucose than in the absence of MnO_2. It also consumes more glucose in the medium than without it (Table 13.4). Results very similar to those with growing cultures of *Bacillus* 29 have also been obtained with another marine culture, coccus 32 (Table 13.4).

Table 13.4 Chemical and Physical Changes During Growth of *Bacillus* 29 and Coccus 32 in the Presence and Absence of MnO_2 [a]

Culture	Time (days)	Glucose consumed (nmol/ml)		Manganese released (nmol/ml)	Pyruvate produced (nmol/ml)		Lactate produced (nmol/ml)		E_h (mv)		pH		Cells/ml ($\times 10^7$)	
		$+MnO_2$	$-MnO_2$	$+MnO_2$	$+MnO_2$	$-MnO_2$	$+MnO_2$	$-MnO_2$	$+MnO_2$	$-MnO_2$	$+MnO_2$	$-MnO_2$	$+MnO_2$	$-MnO_2$
29	0	0	0	0	0	0	0	0	469	474	7.3	7.2	1.6	1.6
	1	1,720	655	25	43	32	461	449	528	482	6.6	6.7	3.8	4.0
	3	3,060	1,239	240	190	153	1,635	1,612	502	512	5.4	5.6	4.5	4.8
	4	3,686	2,342	420	438	330	2,430	1,955	573	514	5.7	5.5	4.5	5.0
	7	6,120	3,635	1,060	705	243	2,080	1,387	615	594	5.7	5.5	4.6	5.2
	8	8,750	5,286	1,260	879	330	2,040	1,705	617	589	5.5	4.9	4.8	5.0
	11	11,893	7,281	1,830	936	225	2,430	1,761	602	589	5.8	5.1	5.0	5.0
32	0	0	0	0	0	0	0	0	426	426	7.7	7.7	2.5	2.5
	1	480	0	0	50	52	0	0	394	419	7.0	6.9	4.6	5.8
	2	1,440	1,440	0	84	75	9	5	519	490	6.8	6.5	6.2	7.0
	3			130	138	79			491	496	6.6	6.4	6.2	7.0
	7	3,000	2,650	490	1,025	575	174	68	484	495	5.7	4.7	6.0	4.8
	9	4,900	3,125	780	1,650	980	250	124	514	549	5.4	4.5	5.8	4.2
	11	5,550	3,600	1,020	2,625	1,120	250	123	475	509	5.6	4.3	5.4	4.8

[a] Culture medium: For *Bacillus* 29, 0.48% glucose and 0.048% peptone in sea water; for coccus 32, 0.60% glucose and 0.048% peptone in sea water. One-gram portions of reagent grade MnO_2 were added to 20 ml of medium, as needed. Incubation was at 25°C. (Source: From Trimble and Ehrlich, 1968; by permission.)

224

MnO_2 reduction has also been studied with resting-cell cultures and with cell extracts of *Bacillus* 29 (Ehrlich, 1966; Trimble and Ehrlich, 1970). Resting cells, previously grown in the absence of added manganese, are able to reduce MnO_2 only in the presence of dilute ferri- or ferrocyanide, whereas resting cells previously grown in the presence of added $MnSO_4$ are able to reduce MnO_2 without added ferri- or ferrocyanide. The ferri- or ferrocyanide substitutes for a missing component in the electron transport system in cells not grown in the presence of $MnSO_4$. Mann and Quastel (1946) had previously observed MnO_2 reduction by whole cells of bacteria when pyocyanin was used as electron carrier. They thought, however, that such extracellular carriers were essential to enzymatic reduction of MnO_2. Since in *Bacillus* 29 inhibition of protein synthesis with chloramphenicol or inhibition of messenger RNA synthesis with actinomycin D prevented the appearance of the missing electron transport component in the presence of added $MnSO_4$, it has been concluded that the missing component in *Bacillus* 29 is induced by added manganese in the medium (Trimble and Ehrlich, 1970). The manganese inducer can be in the form of Mn(II) or Mn(IV), since in growth experiments the bacteria become induced whether in contact with $MnSO_4$ or MnO_2. Reduction of MnO_2 with cell extracts of *Bacillus* 29 proceeds in the same way as with resting cells. Extract from cells previously grown without added $MnSO_4$ do not reduce MnO_2 in the absence of added ferri- or ferrocyanide, whereas extracts from cells previously grown in the presence of added $MnSO_4$ do.

Many other bacterial isolates from the marine environment have since been shown to be able to reduce MnO_2 by a mechanism like that of *Bacillus* 29 (Ehrlich et al., 1972; Ehrlich, 1973). The chemical reactions brought about by the bacteria can be summarized by the following equations:

$$\text{Glucose} \xrightarrow{\text{bacteria}} ne^- + nH^+ + \text{end products} \tag{13.8}$$

$$n/2\ MnO_2 + ne^- + nH^+ \xrightarrow[\substack{\text{uninduced bacteria +}\\ \text{ferri- or ferrocyanide}}]{\text{induced bacteria or}} n/2\ Mn(OH)_2 \tag{13.9}$$

$$n/2\ Mn(OH)_2 + nH^+ \longrightarrow n/2\ Mn^{2+} + nH_2O \tag{13.10}$$

The reason for representing the direct product of reduction of MnO_2 by the bacteria as $Mn(OH)_2$ rather than as Mn^{2+} is that in resting-cell experiments or in experiments with cell-free extracts, both of which have a duration of only 3-4 hr, it is necessary to acidify the reaction mixture to about pH 2 on completion of incubation in order to bring Mn^{2+} into solution. Such acidification is not necessary in growth experiments where acid production from glucose by the bacteria or complexation by medium constituents bring Mn^{2+} into solution (Trimble and Ehrlich, 1968).

Equations 13.8 and 13.9 indicate that MnO_2 serves as a terminal electron acceptor. A study of *Bacillus* 29 has shown the presence of cytochromes of the a, b, and c types, flavoproteins, and possibly nonheme iron protein. Of these, flavoproteins, cytochrome c, and a metalloprotein are probably involved in electron transport from the oxidizable substrate to MnO_2 (Ghiorse and Ehrlich, 1976). Inhibitor studies with other marine MnO_2-reducing bacteria have shown that the electron transport system to MnO_2 may differ in different types of organisms (Ehrlich, unpublished data). The electron transport system appears to be membrane-associated. This would seem to be essential since MnO_2 is insoluble. Hence, MnO_2-reducing cells must be in intimate contact with MnO_2 to reduce it (Ghiorse and Ehrlich, 1976).

Whether MnO_2 can completely replace oxygen as a terminal electron acceptor has not yet been determined. It has been found that MnO_2 in an aerobic system with *Bacillus* 29 decreases oxygen consumption (Trimble, 1967).

Nonenzymatic Reduction of Mn(IV)   Some bacteria and probably all fungi can reduce Mn(IV) oxides such as MnO_2 indirectly (nonenzymatically). A likely mechanism of reaction is the production of metabolic products that are strong enough reductants for Mn(IV) oxides. *Escherichia coli,* for instance, produces formic acid from glucose which is capable of reacting with MnO_2 :

$$3H^+ + HCOO^- + MnO_2 \rightarrow Mn^{2+} + CO_2 + 2H_2O \qquad (13.11)$$

Many fungi produce oxalic acid, which can also reduce MnO_2 :

$$4H^+ + {}^-OOCCOO^- + MnO_2 \rightarrow Mn^{2+} + 2CO_2 + 2H_2O \qquad (13.12)$$

Since the electron transport mechanism in fungi, which are eukaryotic organisms, is located in mitochondrial membranes and not in the plasma membrane, as in prokaryotic cells, fungi cannot be expected to reduce MnO_2 enzymatically (Ehrlich, 1978a). Whether MnO_2 will be reduced in acid solution under anoxic conditions by reversal of the reaction

$$Mn^{2+} + \tfrac{1}{2}O_2 + H_2O \rightarrow MnO_2 + 2H^+ \qquad (13.13)$$

is very questionable, owing to the high energy of activation requirement.

The fact that MnO_2 is reduced indirectly by fungi can be readily demonstrated on a glucose-containing agar medium which has MnO_2 incorporated into it (see, e.g., Schweisfurth, 1968; Tortoriello, 1971). Fungal colonies growing in such a medium develop a halo (clear zone) around them in which MnO_2 has been dissolved (reduced). Since enzymes do not work across a separation between them and their substrate, this can only be explained on the basis that a reducing compound is released which then reacts with the MnO_2 .

13.3 MICROBIAL TRANSFORMATIONS OF MANGANESE IN SOIL

Forms of Manganese in Soil and Their Detection   The foregoing types of interactions of microorganisms with manganese may have profound effects in the biosphere in that they may help or hinder the mobility of manganese and its availability as a micronutrient. From a geological standpoint, these interactions may also have a profound influence on the distribution of manganese in the lithosphere and hydrosphere (i.e., they may contribute to accumulation or dissipation of locally concentrated deposits).

Soil manganese may exist in a mobile or immobile form. The mobile form includes Mn^{2+} ions, complexes of Mn^{2+}, particularly organic complexes such as those of humic and tannic acids, and possibly complexes of $Mn(III)$. Immobile forms include adsorbed Mn^{2+} ions, $Mn(OH)_2$, and insoluble salts of $Mn(II)$, such as $MnCO_3$, $MnSiO_3$, and MnS, and various $Mn(IV)$ oxides. The insolubility of immobile compounds is governed by prevailing pH and E_h conditions. Reducing conditions at near-neutral or alkaline pH are especially favorable for the stability of manganous compounds, while oxidizing conditions at near-neutral or alkaline pH are especially favorable for the stability of $Mn(IV)$ oxides. Agriculturally, the most important forms of insoluble manganese include the oxides of $Mn(IV)$ and mixed oxides such as $MnO \cdot MnO_2$ or Mn_2O_3, or $2MnO \cdot MnO_2$ or Mn_3O_4. Their stability may also be affected by the presence of iron (Collins and Buol, 1970). Soil chemists have distinguished between the various forms of manganese in soil in an empirical fashion through the use of different extraction methods (see, e.g., Robinson, 1929; Sherman et al., 1942; Leeper, 1947; Reid and Miller, 1963). Thus, Sherman et al. (1942) measured, in successive steps, water-soluble manganese by extracting a soil sample with distilled water, then exchangeable manganese by extracting the residue in 1 N NH_4 acetate (pH 7.0), and, finally, easily reducible manganese by extracting the second residue with 1 N NH_4 acetate containing 0.2% hydroquinone. Other investigators have used different extraction reagents to measure exchangeable and easily reducible manganese (Robinson, 1929; Leeper, 1947).

Bromfield and David (1978), using manganese oxide prepared by bacterial oxidation of $MnSO_4$ which they considered to be equivalent to manganese oxides of soils, showed that nonreducing extractants brought manganese into solution in proportion to the amount of oxide in the reaction mixture. Acid extractants brought bound manganous manganese into solution; the more acid, the better the recovery. Reducing agents such as 0.2% hydroquinone completely dissolved any oxide present. Ethylenediaminetetraacetic acid (EDTA) (0.05 M) extractant was most effective at pH values below 6. The extraction of manganese from manganese oxide by NH_4 acetate was enhanced by cations such as Cu^{2+} and Zn^{2+}. Except for the reducing agents, all other reagents probably brought into solution to various degrees mostly the Mn^{2+} that had been adsorbed by the manganese oxide.

The most common mechanisms by which mobile manganese is fixed in soil or sediment are through adsorption to immobile organic or inorganic phases of soil, through precipitation as insoluble salts, or through oxidation to the $Mn(IV)$ state. Of these processes, adsorption and oxidation of Mn^{2+} are the most important mechanisms of immobilization in soil. Oxidation is largely microbial.

Special Manifestations of Manganese Oxides in Soils and On Rocks Nodules or concretions containing manganese and iron oxides have been found in a number of soils (Doherty, 1898; Thresh, 1902; Helbig, 1914; see Robinson, 1929, for these references; and Drosdoff and Nikiforoff, 1940; Roslikova, 1961; Redden and Porter, 1962; see Taylor et al., 1964, for these references). The iron in these formations may occur in significantly larger quantities than the manganese. Besides manganese and iron, they contain traces of Co, Ni, Mo, V, Ga, Pb, Ba, Cr, and Zn (Taylor and McKenzie, 1966). Basing his proposal on suggestions by Murray and Irvine (1894) regarding the origin of marine manganese nodules, Robinson (1929) put forth that soil nodules arise through interaction of manganese with calcium carbonate of calciferous organic remains. It seems not unlikely, however, that microbes are involved in the genesis of soil nodules. Aristovskaya (1963) has implicated *Pedomicrobium* in their formation, and Douka (1977) has implicated a Mn^{2+}-oxidizing *Pseudomonas* sp. and *Citrobacter freundii* in their formation.

Desert varnish is found as a thin brown to black veneer on stones, rocks, and other mineral material in arid regions. It consists of iron and manganese oxides. It was recognized as far back as 1813 (de Roziere: see Krumbein, 1969). Lichens, algae, fungi, and bacteria have been associated with these deposits and may play a role in their formation (Krumbein, 1969). The activity of these microorganisms is probably indirect.

Role of Soil Microbes in Making Manganese Available to Plants Because fixed manganese in soil is not readily available as a plant nutrient, such fixation may lead to a manganese deficiency in plants, such as gray-speck disease. This does not always happen since fixed manganous manganese may be remobilized. Bromfield (1958a,b) has shown that substances released from roots of such plants as oats and vetch may remobilize manganese from insoluble, microbially produced γ-MnO_2 in the from of Mn^{2+} ions, especially at acid pH. Bacteria are apparently not needed to promote $Mn(IV)$-reducing activity of released root substances (Bromfield, 1958a). However, in view of a widespread ability of various types of bacteria to reduce MnO_2, bacterial reduction of oxides of $Mn(IV)$ may be a more important factor in remobilizing fixed manganese in nature.

Microbial Manganese Oxidizing and Reducing Activity in Soil   The ability of microorganisms to oxidize Mn^{2+} to $Mn(IV)$ oxides and to reduce $Mn(IV)$ oxides to Mn^{2+} in soil has been convincingly demonstrated in soil perfusion experiments (Mann and Quastel, 1946). Using percolation columns charged with soil through which 0.02 M $MnSO_4$ were continuously perfused, a buildup of oxides of manganese was noted with time which paralleled the disappearance of manganese from the perfused solution. This buildup of oxides of manganese was attributed to microbial activity because it was found that adding such cell poisons as chloretone, sodium iodoacetate, or sodium azide to an active column caused marked inhibition of further Mn^{2+} oxidation. When glucose solution was added to the soil percolation columns in which oxides of manganese had accumulated, it was found that manganese was being reduced (Mann and Quastel, 1946). Sodium azide inhibited this reduction, although to a lesser extent than it did oxidation. It was inferred that oxides of manganese acted as hydrogen acceptors in a microbially promoted process.

Under some conditions, $Mn(IV)$ oxides in soil can be solubilized as a result of the oxidation of elemental sulfur (S^0) or thiosulfate $(S_2O_3^{2-})$ by *Thiobacillus thiooxidans* (Vavra and Frederick, 1952). This was shown in soil perfusion studies in the laboratory. The production of sulfuric acid was not solely responsible for the solubilization of the MnO_2, because in the presence of acid but in the absence of any reduced sulfur, the MnO_2 was not solubilized. Although the greatest quantity of MnO_2 was reduced when *T. thiooxidans* cells were in contact with MnO_2, slightly more than half of the MnO_2 was reduced under the same conditions when the cells were separated from the MnO_2 by a collodion membrane. This may be a case of simultaneous direct and indirect reduction of MnO_2. A field study confirmed that S^0 and $S_2O_3^{2-}$ can mobilize fixed manganese in an agriculturally manganese deficient soil. This study was performed in Indiana (Garey and Barber, 1952).

The action of *T. thiooxidans* has also been tested in the extraction of manganese from ores (Kazutami and Tano, 1967). The organism leached manganese from an ore consisting of SiO_2 (55%), Fe_2O_3 (25%), MnO_2 (10.6%), MgO (5.23%), and traces of Ca, Al, and S. The ore was treated at a concentration of 3% in this study in a medium containing K_2HPO_4 (0.4%), $MgSO_4$ (0.03%), $CaCl_2$ (0.02%), $FeSO_4$ (0.001%), $(NH_4)_2SO_4$ (0.2%), and S^0 (1%). Addition of FeS or $FeSO_4$ stimulated both the growth of the organism and the solubilization of manganese, while the addition of ZnS stimulated only the solubilization of manganese; $Fe_2(SO_4)_3$ was without effect. Manganese extraction of ores has also been accomplished with *Bacillus* sp. in a medium employing organic nutrients (Perkins and Novielli, 1962). The mechanism of extraction in this case was not examined.

Methods of Detecting Mn^{2+} Oxidizers in Soil   Beijerinck (1913) was the first
to demonstrate manganese-oxidizing bacteria and fungi from soil by culturing
them on agar media containing manganous carbonate or other manganous salts.
He thus identified *Bacillus manganicus* and *Papulospora manganica* as manganese
oxidizers by their deposition of brown Mn(IV) oxide in the colonies or in the
medium. Thiel (1925) also employed this method. Gerretsen (1937) developed
a new method by preparing petri plates of agar mixed with unsterilized soil. He
then removed a central core of agar on the plate and replaced it with a sandy soil
mixture containing 1% MnSO$_4$. As Mn^{2+} ions diffused into the agar, any
developing bacterial or fungal colonies growing on the agar would accumulate a
brown precipitate of manganese oxides in and around themselves if they were
able to oxidize manganese (Fig. 13.10). Most precipitate formed in the pH
range 6.3-7.8 and therefore was of biogenic origin because autoxidation does
not occur readily below pH 8. Plates with antiseptically treated soil did not
develop such a precipitate of manganese oxide. Gerretsen's method was later

Figure 13.10   Gerretsen plate as modified by Leeper and Swaby (1940),
showing manganese oxide deposition (dark halo) as a result of microbial growth
around the central MnSO$_4$-containing agar plug. Initial pH 7.3; final pH 6.7.
[From Leeper and Swaby, 1940. Copyright (1940) The Williams & Wilkins Co.,
Baltimore.]

used by Leeper and Swaby (1940) in a study of Mn^{2+}-oxidizing microorganisms in Australian soils. Others who have observed manganese-oxidizing microorganisms in soil by a variety of methods include Timonin (1950a,b), Timonin et al. (1972); Bromfield and Skerman (1950), Bromfield (1956), Aristovskaya (1961), Aristovskaya and Parinkina (1963), Perfil'ev and Gabe (1969), Schweisfurth (1969), Khak-mun (1967), and others. Perfil'ev introduced a new technique for quasi in situ observation of Mn^{2+}-oxidizing microorganisms through the invention of the **pedoscope**. This apparatus consists of one or more capillaries with optically flat sides for direct microscopic observation of the capillary content (Perfil'ev and Gabe, 1965). Inserting the pedoscope into soil permits development of soil microbes in the capillary lumen under soil conditions. Periodic removal of the pedoscope and examination under the microscope permits a visual assessment of the developmental progress of the organisms (see Chapter 6). On the basis of application of these and other methods, *Athrobacter* or *Corynebacterium*, *Pedomicrobium*, and *Metallogenium* are among the more frequently mentioned Mn^{2+}-oxidizing soil microbes in the literature. It is quite probable, however, that a number of other Mn^{2+}-oxidizing bacteria in soil are important and remain to be identified.

Methods of Detecting Mn(IV)-Reducing Microorganisms in Soil   Schweisfurth (1968) has developed a method for detecting Mn(IV) oxide reducers using a medium consisting of meat extract (0.75 g), glucose (5 g), K_2HPO_4 (0.01 g), Noble agar—Difco (15 g), and tap water (1,000 ml). After solution of the ingredients, the medium is adjusted to pH 6, and then autoclaved at 121°C for 15 min. Before pouring petri plates with this medium, 1-1.2 ml of 2% $KMnO_4$ solution is added to each 100 ml of it, and the pH is readjusted to 6.8. After surface inoculation, plates with this medium are incubated at 20°C for 2-7 days and examined for colonies of active Mn(IV)-reducing microorganisms. Their colonies may be recognized by formation of a colorless halo around them or by flooding the plates with a suitable indicator solution such as benzidine hydrochloride (caution! reagent is carcinogenic) or leukoberbelin blue (Altmann, 1972; Krumbein and Altmann, 1973), which intensifies the halo where the Mn(IV) oxide has been dissolved by coloring the rest of the medium blue. This culture method can only test for organisms capable of reducing Mn(IV) oxides nonenzymatically (i.e., by producing reactive metabolic end products), because the Mn(IV)-reducing enzymes are not diffusible from the cells and thus cannot be responsible for forming the halo around the colonies.

The manganese oxide formed from the reaction of the permanganate in the foregoing medium is very pH-sensitive. Slight acidification is enough to cause reduction of the manganese by glucose. A modification of this medium largely eliminates this problem. The procedure requires a separate preparation of a basal medium containing 0.1% glucose, 0.2% peptone, and 1.5% agar in distilled water, and a capping agar dispensed in 100 X 13 mm test tubes in 2.5-ml amounts consisting of 1.5 ml agar (1.5%) in distilled water, 0.5 ml of 0.01 M potassium ferricyanide, and 1 ml of a suspension of synthetic MnO_2 prepared by the method of Guyard (1864). This MnO_2 is less easily reduced by glucose at an acid pH than that formed from permanganate in the pervious medium. Both basal and capping agars are sterilized by autoclaving. Sterile basal agar is poured in 10-15-ml aliquots into sterile petri plates and allowed to set. Melted capping agar is cooled to about 45°C, inoculated with 1 ml of an appropriate soil suspension, and poured on the basal agar. After setting of the capping agar, the plates are incubated at a desired temperature until maximal development of colonies (7-14 days). As with the previous medium, Mn(IV)-oxide-reducing colonies are identified by clear halos around their perimeter, which may be intensified by use of benzidine or leukoberbelin blue reagents. The added ferricyanide in the capping agar acts as a diffusible electron carrier between the cells of active colonies and the MnO_2 particles and allows thus for detection of potential enzymatic MnO_2 reducers as well as nonenzymatic MnO_2 reducers (Tortoriello, 1971).

13.4 MICROBIAL MANGANESE TRANSFORMATIONS IN FRESH WATER

Most early observations of manganese-oxidizing microorganisms were made in freshwater environments (Neufeld, 1904; Molisch, 1910; Lieske, 1919; as cited by Moese and Brantner, 1966; Thiel, 1925; von Wolzogen-Kuehr, 1927; Zappfe, 1931; Sartory and Meyer, 1947). The organisms were usually detected in sediments or organic debris or in manganiferous crusts. Indeed, the sediments seemed to be where the most intensive transformations of manganese were occurring.

Bacterial Manganese Oxidation and Precipitation in Springs   Manganese precipitation as MnO_2 has been observed in a mineral spring near Komaga-dake on the island of Hokkaido (Hariya and Kikuchi, 1964). Water from this spring contained 4.75 mg of Mn^{2+} per liter but only traces of Fe^{2+} or Fe^{3+}. Concentrations of other constituents were (in milligrams per liter): K^+, 16; Na^+, 128; Ca^{2+}, 101; Mg^{2+}, 52; Cl^-, 156.2; SO_4^{2-}, 481; HCO_3^-, 117.9; and CO_2, 11. The temperature of the water was 23°C, and its pH was 6.8. In experiments with this water, manganese precipitated progressively between 20 and 50 days accompanied by a fall in pH and a gradual rise in E_h after 20 days, which became abrupt after 45 days. Only 1.18 mg of Mn^{2+} per liter were left in solution after 50 days. Manganese-oxidizing bacteria were detected in the mineral water by filtering it through ordindary filter paper and through membrane filters. Sheathed bacteria were recovered from the filter paper and "manganese bacteria" from the membrane filter. The organisms were cultured in two different mineral salts media, one including $Mn(HCO_3)_2$ and the other $MnCl_2 \cdot 4H_2O$ as sole energy sources. Both media contained $(NH_4)_2SO_4$ as sole nitrogen source. The manganese bacteria were relatively large rods ($1 \times 5 \times 8$ μm) which were peritrichously flagellated. They were covered by a manganese hydroxide (sic) precipitate (probably meaning hydrous manganese oxide). This manganese hydroxide recrystallized into pyrolusite and birnessite in the spring.

Bacterial Manganese Deposition in Water Distribution Systems   In 1962, a case of microbial manganese precipitation in a water pipeline connecting a reservoir with the filtration plant of the waterworks of the city of Trier, West Germany, was reported (Schweisfurth and Mertes, 1962). The accumulation of precipitate in the pipes caused a loss of water pressure in the line. The sediment in the pipe had a dark brown to black coloration and was rich in manganese but relatively poor in iron content. Microscopic examination of the sediment revealed the presence of cocci and rods after a removal of MnO_2 with 10% oxalic acid solution. Sheathed bacteria were found only on the rubber-sealed seams, and then always in association with other bacteria. Evidence of fungal mycelia and streptomycetes was also found. In culture experiments, only gram-negative rods and fungi grew up. Chemical examination of the reservoir revealed that the manganese concentration in the bottom water was between 0.25 and 0.5 mg liter^{-1} during most of 1960, except in September-October, when it ran as high as 6 mg liter^{-1}. The peak in manganese concentration in bottom waters was correlated with water temperature, which reached its peak at about the same time as the manganese concentration in the water. The two feedstreams into the reservoir did not contribute large amounts of manganese, only about 0.05 mg liter^{-1}. The major source of the manganese could have been the manganiferous minerals of the reservoir basin and surrounding watershed.

Similar observations were made in some pipelines connecting certain water reservoirs with hydroelectric plants in Tasmania, Australia (Tyler and Marshall, 1967a). In this instance, pipelines from Lake King William were found to have heavy deposits of manganic oxide, whereas those from Great Lake did not. The deposits in pipes leading from Lake King William accumulated to a maximum thickness of 7 mm in 6-12 months. The manganic oxide deposition process was reproduced in the laboratory with a recirculatory apparatus (Tyler and Marshall, 1967a). With water from Lake King William, a deposit of a brown manganic oxide formed at the edge of coverslips after 24 hr and for 6 days thereafter. Subsequent addition of $MnSO_4$ to the water caused further deposition after 6 days. By contrast, only slight traces of deposit developed with Great Lake water in similar experiments. This difference in manganese deposition from the two lake waters was explained in part in terms of the difference in manganese content of the two waters. Lake King William water contained 0.01-0.07 ppm of manganese while Great Lake water contained only 0.001-0.013 ppm. In the laboratory, inoculation of Great Lake water with some Lake King William water did not promote the oxidation unless $MnSO_4$ was also added. It was also found that if Lake King William water was autoclaved or was treated with azide (10^{-3} M final concentration), manganic oxide deposition was prevented, suggesting the participation of a biological agent in the reaction. Inoculation of autoclaved Lake King William water with untreated water caused resumption of the oxidation. The dominant organism involved in the oxidation appeared to be one identified as *Hyphomicrobium* sp. Sheathed bacteria and fungi and possibly *Metallogenium symbioticum* were also encountered in platings from the pipeline deposits but only at low dilutions. It was not resolved whether *Hyphomicrobium* sp. oxidized Mn^{2+} enzymatically or nonenzymatically (Tyler and Marshall, 1967b). Since the publication of the original studies of manganic oxide pipeline deposits in Tasmania, *Hyphomicrobium* has been found in pipeline deposits in other parts of the world (Tyler, 1970). It was not always the only manganese oxidizing organism present, however.

Bacterial Manganese Oxidation in a Karelian Lake   Bacterial manganese deposition has been observed in Lake Punnus-Yarvi on the Karelian peninsula in the U.S.S.R. (Sokolova-Dubinina and Deryugina, 1967a,b). This lake (also described in Chapter 7) is of glacial origin, 7 km long and up to 1.5 km wide, with a maximum depth of 14 km. It is oligotrophic, slightly stratified thermally, and contains significant amounts of dissolved iron (0.7-1.8 mg liter^{-1}) and manganese (0.02-1.4 mg liter^{-1}) only in its deeper waters. The lake is fed by 2 rivers and 24 streams, which drain surrounding swamps. The manganese and iron in the lake water is supplied mainly by surface-and groundwater drainage containing 0.2-0.8 mg liter^{-1} and 0.4-2 mg liter^{-1}, respectively. Most of the manganese and iron in the lake is deposited along the northwestern banks of the Punnus-Ioki Bay situated at the outflow from the lake into Punnus-Ioki River in a deposit 5-7 cm thick at a depth of up to 5-7 m. The deposit is described as including hydrogoethite (nFeO · nH$_2$O), wad (MnO$_2$ · nH$_2$O) and psilomelane (mMnO · MnO$_2$). The iron content of the deposit ranges from 18 to 60% and the manganese content from 10 to 58%. There is also 5-16% SiO$_2$, and a number of elements, such as Al, Ba, and Mg, and the organic matter, in amounts of 0.3-0.7%. *Metallogenium* was found to occur in all parts of the lake. To demonstrate that the formation of the manganese oxide ores in the lake is biogenic, sediment samples were collected from a station in the lake where the ore occurs, and incubated in the laboratory at 8°C (Sokolova-Dubinina and Deryugina, 1967b). After several months, the appearance of dark brown, compact spots that consisted of manganese oxides were noted. No characteristic bacterial structures were found in them on light-microscopic examination.

However, in a later electron microscope study, *Metallogenium* was found in water suspensions from surface material from the ore (Dubinina and Deryugina, 1971). Upon culturing on a manganese acetate agar medium, *Metallogenium* was recovered. Sediment portions from the lake in which microconcretions were not found did not yield any *Metallogenium* on culturing. When thin sections of microconcretions, formed in the laboratory, were compared to the natural lake ores and to manganese oxide ore from the Chiatura manganese deposits, *Metallogenium* was found in all of them. A study of *Metallogenium* in a Perfil'ev peloscope showed progressive encrustation of the organism with manganese oxides as growth proceeded. Ore formation in the lake ore correlated mostly with a high redox potential range of 0.435-0.720 V and a pH range of 6.3-7.1, although it was concluded from data gathered by others that ore formation may begin to occur at an E_h as low as 0.230 V and at a pH of 6.5. At Mn^{2+} concentrations not exceeding 10 mg liter^{-1}, autoxidation was not likely.

It should be commented here that the identity of *Metallogenium* is presently somewhat of a puzzle. Star-shaped coenobia, typical of *Metallogenium,* have now also been recognized in lakes in the Oslo district of southern Norway. However, these coenobia lack identifiable enclosed cellular structures, such as biological membranes or organelles, although threadlike structures are sometimes seen in sections viewed in the electron microscope around which manganese oxide may be deposited (Klaveness, 1977). It has been speculated that the *Metallogenium* organism is actually a coccoid cell at the tip of each arm of the star-shaped coenobium (Klaveness, 1977). The distribution of *Metallogenium* in the Norwegian lakes follows seasonal cycles.

Bacterial Manganese Reduction in Karelian Lakes   Since some of the manganese in Lake Punnus-Yarvi is also deposited as manganous carbonate (rhodochrosite), its origin was examined in relation to possible microbial participation (Sokolova-Dubinina and Deryugina, 1967a). As discussed in Chapter 7, it was concluded that *Bacillus circulans, B. polymyxa,* and an unidentified nonsporeforming rod was in part responsible for the manganous carbonate formation.

Additional studies of the activity of manganese oxide reducing microorganisms from Lake Punnus-Yarvi were made by Troshanov (1968). He examined the dissolution of pulverized lake ores in an agar medium in Perfil'ev's peloscopes inoculated with silt from the lake as a source of the appropriate organisms. After sufficient incubation, he isolated cultures from zones of dissolution in the peloscopes and transferred them to liquid medium. The organisms were tested for their ability to reduce manganese and iron oxides in a medium containing KH_2PO_4, 0.5 g; $MgSO_4$, 0.2 g; $(NH_4)_2SO_4$, 1.0 g; $CaCO_3$, 95 g; sucrose, 5 g; yeast autolysate, 5 ml; ground ore, 30 g; and water, 1,000 ml; the pH of the medium was adjusted to 7.0-7.2. After incubation of these cultures, he measured the amount of Mn^{2+} and Fe^{2+} released from the ore. He found that all isolates of *B. circulans, B. mesentericus, B. mycoides, B. polymyxa, Micrococcus albus,* and others could reduce manganic oxide but that only some could reduce iron. *B. circulans* was the most common. Interestingly, Troshanov found that when he plated silt samples directly without prior enrichment in peloscopes, manganic oxide reducing bacteria were encountered at low frequency. He inferred that not all members of the silt microflora can reduce manganic oxides.

Other lakes on the Karelian peninsula have also been examined for manganese oxide deposition and reduction (Sokolova-Dubinina and Deryugina, 1968; Troshanov, 1968). The studies indicated that for manganese oxidation to occur, a steady supply of Mn^{2+} in runoff and groundwater is needed, together with stable oxidizing conditions. Such conditions are best met only in oligotrophic lakes or in eutrophic lakes with sufficiently high oxidizing conditions. Lake Isk-Yarvi, a shallow lake (maximum depth, 3.5 m) that this highly eutrophic, having a high organic carbon concentration and consequent low redox potential in its sediment, does not support the growth of *Metallogenium* but does support the growth of sulfate-, manganese-, and iron-reducing bacteria in its sediment. Manganese- and iron-reducing bacteria may thus be found even in lakes where no ferromanganese ores accumulate.

Ferromanganese Lake Ores in Various Parts of the World   Lake ores of ferromanganese have been found in a number of water bodies in North America, such as Ship Harbour Lake, Nova Scotia (Kindle, 1932); in Lake Oneida, New York (Gillette, 1961; Dean 1970; Dean and Ghosh, 1978); Great Lake, Nova Scotia, and Mosque Lake, Ontario (Harriss and Troup, 1969, 1970; Lake Ontario (Cronan and Thomas, 1970); Lake George, New York (Schoettle and Friedman, 1971) (Fig. 13.11); and Lake Michigan (Rossman and Callender, 1968). The ores occur as flat, disklike or spherical to elongate structures. Their composition varies widely. Typical analyses are given in Table 13.5. Although Kindle (1932), Gillette (1961), and Dean (1970) thought microbial participation in the formation of these concretions likely, no extensive microbiologic studies were undertaken in any instances. Kindle thought that an indirect mechanism was probably operating in which removal of CO_2 from the lake water through photosynthesis by diatoms caused conversion of Mn^{2+} to MnO_2 as a result of an expected rise in pH (autoxidation). Harriss and Troup (1970), citing a hypothesis of Gorham, also favor the indirect mechanism proposed by Kindle. They explain the alternate iron-rich and manganese-rich layers in their concretions on the basis of the difference in susceptibility to oxidation of iron and manganese and the seasonal variation in oxidizing conditions in the lakes they studied. Gillette (1961) found bacteria in powdered nodules that were able to precipitate iron in laboratory experiments, but these organisms were not

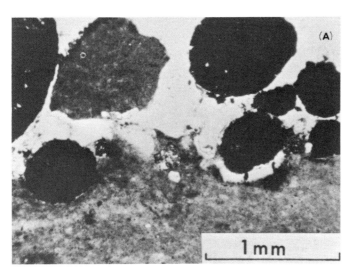

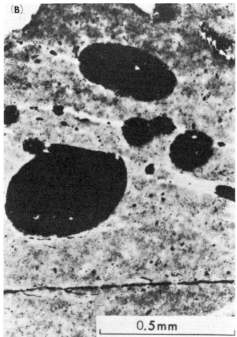

Figure 13.11 Freshwater iron-manganese nodules from Lake George, New York. (A) Nodules sitting on the surface of lake bottom clay. Note one detached particle consisting of a clay aggregate. (B) Nodules in clay near the surface of the lake floor. (From Schoettle and Friedman, 1971; by permission, Geological Society of America.)

Table 13.5 Composition of Some Lake Ores (Values in Percent)

Mn	Fe	Si	Al	Source	Reference
36.08 (MnO_2)	13.74 (Fe_2O_3), 7.70 (FeO)	12.75 (SiO_2)	12.50 (Al_2O_3)	Ship Harbour Lake	Kindle (1932)
13.4-15.4 (Mn)	19.5-27.5 (Fe)	4-10 (Si)	0.7-0.95 (Al)	Oneida Lake	Dean (1970)
31.7-35.9 (Mn)	14.2-20.9	—	—	Great Lake	Harriss and Troup (1970)
15.7 (Mn)	39.8-40.2 (Fe)	—	—	Mosque Lake	Harriss and Troup (1970)
17.0 (Mn)	20.6 (Fe)	—	—	Lake Ontario	Cronan and Thomas (1970)
3.57 (Mn)	33.2	—	—	Lake George	Schoettle and Friedman (1971)
0.89-22.2 (MnO)	1.34-60.8 (FeO)	—	—	Lake Michigan	Rossman and Callender (1968)

typical iron bacteria. Dean (1970) thought that microbial participation in the development of the concretions was possible, but he did not investigate it. Dean and Ghosh (1978) believe that photosynthetic activity by phytoplankton was indirectly responsible for manganese oxidation. They also implicate the algae in concentrating manganese and iron. All workers seem to agree that ferromanganese concretions build up where the sedimentation rate is low. Estimates of the growth rates of the concretions range from 0.015 mm year^{-1} in Lake Ontario to 1.5 mm year^{-1} in Mosque Lake. Much remains to be discovered about the origin of the freshwater ferromanganese concretions. It should be pointed out that not all geologists agree with a microbial explanation of the origin of ferromanganese nodules. Varentsov (1972), for instance, explains the formation of such nodules in Eningi-Lampi Lake, Karelia, U.S.S.R., in terms of chemosorption and autocatalytic oxidation.

The ultimate source of the manganese that is deposited as oxides on and in oxidized sediments in lakes is manganese released from manganese-containing rock minerals in the watershed and carried in runoff and groundwater into rivers feeding the lakes and into the lakes themselves. The oxidized manganese in the sediments may be resolubilized under reducing conditions, as already described, and thus become a part of a manganese cycle. A mathematical model for manganese distribution in sediments of Lake Michigan, based on the assumption that manganese oxidation and precipitation occur under steady-state conditions, has been developed (Robbins and Callender, 1975).

A Manganese Cycle in Lakes   A manganese cycle has been described for Lake Mendota, Wisconsin, and for Round Lake, New York (Delfino and Lee, 1968; Howard and Chisholm, 1975). When the lakes are stratified, Mn^{2+} accumulates in the anoxic hypolimnion as a result of reduction of manganese oxides in the sediments. In the epilimnion at this time, manganese exists as Mn^{2+} and manganese oxides. During turnover, when the thermocline has disappeared and the water is reoxygenated, all forms of manganese in the water column are redistributed and Mn^{2+} is oxidized and then sedimented. Similar observations were made in Tomhannock Reservoir, New York (LaRock, 1969). In this lake, bacterial oxidation and sedimentation of manganese were observed in the shallow waters along parts of the shoreline. This oxidation ceased in response to die-offs of algal blooms, which stimulated microbial reduction of oxidized manganese.

13.5 MICROBIAL TRANSFORMATION OF MANGANESE IN THE MARINE ENVIRONMENT

Mn(II)-oxidizing and Mn(IV)-reducing bacteria have also been found in the marine environment (Ehrlich, 1963b; Krumbein, 1971; Ehrlich, 1975), leading to the belief that microbes also play a role in manganese transformations in that

Table 13.6   Manganese Budget for the Pacific Ocean

Total Mn (as MnO) in sediments	1.4×10^{15} tons
Total Mn (as MnO) in nodules	3.1×10^{11} tons (170 times)[a]
Total Mn (as MnO) in seawater	1.8×10^9 tons
Total Mn (as MnO) in biomass	1×10^7 tons (0.0055 times)[a]

[a] Relative to manganese in seawater.

Source: Based on data from Poldervaart (1955), Merro (1962), and Bowen (1966).

part of the biosphere. Manganese in this environment occurs in greater quantities in the sediments than in seawater and in greater quantities in seawater than in the biomass. Table 13.6 illustrates this for the Pacific Ocean. The average concentration of manganese in seawater is 2 μg liter^{-1}, although this varies within narrow limits of 0.4-8 μg liter^{-1}. The dominant oxidation state of manganese in seawater is +2, despite the alkaline pH of seawater (7.5-8.3) (Park, 1966) and its E_h of +0.43 V (ZoBell, 1946). The stability of the divalent manganese in seawater is attributable to its complexation by the abundant chloride ions (Goldberg and Arrhenius, 1958), by sulfate and bicarbonate ions (Hem, 1963), and by organic substances such as amino acids (Graham, 1959). Although manganese in seawater is important in the nutrition of all living organisms and is present in a sufficient concentration to meet most requirements, it is at too low a concentration as Mn^{2+} to be readily utilized as an energy source by bacteria, as in some fresh waters. Its biological function, when taken directly from seawater, is mainly as enzyme activator inside the cell. It is returned to the seawater environment upon the death of the cell. The primary sources of manganese in seawater are runoff from the continents, submarine volcanism (magma, volcanic exhalations, hydrothermal solutions), and aeolian sources (windborne dust).

Marine Ferromanganese Nodules   Manganese (IV) oxides occur in large quantities in concretions and crusts at the sediment-water interface of the sea in regions where the rate of sedimentation is low (Fig. 13.12) (Margolis and Burns, 1976). They are particularly numerous in the Pacific but occur to significant extents in all the world's oceans. Typical composition of such nodules is given in Table 13.7. The chemical components of the nodule are not evenly distributed throughout its mass (Sorem and Foster, 1972). When examined in cross section, nodules are seen to have developed around a nucleus, which may be a foraminiferal test, pieces of pumice, sharks' teeth, ear bones of whales, and so on.

   The oxidation state of manganese in nodules is mainly +4. It occurs as todorokite, birnessite, and δ-MnO_2 (dissordered birnessite, according to the

Figure 13.12 Photograph of a bed of ferromanganese nodules on the ocean floor. Nodules may range in size from <1 to 25 cm in diameter. Average size has been given as 3 cm. (From Heezen and Hollister, 1971; reproduced by permission from the National Science Foundation.)

Table 13.7 Average Concentration of Some Major Constituents of Manganese Nodules from the Pacific Ocean (percent by weight)

Mn	24.2	Mg	2.7
Fe	14.0	Na	2.6
Co	0.35	Al	2.9
Cu	0.53	Si	9.4
Ni	0.99	L.O.I.	25.8

Source: From Mero (1962).

nomenclature of Burns et al., 1974). The Mn(IV) oxides in the nodules have a strong capacity to scavenge cations (e.g., Crerar and Barnes, 1974; Ehrlich et al., 1973; Loganathan and Burau, 1973; Varentsov and Pronina, 1973), particularly

Mn^{2+}. The nodules thus serve as concentrators of divalent manganese and are therefore able to furnish it in sufficient quantities to serve as an energy source to Mn(II)-oxidizing bacteria. This makes ferromanganese nodules a selective habitat for these microorganisms. Electron microscopic and culture examination have shown the presence of various kinds of bacteria on the surface of and within nodules (Fig. 13.13) (LaRock and Ehrlich, 1975; Ehrlich et al.,1972). Their numbers have been found to range from hundreds to tens of thousands per gram of nodule, as determined by plate counts on seawater-nutrient agar at $14\text{-}18°C$ and atmospheric pressure. These numbers are probably underestimates, since the organisms may grow in clumps or microcolonies on or in the nodule, from where they cannot be easily dislodged. Indeed, it is necessary to plate suspended, crushed nodule material to make the counts.

The microbial population on nodules includes three types of bacteria when considered in terms of their action on manganese compounds. The three types are Mn(II) oxidizers, Mn(IV) reducers, and a group that can neither oxidize Mn(II) nor reduce Mn(IV). In the nodules so far examined, the Mn(IV) reducers were most numerous. These findings, however, must not be interpreted to mean that the nodules examined were undergoing active reduction necessarily. Neither Mn(II) oxidizers nor Mn(IV) reducers need to act on manganese in order to grow. Thus, it is quite conceivable that the nodules were undergoing net manganese accretion at the time of collection. It is a plentiful supply of Mn(II) and oxygen as well as a deficit in organic carbon that is needed to favor Mn(II) oxidation, and it is an adequate supply of oxidizable organic carbon that is needed to favor Mn(IV) reduction. The bacteria which neither oxidize nor reduce manganese may play an important role in keeping the level of oxidizable organic carbon low, thereby favoring the Mn(II)-oxidizing activity by the Mn(II)-oxidizing bacteria.

Most of the organisms found on nodules are gram-negative rods (Ehrlich et al., 1972), although gram-positive bacilli, micrococci, and *Arthrobacter* have also been isolated from them (Ehrlich, 1963b). A curious and as yet unexplained finding has been that a significantly greater number of the isolates recovered from nodules from the central Pacific are unable to grow in freshwater media than from the eastern Pacific.

As discussed in Section 13.1, Mn(II) oxidation by the bacteria from ferromanganese nodules is performed on Mn^{2+} previously bound to Mn(IV) or to certain ferric iron-coated sediment particles (Ehrlich, 1978b). Mn(IV) oxides at neutral to alkaline pH act as scavengers of cations, including Mn^{2+} ions. Since Mn(IV) is the product of Mn(II) oxidation, the oxidation process generates new scavenging sites, and in this way nodules can continue to grow. The scavenging action of Mn(IV) oxides probably explains how other cationic constituents get incorporated into nodules,

$$H_2MnO_3 + X^{2+} \rightarrow XMnO_3 + 2H^+ \tag{13.14}$$

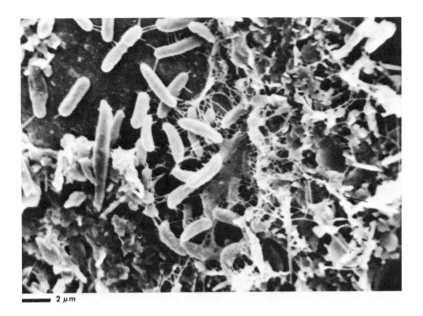

2 µm

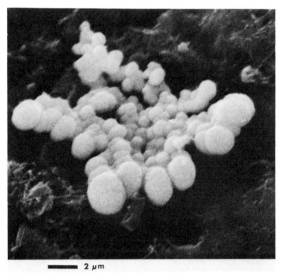

2 µm

Figure 13.13   Scanning electron photomicrographs of bacteria attached to the surface of ferromanganese nodules from Blake Plateau, off the southern Atlantic coast of the United States. Note the slime strands anchoring the rod-shaped bacteria to the nodule surface. (From LaRock and Ehrlich, 1975; by permisson.)

where X^{2+} represents any cation (e.g., Cu^{2+}, Ni^{2+}, Co^{2+}). Iron, if incorporated in this fashion, is probably picked up as $Fe(III)$.

Initially, the $Mn(IV)$ oxide formed in $Mn(II)$ oxidation by the bacteria is probably amorphous. The characteristic mineral assemblages identified in nodules by mineralogists are probably formed subsequently, being the result of slow "aging" processes (crystallization?) involving structural rearrangements of the nodule components.

The rate of growth of manganese nodules in the deep sea is reportedly very slow. Ku and Broecker (1969), for instance, have reported rates ranging from 1 to 10 mm per 10^6 years, based on radioisotope dating methods. Their estimates assume a constant rate of growth. More recently, Heye and Beiersdorf (1973) found variable growth rates for deep sea nodules which they had examined by a fission track method. They reported rates varying from 0 to 15.1 mm per 10^6 years. In other words, the nodules these investigators examined did not grow at constant rates. This suggests that conditions must not be continually favorable for nodule growth and that quiescent periods and, perhaps, even periods of nodule degradation very likely helped by the $Mn(IV)$-reducing bacteria, must intervene between growth periods.

Other biogenic mechanisms of nodule growth have been reported. Butkevitch (1928) tried to explain the growth in terms of iron precipitation by *Gallionella* and *Persius marinus*, the latter newly isolated by him but not otherwise known, which he found associated with brown deposits of the Petchora and White seas. However, Sorokin (1971) has recently shown that *Gallionella* does not occur in deep-sea nodules, nor could he find evidence for the presence of *Metallogenium*. The only types of bacteria he found were heterotrophs. The Petchora and White seas are, of course, marginal seas and untypical of the open ocean, which probably explains Butkevitch's findings. Graham (1959) proposed that manganese nodules form as a result of bacterial destruction of organic complexes of $Mn(II)$ in seawater liberating Mn^{2+} ions, which then autoxidize and precipitate, the oxides collecting around foci such as foraminiferal tests. According to Graham, other trace metals could be deposited in a similar manner. He felt that amino acids or peptides were the complexing agents attacked. He detected amino acid-like material in nodules from Blake Plateau. Graham and Cooper (1959) were able to recover foraminiferal tests from Blake Plateau, among which arenaceous ones were coated with a veneer of manganese-rich material containing Cu, Ni, Co, and Fe. Since the manganese-rich coating was on the surface of the tests, the authors reasoned that the manganese was deposited by the foraminifera, not while alive, but after their death. Calcareous tests were free of this coating. Graham and Cooper implicated a protein-rich coating on the arenaceous tests as responsible for providing a habitat for organisms (bacteria?) which remove trace-element chelates from seawater.

Kalinenko et al. (1962) visualized bacterial colonies growing on ooze particles where they mineralize organic matter coating the particles. In the process, manganese and iron oxides are supposed to be formed and deposited together with other trace elements on the colonies. Slime formed by the bacteria is assumed to cause the deposits to agglutinate, producing micronodules. The investigators made attempts to demonstrate this in laboratory experiments by watching bacterial development on glass slides introduced into oozes from the bottom of several Indian Ocean stations. Kalinenko et al. have visualized the bacteria as the living cement that holds the nodules together.

Greenslate (1974a) has observed manganese deposition in microcavities of planktonic debris, especially diatoms, and has proposed such deposition as being the beginning of nodule growth. He also found remains of shelter-building organisms such as benthic foraminifera on nodules, which became encrusted and ultimately buried in the nodule structure and proposed that the skeletal remains provide a framework on which manganese and other nodule components may be deposited, perhaps with the help of bacterial action (Greenslate, 1974b). Others have since reported evidence of traces of such organisms on nodules (Fredericks-Jantzen et al., 1975; Bignot and Dangeard, 1976; Dugolinsky et al., 1977; Harada, 1978) (Fig. 13.14). The finding that benthic foraminifera and other protozoans have grown and may presently be growing on ferromanganese nodules is of significance in explaining the role of $Mn(II)$-oxidizing bacteria in nodule growth, not previously recognized by marine geologists. Since these foraminifera and other protozoans are **phagotrophic** (i.e., they live at the expense of bacterial cells), they probably feed on the $Mn(II)$-oxidizing bacteria, among others, on the nodules. To maintain this food supply, uneaten $Mn(II)$-oxidizing bacteria must, therefore, continue to multiply. Thus, $Mn(II)$-oxidizing bacteria on nodules may play a dual role: (1) to aid in manganese accretion to nodules and (2) to serve as food for phagotrophic protozoans.

Although the hypotheses of Graham, Greenslate, and to some extent of Kalinenko et al. may well have a bearing on the initiation of nodule formation, it seems doubtful that they apply to the major growth phase of nodules. Graham and Kalinenko et al. assume that most of the Mn^{2+} in seawater is organically complexed, and that it is as preformed $Mn(IV)$ oxide that manganese is incorporated into the growing nodule. However, most Mn^{2+}, as already pointed out, exists as inorganic complexes in seawater from which Mn^{2+} is readily adsorbed by nodules.

As discussed in this section, $Mn(IV)$-reducing bacteria are associated with nodules in significant numbers. Biochemically, all those so far studied bring about the reduction of $Mn(IV)$ by an inducible MnO_2-reductase system (see Section 13.2). Laboratory experiments have shown that when they reduce and thereby solubilize the $Mn(IV)$ oxides of nodules, they cause the simultaneous

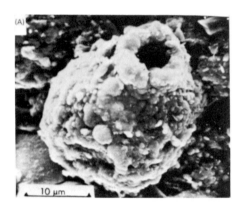

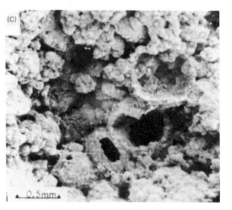

Figure 13.14 Benthic, test-forming protozoans inhabiting the surface of ferromanganese nodules. (A) Fresh remains of chambered encrusting protozoans on a nodule surface, showing siliceous biogenic material used in test construction. (B) Surface of a nodule with partial test of *Saccorhiza ramosa,* the most common and longest of any tubular agglutinating foraminifera yet found on nodules. (C) Test of unidentified form composed almost entirely of manganese micronodules. [From Dugolinsky et al., *J. Sediment Petrol.* 47:428-445. Copyright 1977, The Society of Economic Paleontologists and Mineralogists.]

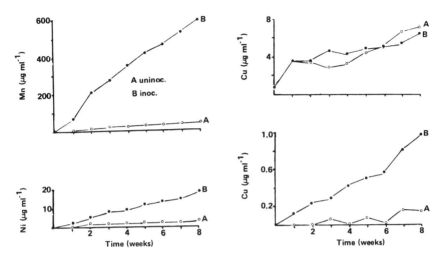

Figure 13.15 Mangense, copper, nickel, and cobalt release from ferromanganese nodule substance by *Bacillus* GJ 33 in seawater containing 1% glucose and 0.05% peptone. (From Ehrlich et al., 1973; by permission.)

release of Cu, Ni, and Co (Fig. 13.15) (Ehrlich et al., 1973). In the laboratory experiments, the Ni and Co release showed an absolute dependence on Mn(IV) reduction. Cu release, on the other hand, showed only partial dependence on it, being initially solubilized by complexation with peptone in the medium, which was independent of the presence of bacteria. Only in later stages did Cu solubilization show a direct dependence on bacterial action. Cu appears to be more loosely bound in the nodule structure than Ni or Co. The need of bacterial action for Cu release appears to arise only when it is "covered up" by Mn(IV) oxides and not in direct contact with the solvent. Only negligible amounts of iron were solubilized in these experiments, even though the Mn(IV)-reducing organisms used in these studies also had the capacity to reduce limonite and goethite. Whether this apparent inability to solubilize iron is due to an inability to reduce it in nodules or due to immediate autoxidation and precipitation after microbial attack is not known.

The reduction of Mn(IV) oxides by the marine bacteria can take place only if a suitable reducing substance is present. In the laboratory, glucose and some other sugars have been found to serve this purpose, but organic nitrogen compounds, specifically peptones and other amino acid polymers, have not. These are rather restrictive conditions. The only sigificant source of reducing substrates naturally present may be bacterial slimes, and even those may not be available to the Mn(IV)-reducing bacteria if they cannot break them down. Mn(IV) reduction on nodules is probably a relatively rare event, despite the large number of organisms having this capacity which are found on nodules. These organisms are also found in significant numbers in sediments. Here they may play a more important role in keeping migrating manganese mobile and available for incorporation into nodules.

13.6 SUMMARY

Manganous manganese can be bacterially oxidized by enzymatic catalysis or by metabolic end-product reaction. Free Mn^{2+} ions may be oxidized by O_2 and catalyzed by an inducible or a constitutive oxidase, or they may be oxidized by metabolic H_2O_2 and catalyzed by catalase. Prebound Mn^{2+} is oxidized by O_2 and catalyzed by a constitutive oxidase. At least some of the oxidase-catalyzed reactions yield useful energy to the cell. Oxidation of free Mn^{2+} ions seems to occur in soil and freshwater environments, whereas oxidation of prebound Mn^{2+} ions seems to occur in the marine environment.

Nonenzymatic manganese oxidation may be promoted by utilization of hydroxycarboxylic acids or by raising the pH and E_h of the environment so that autoxidation is favored. Both bacteria and fungi may be active in this way.

Preformed oxidized manganese may be precipitated by some microbes through adsorption to a surface structure or to extracellular slime. It has been observed with bacteria and fungi.

Mn(IV) oxides may be enzymatically reduced by bacteria to manganous manganese [Mn(II)] with suitable electron donors. This had been especially demonstrated with some bacteria from soil and fresh water, but in greatest detail with some bacteria from marine sources. Mn(IV) oxides may also be nonenzymatically reduced by metabolic end products of bacteria and fungi. Such products include formate and oxalate.

Manganese-reducing and -oxidizing microorganisms play an important role in the mobilization and immobilization of manganese in soil. Fixed manganese in soil may be concentrated in nodules or in arid environments as desert varnish. Since manganese is required in plant nutrition, this microbial activity is ecologically very significant. Manganese-reducing and -oxidizing microorganisms can also play a significant role in the manganese cycle in freshwater and marine environments. The manganese-oxidizing microorganisms may contribute to the accumulation of manganese oxides on and in sediments. The oxides may sometimes be accumulated as concretions formed around a nucleus, such as a sediment grain, a pebble, or a dead biological structure (e.g., a mollusk shell, coral fragment, or other debris). Conversely, manganese-reducing microorganisms may mobilize the oxidized or fixed manganese, releasing it into the water column. In reducing freshwater environments, in the presence of abundant plant debris, the microbial reduction of manganese oxides may lead to manganous carbonate formation, a different form of fixed manganese.

Ferromanganese nodules on parts of the ocean floor are inhabited by bacteria; some of these bacteria can oxidize and others can reduce manganese. Together with benthic foraminifera, which can be expected to feed on the bacteria and which accumulate manganese oxides on their tests and tubes, they appear to constitute a biological system that contributes to nodule formation.

14

Geomicrobial Transformations of Sulfur

Sulfur is an important element for life. In the cell it is especially important in stabilizing protein structure and in the transfer of hydrogen by enzymes in redox metabolism. For some prokaryotes, reduced forms of sulfur can serve as energy sources, and oxidized forms, especially sulfate, can serve as electron acceptors. It is in the context of microbial sulfur oxidation and reduction that the element will be mainly considered in this chapter.

Sulfur is one of the more abundant elements of our planet, having a concentration of 520 ppm in the earth's crust (Goldschmidt, 1954). Inorganic sulfur can exist in the $-2, 0, +2, +4,$ and $+6$ oxidation states. Table 14.1 lists geomicrobially important forms at the various oxidation states. In nature the $-2, 0,$ and $+6$ oxidation states are most common, existing in the form of sulfide, elemental sulfur, and sulfate, respectively. In field soils in humid, temperate regions, the total sulfur concentration may range from 100 to 1,500 ppm, of which 50-500 ppm is soluble in weak acid or water (Lawton, 1955). Most of the sulfur in soil of pastureland in humid to semiarid climates is organic, whereas that in drier soils is contained in gypsum ($CaSO_4 \cdot 2H_2O$), epsomite ($MgSO_4 \cdot 7H_2O$), and in lesser amounts of sphalerite (ZnS), chalcopyrite ($CuFeS_2$), and pyrite or marcasite (FeS_2) (Freney, 1967). Metal sulfides in general are relatively water insoluble, and sulfide-sulfur can thus be considered an immobilizing agent of metal ions (see Chapter 15). Sulfate salts of most metals are generally water-soluble, a major exception being ferric sulfate. The sulfate salts of such alkaline earths as Ca, Ba, and Sr also tend to be insoluble in water, and thus sulfate can serve to immobilize them too.

Elemental sulfur may react with sulfite to form thiosulfate (Roy and Trudinger, 1970):

$$S_8^0 + 8SO_3^{2-} \rightarrow 8S_2O_3^{2-} \tag{14.1}$$

The reaction is reversible, the forward reaction being favored by neutral to alkaline pH, whereas the back reaction is favored by acid pH.

251

Table 14.1   Geomicrobially Important Forms of Sulfur and Their Oxidation
States

Compound	Formula	Oxidation state(s) of sulfur
Sulfide	S^{2-}	-2
Sulfur[a]	S_8^{0}	0
Hyposulfite (dithionite)	$S_2O_4^{2-}$	$+2$
Sulfite	SO_3^{2-}	$+4$
Thiosulfate[b]	$S_2O_3^{2-}$	$0, +4$
Dithionate	$S_2O_6^{2-}$	$+4$
Trithionate	$S_3O_6^{2-}$	$0, +4$
Tetrathionate	$S_4O_6^{2-}$	$0, +4$
Pentathionate	S_5O_6	$0, +4$
Sulfate	SO_4^{2-}	$+6$

[a] Occurs in an octagonal ring in crystalline form.

[b] Outer S has a valence of 0; inner S has a valence of +4.

Elemental sulfur also reacts with sulfide, forming polysulfanes (Roy and
Trudinger, 1970):

$$S_8^0 \xrightarrow{\ HS^-\ } HS_n^-$$
(14.2)

The value of n may equal 2, 3, and so on, and is related to the sulfide concentration. Polythionates, starting with trithionate ($S_3O_6^{2-}$), may be viewed as
disulfonic acid derivatives of sulfanes (Roy and Trudinger, 1970).

14.1   MICROBIAL TRANSFORMATIONS OF ORGANIC SULFUR COMPOUNDS

Microbial Degradation of Organic Sulfur Compounds   Microbes degrade organic
sulfur compounds, examples of which are the amino acids cysteine and methionine, agar-agar (a sulfuric acid ester of a linear galactan), tyrosine-O-sulfate

$$HO_3SO\text{—}\!\!\!\bigcirc\!\!\!\text{—}CH_2CHCOOH \atop \qquad\qquad\ \ NH_2$$, and so on. Desulfuration of cysteine by bacteria may

occur anaerobically by the reaction (Freney, 1967)

$$HSCH_2\underset{\underset{NH_2}{|}}{C}HCOOH \xrightarrow[\text{cysteine desulfhydrase}]{+ H_2O} H_2S + NH_3 + CH_3COCOOH \quad (14.3)$$

or by the reaction (Freney, 1967)

$$HSCH_2\underset{\underset{NH_2}{|}}{C}HCOOH \xrightleftharpoons[\underset{\text{serine sulfhydrase}}{- H_2O}]{+ H_2O} HOCH_2\underset{\underset{NH_2}{|}}{C}HCOOH + H_2S \quad (14.4)$$

The sulfur of cysteine may also be released aerobically. The reaction sequence in that instance is not completely certain and may differ with different types of organisms (Freney, 1967; Roy and Trudinger, 1970). Although alanine-3-sulfinic acid ($HO_2S-CH_2\underset{\underset{NH_2}{|}}{C}HCOOH$) has been postulated as a key intermediate by some workers, others have questioned it, at least for rat liver mitochondria (Wainer, 1964, 1967).

Methione is decomposed by extracts of *Clostridium sporogenes* or *Pseudomonas* sp. as follows (Freney, 1967):

$$\text{Methionine} \longrightarrow \alpha\text{-ketobutyrate} + \underset{\text{mercaptan}}{CH_3SH} + NH_3 \quad (14.5)$$

Microbial Assimilation of Inorganic Sulfur Compounds   Inorganic sulfur is usually assimilated as sulfate by appropriate microorganisms and higher forms of life. One possible pathway may be the reduction of sulfate to sulfide and its subsequent reaction with serine to form cysteine, as in *S. typhimurium* (see Freney, 1967, p. 239):*

$$SO_4{}^{2-} \xrightarrow{\text{ATP sulfurylase}} APS + PP \quad (14.6)$$

$$APS + ATP \xrightarrow{\text{APS kinase}} PAPS + ADP \quad (14.7)$$

*APS, adenosine 5'-sulfatophosphate; PAPA, adenosine 3'-phosphate-5'-sulfatophosphate; PP, inorganic pyrophosphate; PAP, adenosine 3',5'-diphosphate.

$$\text{PAPS} + 2e^- \xrightarrow[\text{NADH}]{\text{PAPS reductase}} SO_3^{2-} + \text{PAP} \qquad (14.8)$$

$$SO_3^{2-} + 6H^+ + 6e^- \xrightarrow[\text{NADH}]{SO_3^{2-} \text{ reductase}} S^{2-} + 3H_2O \qquad (14.9)$$

$$S^{2-} + \text{serine} \xrightarrow{\text{cysteine synthase}} \text{cysteine} + H_2O \qquad (14.10)$$

This sequence has also been noted in *Bacillus subtilis, Aspergillus niger, Micrococcus aureus,* and *Aerobacter aerogenes* (Roy and Trudinger, 1970). Reaction 14.10 may be replaced by the following sequence:

$$\text{Serine} + \text{acetyl-CoA} \longrightarrow \underset{\substack{| \\ \text{CHNH}_2 \\ | \\ \text{COOH} \\ \text{O-acetylserine}}}{\text{CH}_2\text{OOCCH}_3} + \text{CoASH} \qquad (14.11)$$

$$\text{O-Acetylserine} + H_2S \longrightarrow \text{cysteine} + \text{acetate} \qquad (14.12)$$

This latter sequence has been noted in *Escherichia coli* and *Salmonella typhimurium* (Roy and Trudinger, 1970). The reduction of sulfate to "active thiosulfate" and its incorporation into serine to form cysteine is also possible for some organisms, such as *E. coli* (Freney, 1967).

14.2 MICROBIAL OXIDATION OF INORGANIC SULFUR COMPOUNDS

Aerobic Microbial Sulfide Oxidation   Hydrogen sulfide may be oxidized by aerobic and anaerobic bacteria. The aerobic bacteria are often autotrophic and derive useful energy and reducing power for CO_2 assimilation from the process. *Thiobacillus thioparus* can oxidize soluble sulfide to elemental sulfur at an initially alkaline pH (about 8) and an rH_2 value of about 12. As a result of bacterial action on the sulfide, the pH falls, after an initial slight increase, to about 7.5 in 4 days, and the rH_2 rises to almost 20 (Sokolova and Karavaiko, 1968). The reaction can be summarized as follows:

$$S^{2-} + \tfrac{1}{2}O_2 + 2H^+ \rightarrow S^0 + H_2O \qquad (14.13)$$

Under conditions of greater aeration, *T. thioparus* will oxidize soluble sulfide all the way to sulfate (London and Rittenberg, 1964; Sokolova and Karavaiko,

1968). The mechanism proposed by London and Rittenberg (1964) for this oxidation is

$$4S^{2-} \rightarrow 2S_2O_3^{2-} \rightarrow S_4O_6^{2-} \rightarrow SO_3^{2-} + S_3O_6^{2-} \rightarrow 4SO_3^{2-} \rightarrow 4SO_4^{2-} \quad (14.14)$$

If this mechanism correctly explains the reaction sequence, it raises a question of whether elemental sulfur is ever a direct intermediate in sulfide oxidation by this organism.

T. thiooxidans has also been shown to be able to oxidize sulfide aerobically, but it always produces sulfate from it (London and Rittenberg, 1964). Sorokin (1970) has questioned the sulfide-oxidizing ability of thiobacilli, believing that they oxidize only thiosulfate formed from sulfide by chemical oxidation with oxygen, and that any elemental sulfur formed by them is due to the chemical interaction of bacterial oxidation products with S^{2-} and $S_2O_3^{2-}$, as previously proposed by Nathansohn (1902) and Vishniac (1952). This view is not generally accepted today.

Recently, *Thiovulum* sp. was shown to oxidize sulfide to sulfur or sulfate under lowered oxygen tension (Wirsen and Jannasch, 1978). The organism appears to be an autotroph. It is found in marine and freshwater environments. It has not yet been grown in pure culture.

Anaerobic Microbial Oxidation of Sulfide   Most bacteria that oxidize sulfide anaerobically are photoautotrophs, but one is a chemoautotroph. The photo-autotrophs include primarily the purple sulfur bacteria (Chromatiaceae) (Fig. 14.1) and green sulfur bacteria (Chlorobiaceae), but also some purple nonsulfur bacteria (Rhodospirillaceae) (Pfennig, 1977), and some cyanobacteria. For them, sulfide serves as electron donor for photosynthesis. They need this reducing power for the reduction of fixed CO_2. The product of their sulfide oxidation is usually elemental sulfur which is deposited within cells by the purple sulfur bacteria and outside cells by the green sulfur bacteria and cyanobacteria. Under conditions of H_2S limitation, purple sulfur bacteria may oxidize accumulated sulfur further to sulfuric acid. The purple nonsulfur bacteria, *Rhodopseudomonas palustris* and *R. sulfidophila*, oxidize H_2S directly to sulfate without intermediate formation of elemental sulfur (Hansen and van Gemerden, 1972; Hansen and Veldkamp, 1973). *Rhodospirillum rubrum*, *Rhodopseudomonas capsulata*, and *R. spheroides*, on the other hand, oxidize H_2S to elemental sulfur, which they deposit extracellularly (Hansen and van Gemerden, 1972). Except for *R. sulfidophila*, they differ from purple sulfur bacteria in that they do not tolerate as high a concentration of H_2S as the latter. A light-dependent dismutation of elemental sulfur into H_2S and sulfate by *Chlorobium limicola* forma *thiosulfatophilum* has been demonstrated by Paschinger et al. 1974).

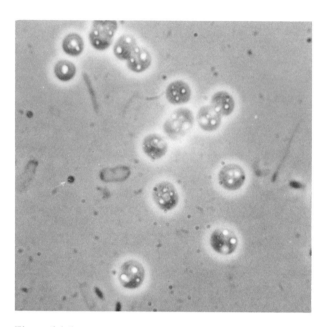

Figure 14.1   Unidentified purple sulfur bacteria (probably *Chromatium* sp.) in an enrichment (5,385X). Note the conspicuous sulfur granules in the spherical cells.

Some cyanobacteria have recently been shown to be able to use H_2S as a source of reducing power anaerobically as photosynthetic bacteria do (Castenholz, 1976, 1977; Cohen et al., 1975; Garlick et al., 1977; Padan, 1979). Under these conditions they use only photosystem I and produce elemental sulfur. This ability suggests an evolutionary relationship to photosynthetic bacteria. *Oscillatoria limnetica* can even reduce elemental sulfur to H_2S anaerobically in the dark, using its intracellular store of polyglucose as the source of reducing power (Oren and Shilo, 1979).

One of the chemoautotrophic thiobacilli, the facultative *Thiobacillus denitrificans,* can oxidize sulfide, sulfite, and thiosulfate anaerobically with nitrate as the electron acceptor (Baalsrud and Baalsrud, 1954; Milhaud et al., 1958; Peeters and Aleem, 1970; Aminuddin and Nicholas, 1973). The nitrate is reduced to N_2 and the nitrite to NO, N_2O, and N_2 in the process.

Oxidation of Sulfide by Mixotrophs and Heterotrophs   Hydrogen sulfide oxidation is not limited to autotrophs. *Beggiatoa* (Fig. 14.2) can grow mixotrophically on H_2S at low oxygen tension (*Bergey's Manual,* 1974). It deposits sulfur

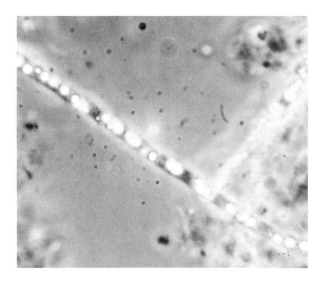

Figure 14.2   Trichome of *Beggiatoa* with sulfur granules in a pond water enrichment (5,240X). (Courtesy of E. J. Arcuri.)

granules in its cells external to the cytoplasmic membrane in invaginated, double-layered membrane pockets (Strohl and Larkin, 1978). The sulfur granules are the product of H_2S oxidation. The sulfur can be further oxidized to sulfate when H_2S is limiting (Pringsheim, 1967). A few strains of *Beggiatoa* may be able to grow autotrophically, but even those grow better mixotrophically (Kowallik and Pringsheim, 1967). The heterotrophs *Sphaerotilus natans*, *Alternaria*, and yeast have also been reported to oxidize H_2S to elemental sulfur (Skerman et al., 1957a; Skerman et al., 1957b). It has not been established whether these organisms derive useful energy from this oxidation.

There may be more than one enzymatic mechanism of sulfide oxidation since some organisms form elemental sulfur as an end product or as an intermediate in sulfate production, whereas others do not. Roy and Trudinger (1970) have proposed the following pathway for organisms that form sulfate from sulfide:

$$S^{2-} \rightarrow X \rightarrow SO_3^{2-} \rightarrow SO_4^{2-} \qquad (14.15)$$
$$\updownarrow$$
$$S_8$$

Here X and not S_8 is the true intermediate in the oxidation of sulfide to sulfite. Roy and Trudinger visualize X as a derivative of glutathione or a membrane-bound thiol. Their scheme also accounts for the observation that some organisms draw on S_8 (elemental sulfur) as an energy source. The other mechanism for sulfide oxidation to sulfate is the one of London and Rittenberg (1974) (see equation 14.14) (see also Vishniac and Santer, 1957). Here thiosulfate and polythionates are the intermediates.

Oxidation of Elemental Sulfur by Autotrophs   Elemental sulfur may be biologically oxidized to sulfuric acid:

$$S^0 + \tfrac{3}{2} O_2 + H_2O \rightarrow H_2SO_4 \qquad\qquad (14.16)$$

At a sufficient oxygen tension, *T. thioparus* will oxidize S^0 slowly to sulfuric acid, but *T. thiooxidans* or *T. ferrooxidans* are more commonly involved. *T. thioparus* does not grow below about pH 4.5, its range for growth being pH 10-4.5, whereas *T. thiooxidans* or *T. ferrooxidans* grow on elemental sulfur at pH values between 1 and 7 (*Bergey's Manual*, 1974).

Recently, a thermophilic and acidophilic bacterium, *Sulfolobus acido-caldarius*, was discovered which is capable of oxidizing elemental sulfur aerobically to sulfuric acid at temperatures between 55 and 85°C (70-75°C optimum) in the pH range 0.9-5.8 (pH 2-3, optimum) (Brock et al., 1972). The organism is a facultative autotroph which consists of spherical cells with lobes. A similar organism that is also capable of oxidizing iron and some metal sulfides has also been discovered (Brierley and Brierley, 1973) (see Fig. 12.5).

Other thermophilic bacteria capable of oxidizing reduced forms of sulfur have also been described recently. They are mostly incompletely characterized. Some were isolated from sulfurous hot springs, others from ore deposits. One of these has been described as a motile rod capable of forming endospores in plectridia. It is a facultative autotroph capable of oxidizing various sulfides and organic compounds. It has been named *Thiobacillus thermophilica* Imshenetskii (Egorova and Deryugina, 1963). Another is an aerobic, facultative thermophile, capable of sporulation, which is capable not only of oxidizing elemental sulfur but also Fe^{2+} and metal sulfides. It has been named *Sulfobacillus thermosulfidooxidans* (Golovacheva and Karavaiko, 1978). Still another is a *Thiobacillus* which is a gram-negative, facultative autotroph capable of growth at 50 and 55°C with a pH optimum of 5.6 (pH range 4.8-8.0) (Williams and Hoare, 1972). Other thermophilic thiobacillus-like bacteria have been isolated which can grow on thiosulfate at 60 and 75°C at pH 7.5, and still others which can grow at 60 or 75°C and a pH of 4.8 (LeRoux et al., 1977).

Oxidation of Elemental Sulfur by Heterotrophs   A number of heterotrophs, including bacteria and fungi, are capable of oxidizing S^0 and $S_2O_3^{2-}$. Guittoneau and coworkers (see Roy and Trudinger, 1970, pp. 248-249) reported that most cultures either oxidized S^0 to $S_2O_3^{2-}$.

$$2S^0 + O_2 + OH^- \rightarrow S_2O_3^{2-} + H^+ \qquad\qquad (14.17)$$

or they oxidized $S_2O_3^{2-}$ to sulfate,

$$S_2O_3^{2-} + 2O_2 + OH^- \rightarrow 2SO_4^{2-} + H^+ \qquad\qquad (14.18)$$

so that mixed heterotrophic cultures are needed to oxidize S^0 to SO_4^{2-}. These oxidations may yield useful energy to the cells (see, e.g., Tuttle et al., 1974).

Biochemistry of Sulfide and Elemental Sulfur Oxidation   The enzymes involved in sulfide oxidation have not as yet been clearly identified. However, Moriarty and Nichols (1969, 1970a) have prepared a cell membrane fraction that oxidizes sulfide to a membrane-bound polysulfur whose production depends on electron transfer to oxygen through the cytochrome system in the membrane. Oxidation of S^0 has been obtained with cell-free extract of *T. thiooxidans* to which catalytic amounts of glutathione had been added (Suzuki, 1965). The product of the oxidation was apparently $S_2O_3^{2-}$, but it was subsequently shown to be SO_3^{2-} (Suzuki and Silver, 1966). A sulfur-oxidizing enzyme was isolated from *T. thioparus* which oxidized S^0 to SO_3^{2-} with glutathione as cofactor (Suzuki and Silver, 1966). The enzyme contained nonheme iron and was described as an oxygenase. A similar enzyme system appears to exist in *T. thiooxidans*. Thiosulfate, found as oxidation product in the earlier studies, was shown to have really been the result of nonbiological interaction of residual S^0 and SO_3^{2-} from the microbial oxidation of S^0.

Biochemistry of Sulfite Oxidation   Sulfite may be oxidized by two different mechanisms, one of which involves substrate-level phosphorylation while the other does not, although both can yield useful energy through oxidative phosphorylation in the intact cell. In the substrate-level phosphorylation mechanism, sulfite reacts oxidatively with AMP to give APS:

$$SO_3^{2-} + AMP \xrightarrow{\text{APS reductase}} APS + 2e^- \qquad\qquad (14.19)$$

The sulfate of APS is then exchanged for phosphate:

$$APS + P_i \xrightarrow{\text{ADP sulfurylase}} ADP + SO_4^{2-} \qquad\qquad (14.20)$$

ADP can be converted to ATP as follows:

$$2\,ADP \xrightarrow{\text{adenylate kinase}} ATP + AMP \qquad (14.21)$$

Hence, the oxidation of 1 mol of sulfite yields 0.5 mol of ATP from substrate-level phosphorylation. However, additional energy as ATP may be gained from shuttling the electrons in equation 14.19 through the cytochrome system to oxygen (Davis and Johnson, 1967).

A number of thiobacilli appear to use an AMP-independent sulfite oxidase system (Roy and Trudinger, 1970, p. 214). These systems do not all seem to be alike. The AMP-independent sulfite oxidase of autotrophically grown *T. novellus* may be pictured as follows (Charles and Suzuki, 1966):

$$SO_3^{2-} \rightarrow \text{cytochrome c} \rightarrow \text{cytochrome oxidase} \rightarrow O_2 \qquad (14.22)$$

The sulfite oxidase of *T. neapolitanus*, on the other hand, may be pictured as a single enzyme that may react with sulfite and either with AMP to give APS or water to form sulfate (Roy and Trudinger, 1970). The enzyme reacts directly with oxygen. The sulfite-oxidizing enzymes which do not require the presence of AMP have also been detected in *T. thiooxidans, T. denitrificans,* and *T. thioparus. T. concretivorus,* now considered as a strain of *T. thiooxidans,* has been reported to shuttle electrons from SO_3^{2-} oxidation via the following pathway to oxygen (Moriarty and Nicholas, 1970b):

$$SO_3^{2-} \rightarrow \text{(flavin?)} \rightarrow \text{coenzyme } Q_8 \rightarrow \text{cytochrome b} \rightarrow$$
$$\text{cytochrome c} \rightarrow \text{cytochrome } a_1 \rightarrow O_2 \qquad (14.23)$$

Reduced Sulfur Compounds as Sources of Metabolic Reducing Power   For chemoautotrophs, reduced sulfur substrates are not only a possible source of energy but also a source of reducing power needed in CO_2 assimilation (see Chapter 5). However, in these organisms, reduced sulfur compounds cannot, for the most part, reduce pyridine nucleotides without the expenditure of energy because their redox potential is higher than that of pyridine nucleotides. This has been shown for autotrophically grown *T. novellus* and for *T. neapolitanus* (Aleem, 1966a,b; 1969; Roth et al., 1973). The energy for pyridine nucleotide reduction is obtained from ATP. Thus, some of the ATP generated in the oxidation of reduced sulfur compounds is employed for the production of reduced pyridine nucleotides. In photoautotrophs, the reduction of pyridine nucleotides by thiosulfate requires light energy in a process called noncyclic photophosphorylation (see Roy and Trudinger, 1970, p. 247; Arnon et al., 1961). The reduction of pyridine nucleotides by sulfide in these organisms also probably requires light energy or ATP (Doetsch and Cook, 1973, p. 278; Gest, 1972).

Reduced Sulfur Compounds as Energy Sources for Mixotrophs   Although it has been long recognized that chemolithotrophs can derive useful energy from the oxidation of reduced sulfur compounds, it is only recently that limited evidence has been obtained that some mixotrophs can also benefit energetically from the oxidation of inorganic sulfur compounds. Growth of *T. intermedius*, which grows poorly in a thiosulfate-mineral salts medium, is greatly stimulated by the addition of yeast extract, glucose, glutamate, and other organic supplements (London, 1963; London and Rittenberg, 1966). The organic matter seems to repress the CO_2 assimilating mechanism but not the energy-generating one (London and Rittenberg, 1966). The organism can grow heterotrophically on yeast extract and glucose or glutamate but not in a glucose-mineral salts medium without $S_2O_3^{2-}$ (London and Rittenberg, 1966). It needs $S_2O_3^{2-}$ or organic sulfur compounds because it cannot assimilate sulfate (Smith and Rittenberg, 1974). *T. perometabolis* cannot grow at all in thiosulfate-mineral salts medium but requires the addition of yeast extract, casein hydrolysate, or an appropriate single organic compound in order to utilize thiosulfate as an energy source (London and Rittenberg, 1967). Growth on yeast extract or casein hydrolysate is much less luxuriant without thiosulfate. *Thiobacillus organoparus*, a facultative, heterotrophic bacterium, first isolated from acid mine water in copper deposits in Alaverdi (Armenian S.S.R.), can grow autotrophically and mixotrophically with reduced sulfur compounds (Markosyan, 1973). It is acidophilic.

Three marine pseudomonads have recently been shown to be able to grow mixotrophically on reduced sulfur compounds (Tuttle et al., 1974). Growth of these cultures on yeast extract was stimulated by the additon of thiosulfate. The thiosulfate was oxidized to tetrathionate and caused a greater fraction of organic carbon to be assimilated than oxidized as compared to growth without added thiosulfate. As mentioned previously, a number of other heterotrophic bacteria, actinomycetes, and filamentous fungi have also been reported able to oxidize sulfur or thiosulfate to tetrathionate (Trautwein, 1921; Starkey, 1934; Guittoneau, 1927; Guittoneau and Keilling, 1927). It is not known, however, whether these organisms can gain energy from these oxidations, but they are probably important in promoting the sulfur cycle in most soils (Vishniac and Santer, 1957).

14.3 MICROBIAL REDUCTION OF INORGANIC SULFUR COMPOUNDS

Reduction of Elemental Sulfur   Elemental sulfur can be reduced anaerobically to H_2S by *Desulfuromonas acetoxidans* n.gen., n.sp. (Pfennig and Biebl, 1976), using acetate as reductant. The organism is a motile, gram-negative rod (0.4-0.7 $\times$ 1.4 μm) that forms pink to ochre colonies in agar medium. It is unable to ferment organic compounds. It derives its energy by anaerobic respiration on acetate, ethanol, or propanol using elemental sulfur as the terminal electron acceptor. It cannot substitute sulfate, sulfite, or thiosulfate as alternative electron acceptors, although it can use organic disulfides, malate, or fumarate for this purpose. It assimilates acetate. The organism has been isolated from H_2S-containing muds in enrichments with seawater, brackish water, and fresh water.

Elemental sulfur was shown to serve as terminal electron acceptor for *Desulfovibrio gigas* and sulfate-reducing bacterial strains 9974, 4474, and 5174 with ethanol as the electron donor (Biebl and Pfennig, 1977). Growth on elemental sulfur by *D. gigas* was slow. Type strains of *D. desulfuricans, D. vulgaris, Desulfotomaculum nigrificans,* and *Desulfomonas pigra* strain 11112 were unable to grow with elemental sulfur as the terminal electron acceptor (Biebl and Pfennig, 1977). Elemental sulfur has also been shown to be reduced by strains of *Rhodotorula* and *Trichosporon* growing in the presence of glucose (Ehrlich and Fox, 1967a).

Assimilatory Sulfate Reduction   Sulfate has been known for some time to serve as the terminal electron acceptor for a special group of anaerobic bacteria. However, since it is also reduced in assimilation, a distinction must be made between assimilatory and dissimilatory processes. The assimilatory process is carried out on a limited scale by a wide variety of organisms for the purpose of converting inorganic sulfate into sulfur amino acids. Sulfide does not accumulate in this case. Its reaction may be summarized as follows:

$$SO_4^{2-} + ATP \xrightarrow{\text{ATP sulfurylase}} APS + PP \qquad (14.24)$$

$$APS + ATP \xrightarrow{\text{APS kinase}} PAPS + ADP \qquad (14.25)$$

$$NADPH + H^+ + PAPS \xrightarrow{\text{PAPS reductase}} PAP + SO_3^{2-} + NADP^+ \quad (14.26)$$

$$3NADPH + 3H^+ + SO_3^{2-} + 2H^+ \xrightarrow{SO_3^{2-} \text{ reductase}} H_2S + 3H_2O + 3NADP^+$$
$$(14.27)$$

$$H_2S + serine \xrightarrow{\text{CoAS acetate}} cysteine + H_2O \qquad (14.28)$$

The sulfite in these reactions may never appear free in the cells but be enzyme-bound for reduction to H_2S (see Roy and Trudinger, 1970). Assimilatory sulfate reduction does not require anaerobic growth conditions.

Dissimilatory Sulfate Reduction   Of greater geochemical interest is dissimilatory sulfate reduction. It is an important form of respiration in anaerobic marine environments (Jørgensen and Fenchel, 1974; Jørgensen, 1977a,b; Sørensen et al., 1979). It is mainly carried out by *Desulfovibrio* spp., *Desulfotomaculum* spp. (LeGall and Postgate, 1973), and *Desulfomonas pigra* (Moore et al., 1976) (Fig. 14.3). Recently, another sulfate-reducing organism has been isolated from sheep rumen which resembles *Desulfovibrio* (Huisingh et al., 1974). *Bacillus megaterium* and *Pseudomonas zelinskii* have also been reported to produce free H_2S from sulfate (Bromfield, 1953; Shturm, 1948). However, the physiology of sulfate reduction by these two organisms should not be taken in common with the other three genera of sulfate reducers. *Desulfovibrio*, *Desulfotomaculum*, and *Desulfomonas* are strict anaerobes and use sulfate or some other partially oxidized sulfur compounds as terminal electron acceptors in anaerobic respiration (Postgate, 1951; Moore et al., 1976). *Desulfovibrio* is a gram-negative rod with polar flagella. It is a heterotroph which oxidizes lactate, pyruvate, and malate to acetate and CO_2 but rarely attacks carbohydrates. *Desulfovibrio* characteristically contains cytochrome c_3 and desulfoviridin

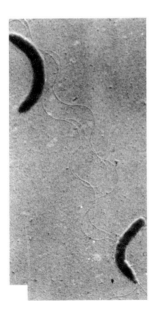

Figure 14.3   Electron micrograph of the Benghazi strain of *Desulfovibrio africanus*. Gold-shadow-cast (6,000X). (From Campbell et al., 1966; by permission.)

pigments. Some strains contain additional cytochromes, such as cytochrome c_{553}, cytochrome cc_3, and cytochromes b and d (LeGall and Postgate, 1973). It grows between 0 and 44°C (25-30°C optimum). *Desulfotomaculum* is classed as a straight or curved gram-negative rod capable of forming endospores. It preferably oxidizes lactate and pyruvate to acetate and CO_2, and attacks carbohydrate only rarely. One species, *D. acetoxidans,* was recently reported to oxidize acetate stoichiometrically with sulfate, forming CO_2 and H_2S (Widdel and Pfennig, 1977). *Desulfotomaculum* contains a protoheme type of cytochrome (cytochrome b) and lacks cytochrome c_3 and desulfoviridin (*Bergey's Manual,* 1974; LeGall and Postgate, 1973). It grows between 30 and 70°C (35-55°C optimum). *Desulfomonas* is a straight, nonmotile, nonsporeforming, gram-negative rod which can reduce SO_4^{2-} with lactate or pyruvate, producing acetate and CO_2. It possesses desulfoviridin and shows absorption maxima in dithionite-reduced minus oxidized difference spectra at 425, 527, and 556 nm, indicating the presence of cytochrome. It grows best at 37°C.

Although sulfate is the usual terminal electron acceptor for *Desulfovibrio,* some strains can grow on pyruvate or fumarate in the absence of added sulfate (Postgate, 1952, 1963). *Desulfomonas* can grow in the absence of sulfate when using lactate or pyruvate as an energy source by disposing of excess reducing power as H_2. However, if H_2 is allowed to accumulate, it is inhibitory to growth.

Desulfovibrio has been reported to grow mixotrophically (Mechalas and Rittenberg, 1960; Sorokin, 1966a,b,c,d; Badziong and Thauer, 1978; Badziong et al., 1978). It can grow on hydrogen as an energy source in the presence of yeast extract. It can also grow on hydrogen and formate as energy sources in the presence of CO_2 and some acetate. It cannot grow autotrophically because it is unable to derive its carbon exclusively from CO_2. A strain used by Sorokin (1966a) is unusual as a heterotroph, however, because when growing on hydrogen plus some acetate, it can derive as much as 50% of its carbon from CO_2, whereas on lactate it can derive as much as 30% of its carbon from CO_2. Badziong et al. (1978) found a strain that on H_2 and acetate derived 30% of its carbon from CO_2.

The reduction of sulfate to hydrogen sulfide requires a sequence of biochemical steps. As in assimilatory sulfate reduction, the sulfate ion must first be energized before it is reduced:

$$SO_4^{2-} + ATP \xrightarrow{\text{ATP sulfurylase}} APS + PP \tag{14.29}$$

The reaction is pulled in the direction of APS because of the hydrolysis of pyrophosphate:

$$PP + H_2O \xrightarrow{\text{pyrophosphatase}} 2P_i \tag{14.30}$$

Unlike in assimilatory sulfate reduction, no further phosphorylation of APS is required before reduction. APS is reduced directly to sulfite:

$$\text{APS} + 2e^- \xrightarrow{\text{APS reductase}} \text{AMP} + SO_3^{2-} \quad (14.31)$$

The APS reductase, unlike PAPS reductase, does not require NADP as a cofactor, but, like PAPS reductase, contains bound FAD and iron (for further discussion, see, e.g., Roy and Trudinger, 1970).

How the sulfite is reduced to H_2S is still not fully resolved. One line of experimental evidence suggests a multistep process involving trithionate and thiosulfate as intermediates (Kobiyashi et al., 1969; modified by Akagi et al., 1974; Drake and Akagi, 1978):

$$3HSO_3^- + H_2 \xrightarrow{\text{bisulfite reductase}} S_3O_6^{2-} + 2H_2O + OH^- \quad (14.32)$$

$$S_3O_6^{2-} + H_2 \xrightarrow{\text{trithionate reductase}} S_2O_3^{2-} + HSO_3^- + H^+ \quad (14.33)$$

$$S_2O_3^{2-} + H_2 \xrightarrow{\text{thiosulfate reductase}} HS^- + HSO_3^- \quad (14.34)$$

In most *Desulfovibrio* cultures, the bisulfite reductase seems to be identical to desulfoviridin (Kobyashi et al., 1972; Lee and Peck, 1971). However, in *D. desulfuricans* strain Norway, which lacks desulfoviridin, desulforubidin appears to be the bisulfite reductase (Lee et al., 1973). In *Desulfotomaculum nigrificans*, a carbon monoxide-binding pigment, called P_{582} by Trudinger (1970), accounts for the bisulfite reductase activity, which according to Akagi et al. (1974), leads to the formation of trithionate, with thiosulfate and sulfide accumulating as endogenous side products.

Chambers and Trudinger (1975) have questioned whether the trithionate pathway of sulfate reduction is the major pathway of *Desulfovibrio* spp. They found that results of experiments with isotopically labeled $^{35}SO_3^{2-}$, $^{35}SSO_3^{2-}$, and $S^{35}SO_3^{2-}$ could not be reconciled with the trithionate pathway, but were more consistent with a pathway involving the assimilatory kind of sulfite reductase that produces sulfide without any intermediates.

Sulfur Isotope Fractionation Sulfate-reducing bacteria can distinguish between ^{32}S and ^{34}S isotopes of sulfur (i.e., they can bring about isotope fractionation) (Harrison and Thode, 1957; Jones and Starkey, 1957). The ^{32}S isotope of sulfur is the most abundant (average 95.1%) and the stable ^{34}S isotope is the next most abundant (average 4.2%). The $^{32}S/^{34}S$ ratio of natural sulfur compounds ranges between 21.3 and 23.2. Meteoritic sulfur has an $^{32}S/^{34}S$ ratio of 22.22. Since this ratio appears to be relatively constant from sample to sample, it is often used as a reference standard against which to compare sulfur isotope ratios of other materials, which may be either enriched or depleted in ^{34}S. Under conditions of slow growth (i.e., slow reduction of sulfate), the sulfate reducers attack $^{32}SO_4^{2-}$ more readily than $^{34}SO_4^{2-}$ (Jones and Starkey, 1957, 1962) (Table 14.2). The nature of the electron donor may also affect the degree of isotope fractionation (Kemp and Thode, 1968). The degree of sulfur isotope fractionation is calculated in terms of δ^{34}S values expressed in parts per thousand (‰):

$$\delta^{34}S = \frac{^{34}S/^{32}S \text{ sample} - {^{34}S/^{32}S} \text{ meteoritic or standard}}{^{34}S/^{32}S \text{ meteoritic or standard}} \times 1,000 \quad (14.35)$$

Harrison and Thode (1957) proposed that it was the rate-controlling S—O bond breakage in bacterial sulfate reduction (i.e., the reduction of APS to sulfite and AMP) that is reponsible for the isotope fractionation phenomenon.

Dissimilatory sulfate reduction is not the only process that may lead to sulfur isotope fractionation. Sulfite reduction by *Desulfovibrio* and *Saccharomyces cerevisiae* (Kapalan and Rittenberg, 1962, p. 81), sulfide release from cysteine by *Proteus vulgaris* (Kaplan and Rittenberg, 1962, 1964), assimilatory sulfate reduction by *E. coli* and *S. cerevisiae* (Kaplan and Rittenberg, 1962), and H_2S oxidation by chemosynthetic and photosynthetic autotrophs such as *T. concretivorus* and *Chromatium* (Kaplan and Rittenberg, 1962) can lead to sulfur isotope fractionation.

Isotopic anlaysis of sulfur minerals in nature has helped in deciding whether biogenesis was involved in their accumulation. Any given deposit must, however, be sampled at a number of locations since isotope enrichment values (δ^{34}S) generally fall in a narrow or a wide range. Nonbiogenic δ^{34}S values generally tend to fall in a narrow range and have a positive sign, whereas biogenic values tend to fall in a wide range and have a negative sign.

Natural Sulfur Accumulations   Local accumulations of sulfur may occur in the form of sulfates [e.g., anhydrite ($CaSO_4$), gypsum ($CaSO_4 \cdot 2H_2O$), epsomite ($MgSO_4 \cdot 7H_2O$), jarosite [$KFe_3(OH)_6(SO_4)_2$], celestite ($SrSO_4$)], as elemental sulfur, and as sulfide [e.g., hydrotroilite ($FeS \cdot nH_2O$), pyrite or marcasite (FeS_2), pyrrhotite (Fe_nS_{n-1}), chalcocite (Cu_2S), covellite (CuS), chalcopyrite ($CuFeS_2$), sphalerite (ZnS), galena (PbS), cinnabar (HgS), orpiment (As_2S_3)]. Some of these minerals can be formed biogenically, and many can also be degraded by biological agents (see Chapter 15). Many of the simple sulfates such as those of calcium and magnesium are associated with evaporites, deposits that have resulted from the evaporation of seawater or other saline waters that contain significant concentrations of the ionic constituents of those salts.

Table 14.2 Sulfide Production and Fractionation of Stable Isotopes of Sulfur by *Desulfovibrio desulfuricans*

		A.	Cultivated at 28°C (rapid growth)		
Sample number	Incubation period (hr)[a]	Sulfide S in PbS (mg)	Sulfate reduced[b] (%)	Number of isotope determinations	δ^{34}S
1	44	996	6.3	4	−5.4
2	8	2168	20.0	4	−4.9
3	4	1931	32.2	2	−3.1
4	5	1448	41.4	2	−3.1
5	4	1394	50.2	2	−5.4
6	3	1248	58.1	2	−5.4
7	9	317	60.1	2	−6.7
8	14	191	61.3	2	−8.9
9	41	103	62.0	2	−9.8
10	68	115	62.7	2	−12.9
11	59.5	387	65.2	2	−7.2
12	25.5	901	70.9	2	−3.1
13	7	615	74.8	2	−0.5
14	6	474	77.8	2	+0.9
15	24	856	83.2	2	+0.5
16	43	106	83.9	2	−4.9

	B.	Cultivated at low temperatures (slow growth)		
Sample number	Incubation period (hr)[c]	Sulfide S in PbS (mg)	Sulfate reduced (%)	δ^{34}S
17	200	21.2	4.4	−22.1
18	142	65.8	4.5	−25.9
19	120	112.7	8.3	−25.9
20	120	167.8	10.0	−24.2
21	96	174.6	13.1	−24.2
22	120	180.8	16.0	−22.9
23	144	134.5	16.8	−21.6
24	120	102.0	18.4	−19.5

[a] Periods were calculated from the time that sulfide first appeared in the culture substrate. This was 60 hr after the medium was inoculated.

[b] The initial sulfate S in the substrate was 3943 ppm.

[c] Periods were calculated from the time that sulfide first appeared in the culture substrate; this was 18 hr after the medium was inoculated.

Source: Adapted from Jones and Starkey (1957); by permission.

14.4 SYNGENETIC SULFUR DEPOSITS

Elemental sulfur deposits, whether formed abiogenically or biogenically, result from the oxidation of H_2S:

$$H_2S + \tfrac{1}{2}O_2 \rightarrow S^0 + H_2O \tag{14.36}$$

However, in some fumaroles, sulfur may also form abiogenically through the interaction of H_2S and SO_2:

$$H_2S + SO_2 \rightarrow 3S^0 + 2H_2O \tag{14.37}$$

Most known sulfur deposits are not of volcanogenic origin. Indeed, of known reserves, only 5% are the result of volcanism (Ivanov, 1968, p. 139). Biogenic sulfur may form **syngenetically** or **epigenetically**. In **syngenetic** formation, sulfur is deposited contemporaneously with the enclosing host rock in a sedimentary process. In **epigenetic** formation, sulfur is laid down in preformed host rock. Its formation may involve a diagenetic process in which a sulfate component of the host rock is converted to sulfur, or it may involve the conversion of sulfur of dissolved sulfate or sulfide in a solution percolating through cracks and fissures of host rock. Syngenetic sulfur deposits are generally formed in limnetic environments, whereas epigenetic sulfur deposits tend to form in terrestrial environments.

Cyrenaican Lakes   Typical examples of contemporary syngenetic sulfur deposits are found in the sediments of the Cyrenaican lakes in Libya, North Africa; Sulfur Lake (Lake Sernoye) in the Kuibyshev region of the U.S.S.R.; and Lake Eyre in Australia. The origin of the sulfur in the Cyrenaican lakes was first studied by Butlin and Postgate (Butlin and Postgate, 1952; Butlin, 1953). Of four lakes that they have examined, Aïn ez Zauia, Aïn el Rabaiba, and Aïn el Braghi contain extensive native sulfur in their sediments, making up as much as half of the silt. The waters of these lakes have a strong odor of hydrogen sulfide and are opalescent, owing to a fine suspension of sulfur crystals. A fourth lake, Aïn amm el Gelud, also contains sulfuretted water but shows no evidence of sulfur in its sediment. Aïn es Zauia was most thoroughly studied. It is made up of two adjacent basins, 55 × 30 m and 90 × 70 m in expanse, respectively, and no deeper than 1.5 m. Other characteristics of the lake are summarized in Tables 14.3 and 14.4. The water in the lake is introduced by warm springs (Butlin, 1953). The border of Aïn ez Zauia, as well as those of the other two sulfur-producing lakes, were found to exhibit a characteristic red-colored, carpetlike, gelatinous material which extended several yards into shallow water in some places. The underside of this red, gelantinous material showed a green and black material. Some of the red material was found floating

Table 14.3   Physical Characteristics of Lake Aïn ez Zauia

Surface area	$7,950 \text{ m}^2$
Max. depth	1.5 m
Surface temp.	$30°C$
Bottom temp.	$32°C$
Air temp.	$16°C$
Sulfur production per year	100 tons

Table 14.4   Chemical Composition of the Waters of Lake Aïn ez Zauia

H_2S in surface water	15-20 mg liter^{-1}		
H_2S in bottom water	108 mg liter^{-1}		
Total solids	25.25 g liter^{-1}		
Ca	1,179 mg liter^{-1}	Cl	13,520 mg liter^{-1}
Mg	336 mg liter^{-1}	HCO_3	145 mg liter^{-1}
Na	7,636 mg liter^{-1}	SO_4	1,848 mg liter^{-1}
K	320 mg liter^{-1}	NO_3	3 mg liter^{-1}
NH_3	8 mg liter^{-1}	SiO_2	70 mg liter^{-1}

in the form of red, bulbous formations. The red-colored material was massive growth of the photosynthetic purple bacterium *Chromatium* and the green material consisted of growth of the green photosynthetic bacterium *Chlorobium*. Many sulfate-reducing bacteria wers also found in the lakes. From these observations it was concluded that the sulfate reducers were responsible for reducing the sulfate in the water to hydrogen sulfide, utilizing some of the carbon fixed by the autotrophic bacteria as carbon and energy source, whereas the photosynthetic bacteria oxidized the hydrogen sulfide produced by the sulfate-reducing bacteria to elemental sulfur and in the process assimilated CO_2 photosynthetically (Butlin and Postgate, 1952). Although it was recognized that some of the hydrogen sulfide in the lake could undergo autoxidation, the importance of this process was discounted because Aïn amm el Gelud, which contains sulferetted waters, exhibits no significant sulfur deposit in its sediment and also lacks a noticeable growth of photosynthetic bacteria. Butlin and Postgate were able to reconstruct an artificial system in the laboratory with pure and mixed cultures of sulfate-reducing and photosynthetic sulfur bacteria which reproduced the process they postulated for sulfur deposition in the Cyrenaican lakes. Significantly, however, they supplemented their artificial lake water with 0.1% sodium malate.

Ivanov (1968) has criticized the model of Butlin and Postgate for sulfur biogenesis in the Cyrenaican lakes by interaction of sulfate-reducing bacteria and photosynthetic bacteria on the basis of a deficiency of organic carbon. Ivanov argued that in a cycle in which the photosynthetic sulfur bacteria generate the organic carbon with which the sulfate-reducing bacteria reduce sulfate, each turn of the cycle produces one-fourth or less of the hydrogen sulfide that was produced in the previous cycle. This is best illustrated by the following two reactions:

$$2CO_2 + 4H_2S \rightarrow 2(CH_2O) + 4S^0 + 2H_2O \tag{14.38}$$

$$2(CH_2O) + SO_4^{2-} + 2H^+ \rightarrow H_2S + 2H_2O + 2CO_2 \tag{14.39}$$

The first of these equations illustrates the photosynthetic reaction and the second illustrates sulfate reduction. It is seen that to produce the organic carbon (CH_2O) needed to reduce the sulfate, four times as much H_2S is consumed as is produced in sulfate reduction. Ivanov (1968) therefore argued that most of the sulfide turned into sulfur by the photosynthetic bacteria is introduced into the lake by warm springs and does not result from sulfate reduction. He noted that Butlin and Postgate (Butlin, 1953) had actually demonstrated that many artesian wells in the area contained sulfuretted waters with sulfate-reducing bacteria. Ivanov, however, did not consider the possibility that these wells might also inject into the lakes H_2 which the sulfate reducers could employ as an alternative energy source and reductant of sulfate in a carbon-sparing action.

Ivanov also suggested that a portion of the sulfur in the lake may be produced by nonphotosynthetic sulfur bacteria and by autoxidation. No matter what the source of the H_2S, biogenesis of the sulfur in the Cyrenaican lakes has been confirmed on the basis of isotope analysis (Macnamara and Thode, 1951; Harrison and Thode, 1958; Kaplan et al., 1960).

Lake Sernoye   This lake is located in the U.S.S.R. It is an artificial, relatively shallow reservoir fed by the Sergievsk sulfuretted springs (Ivanov, 1968). The water output of these springs is around 6,000 m^3 day^{-1}. The waters contain 83-86 mg of H_2S per liter and have a pH of 6.7. The water temperature in summer ranges around 8°C. The waters which flow out of the lake into Molochni Creek are reported to be turbid and opalescent, owing to suspended native sulfur in them which results from the oxidation of the H_2S in the lake. Much of the lake sediment contains about 0.5% native sulfur, but some sediment contains as much as 2-5%. The lake freezes over in winter, at which time no significant oxidation of H_2S occurs. This fact is reflected by the stratified occurrence of sulfur in the lake sediment. Pure sulfur crystals paragenetic with calcite crystals have been found in some sediment cores (Sokolova, 1962). Most sulfur in the lake is deposited around the sulfuretted springs. At these locations purple and green sulfur bacteria are seen en masse. Impression smears show the presence of *Chromatium* and large numbers of rod-shaped bacteria which in culture reveal themselves to be mostly thiobacilli. A radiotracer study of Lake Sernoye water from these stations, which measured chemical and biological dark oxidation and biological light-dependent oxidation of added $Na_2{}^{35}S$, revealed that the microflora of the lake made a significant contribution (more than 50%) to the H_2S oxidation in the lake. About the same amount of native sulfur was precipitated in the dark and in the light, but more sulfate was formed in the light. These results were interpreted to mean that most of the H_2S in the lake which is biologically oxidized to native sulfur is attacked by the thiobacilli. The photosynthetic bacteria appear to oxidize H_2S for the most part directly to sulfate. They are of a type that is physiologically like *Chlorobium thiosulfatophilum*. An average dark production of sulfur during the summer months has been estimated to be 150 kg (Ivanov, 1968).

Lake Eyre   This Australian lake represents another locality in which evidence of syngenetic sulfur deposition has been noted. In shallow water on the southern bank of this lake, sulfur nodules have been found by Bonython (see Ivanov, 1968, pp. 146-150). The nodules are oval to spherical and usually covered with crusts of crystalline gypsum on the outside while being cavernous on the inside (Baas Becking and Kaplan, 1956). Their composition includes (in percent by dry weight): $CaSO_4$, 34.8; S^0, 62-63; NaCl, 0.8; Fe_2O_3, 0.45; $CaCO_3$, 0.32; organic carbon, 1.8; and moisture, 7.54 (Baas Becking and Kaplan, 1956). Most nodules contain active sulfate-reducing bacteria and thiobacilli, as do the water and muds of the lake (Baas Becking and Kaplan, 1956). Flagellates and cellulytic, methane-forming, and other bacteria abound also. Baas Becking and Kaplan (1956) at first proposed that the nodules were forming at the present time, with the photosynthetic flagellates providing organic carbon which cellulytic bacteria convert into a form utilizable by sulfate-reducing bacteria for the reduction of sulfate of gypsum in the surrounding sedimentary rock. The H_2S was then visualized as being subjected to chemical and biological (thiobacilli) oxidation to native sulfur. The nodule structure was seen to result from the original dispersion of the gypsum in septaria in which gypsum is replaced by sulfur. A difficulty with this model, as Ivanov (1968) points out, is that the present oxidation-reduction potential of the ecosystem is 250-350 mV, which is too high for intense sulfate reduction, which requires an oxidation-reduction potential no higher than around -110 mV.

Radiodating of nodules has shown them to be 19,600 years old (Baas Becking and Kaplan, 1956). The $^{32}S/^{34}S$ ratio of the sulfur and of the gypsum of the outer crust of the nodules is remarkably similar (22.40-22.56 and 22.31-22.53, respectively), whereas that of the gypsum of the surrounding rock is 22.11. This clearly suggests that the gypsum of the nodule crust is a secondary formation, biogenically produced through the oxidation of the nodule sulfur which was itself biogenically produced, but in Quaternary time, the age of the surrounding sedimentary deposit.

Solar Lake   An example of a lake in which sulfur is produced biogenically but not permanently deposited is found in the Sinai on the western shore of the Gulf of Aqaba. It is a small body of water called Solar Lake, which has undergone extensive limnological study (Cohen et al., 1977a,b,c). It is tropical and hypersaline, with a chemocline (O_2/H_2S interface) and a thermocline, which in winter is inverted (i.e., the hypolimnion is warmer than the epilimnion). The chemocline, which is 0-10 cm thick and located at a depth of 2-4 m, undergoes diurnal migration over a distance of 20-30 cm. The chief cause of this migration is the activity of the cyanobacteria *Oscillatoria limnetica* and *Microcoleus* sp., whose growth extends from the epilimnion into the hypolimnion. Sulfate-reducing bacteria, including a *Desulfotomaculum acetoxidans* type, near the bottom in the anoxic hypolimnion generate H_2S from the SO_4^{2-} in the lake water. Some of this H_2S migrates upward to the chemocline. During the early daylight hours, H_2S in the chemocline and below it is oxidized to elemental sulfur by anaerobic photosynthesis of the cyanobacteria. After H_2S is depleted, the cyanobacteria switch to aerobic photosynthesis, generating O_2. Thus, during the daylight hours the chemocline gradually drops. After dark, when all photosynthesis by the cyanobacteria has ceased, H_2S generated by the sulfate reducers builds up in the chemocline, together with H_2S generated by the cyanobacteria from the S^0 they formed earlier and by bacteria such as *Desulfuromonas acetoxidans*. Thus, during dark hours, the chemocline rises. The cycle is repeated with the break of day. Some thiosulfate is found in the chemocline during daylight hours, primarily as a result of chemical oxidation of sulfide. This thiosulfate is reduced in the night hours by biological and chemical means. Sulfur thus undergoes cyclic transformations in this lake such that elemental sulfur does not acucmulate to a significant extent. [See Jørgensen et al. 1979a,b) for further details of the sulfur cycle in this lake].

Thermal Lakes and Springs   An example of syngenetic sulfur deposition in a high-temperature environment is Lake Ixpaca in Guatemala. It is a crater lake that is supplied with H_2S form solfataras (fumarolic hot springs that yield sulfuretted waters) (Ljunggren, 1960). The H_2S concentration of the lake water was reported to be 0.10-0.18 g liter^{-1}. This H_2S is oxidized to native sulfur, rendering the water of the lake opalescent. A portion of the sulfur settles out and is incorporated into the sediment. Another lesser portion of the sulfur is oxidized to sulfuric acid, lowering the pH of the lake to 2.27. The sulfate content of the lake water ranges from 0.46 to 1.17 g liter^{-1}. The water from the solfataras is very hot (87-95°C), whereas that of the lake ranges from 29 to 32°C. The sulfuric acid of the lake water is very corrosive. It decomposes igneous minerals such as pyroxenes and feldspars into clay minerals (e.g., pickingerite). Ljunggren (1960) found an extensive presence of *Beggiatoa* in the waters and associated with the sediments of this lake. He implicated this organism in the conversion of H_2S into native sulfur. He apparently did not investigate whether the sulfuric acid might not also be biogenically formed, at least in part, nor did he consider the possibility that other organisms in addition to *Beggiatoa* might be involved in the genesis of S^0 from H_2S. More recent studies of hot springs in the United States have revealed the active participation of microbes in sulfur oxidation to sulfuric acid (Brock, 1978; Ehrlich and Schoen, 1967; Schoen and Ehrlich, 1968; Schoen and Rye, 1970). At most locations in these hot-spring areas, the H_2S of the solfataras appeared to be chemically oxidized to native sulfur. However, in the case of Mammoth Hot Springs in Yellowstone National Park, H_2S appeared to be biochemically oxidized to native sulfur, as concluded from sulfur isotope fractionation studies. Physiological evidence of bacterial H_2S oxidation at temperatures as high as 93°C (80-90°C optimum) in Boulder Spring of Yellowstone National Park (an alkaline hot spring, pH 8-9) has also been reported recently (Brock et al.,1971). The bacteria in this instance are mixotrophic, being able to use H_2S or other reduced sulfur compounds as an energy source and organic matter as a carbon source. A more detailed physiological study of elemental sulfur oxidation to sulfuric acid in hot acid soils and hot springs of Yellowstone National Park revealed that *Thiobacillus thiooxidans* is active at temperatures below 55°C and *Sulfolobus acidocaldarius* at temperatures between 55 and 85°C (Fliermans and Brock, 1972; Mosser et al.,1973). Almost all of the sulfur oxidation in the hot acid soils and hot springs is biochemical in origin since sulfur appears to be stable in the absence of bacterial activity (Mosser et al.,1973). *S. acidocaldarius* from these sources consists of spherical cells that form frequent lobes and lack peptidoglycan in the cell wall (Brierley, 1966; Brock et al., 1972). The organism is acidophilic (optimum pH 2-3; range pH 0.9-5.8) and thermophilic (temperature optimum 70-75°C; range 55-80°C). It has a guanine-cytosine content of 60-68 mol%. In growing cultures in the laboratory, the growth rate parallels the oxidation rate on elemental sulfur (Shivvers and Brock, 1973). The organisms attach themselves to sulfur crystals. The presence of yeast extract in the medium partially inhibits sulfur oxidation but not growth. The growth rates of *S. acidocaldarius* in several hot springs in Yellowstone National Park exhibit steady-state doubling times on the order of 10-20 hr in the water of small springs having volumes for 20 to 2,000 liters, and on the order of 30 days in large springs having 1×10^6-liter volumes. Exponential doubling times measured in the water of artificially drained springs are on the order of a few hours (Mosser et al., 1974).

In alkaline hot springs, a bacterium called *Chloroflexus aurantiacus* gen. nov., sp. nov. has also been found (Brock, 1978; Pierson and Castenholz, 1974). It is characterized as a gliding, filamentous (0.5-0.7 μm in width, variable in length), phototrophic bacterium with a tendency to form orange mats below, and to a lesser extent above thin layers of cyanobacteria (e.g., *Synechococcus*). Its photosynthetic pigments include bacteriochlorophylls a and c, and β- and γ-carotene. The pigments occur in *Chlorobium* vesicles. Anaerobically, the organism is capable of photoautotrophic growth in the presence of sulfide and bicarbonate (Madigan and Brock, 1975), and of photoheterotrophic growth with yeast extract and certain other organic supplements. Aerobically, the organism is capable of heterotrophic growth in the dark. Although showing some resemblance to Rhodospirillaceae, it is considered to be more closely related to the Chlorobiaceae. *Chloroflexus* has been reported to be a major component of laminated mats in hot springs in Yellowstone National Park in the temperature range 55-70°C (Doemel and Brock, 1974, 1977). The minor component of the mats was reported to be the nonmotile, unicellular cyanobacterium *Synechococcus*. The mats often incorporate detrital silica in the form of siliceous sinter from the geyser basins and in time become transformed into stromatolites. Sulfate in the mats can be reduced to sulfide below the upper 3 mm, and the sulfide can be converted to elemental sulfur by *Chloroflexus* in the mats (Doemel and Brock, 1976). Here, then, is another example where below 70°C at least some elemental sulfur in hot springs or their effluent may be of biogenic origin.

14.5   EPIGENETIC SULFUR DEPOSITS

Sicilian Deposits   Examples of epigenetic sedimentary sulfur deposits are those on the volcanic island of Sicily. Isotopic study (Jensen, 1962) has shown that the sulfur in these deposits, being significantly enriched in ^{32}S relative to the associated sulfate, is not of volcanic origin but of biological origin. Sulfate-reducing bacteria are presumed to have reduced sulfates in the sediment beds to sulfide, and the sulfide was then oxidized chemically or biologically to sulfur. The organic carbon used in the reduction of the sulfate presumably came from organic detritus in the sediment (algal and other remains).

Salt Domes   Another example of biogenic native sulfur of epigenetic origin in a sedimentary environment is that associated with salt domes (Fig. 14.4) as found on the gulf coast of the United States (the northern and western shores of

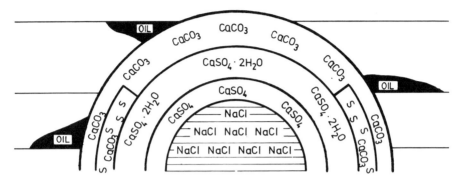

Figure 14.4 Diagrammatic representation of a salt dome. (After Ivanov, 1968).

the gulf coast, including those of Texas, Louisiana, and Mexico) (see, e.g., the description by Ivanov, 1968, pp. 92ff). Such salt domes reside over a central plug of 90-95% rock salt (NaCl) and 5-10% anhydrite ($CaSO_4$) and traces of dolomite $CaMg(CO_3)_2$, barite ($BaSO_4$), and celestite ($SrSO_4$). Petroleum may be entrapped in peripheral deformations. The salt domes contain mainly anhydrite located directly over the salt plug, and are topped by calcite which may have commercially exploitable petroleum associated with it. Between the calcite and the anhydrite exists a zone containing gypsum ($CaSO_4 \cdot 2H_2O$), calcite, and anhydrite relicts. Sulfur is associated with calcite and the intermediate zones. The salt domes originated from evaporites formed in the Jurassic (Middle Mesozoic 135-181 million years ago) to late Paleozoic (230-280 million years ago). A current theory of the origin of salt domes (see Strahler, 1977) is that they began as beds of evaporites along continental margins of newly emergent oceans, such as the Atlantic Ocean about 180 million years ago. The evaporite derived from hypersaline waters with the aid of the heat emanating from the interior of the earth in these tectonically active areas. As the ocean basins broadened due to continental drift, and as the continental margins became more defined, turbidity currents began to bury the evaporite beds under ever-thicker layers of sediment. Ultimately, these sediment layers became so heavy that they forced portions of the evaporite, which has plastic properties, upward as salt plugs through ever younger sediment stata. As these plugs intruded into the groundwater zone, they lost their more water-soluble constituents, particularly their rock salt, leaving behind relatively insoluble anhydrite which became cap rock. In time, some of the anhydrite was converted to the more-soluble gypsum and some was dissolved away. At that point, bacterial sulfate reduction is thought to have occurred, lasting perhaps for a period of 1 million years. This reduction process is thought to have utilized the organic carbon from adjacent petroleum deposits. Although some Russian workers have claimed that sulfate-reducing bacteria can use hydrocarbons as their source of reducing power in sulfate reduction, it is more likely that other heterotrophic bacteria first converted these compounds into substances that sulfate-reducing bacteria can utilize for sulfate reduction. On the other hand, recent studies have revealed a sulfate-reducing bacterium that can use methane (CH_4) as a source of reducing power (Panganiban and Hanson, 1976; Panganiban et al., 1979). During the sulfate-reducing process, not only was H_2S generated in the evolving salt dome, but also CO_2. The H_2S became oxidized biologically or chemically to native sulfur, whereas the CO_2 was extensively precipitated as carbonate.

$$CaSO_4 + 2C_{org} \rightarrow 2CO_2 + CaS \tag{14.40}$$

$$CaS + CO_2 + H_2O \rightarrow CaCO_3 + H_2S \tag{14.41}$$

$$H_2S + \tfrac{1}{2}O_2 \rightarrow S^0 + H_2O \tag{14.42}$$

The sulfur and secondary calcite are physically associated in the cap rock, and their isotopic enrichment values indicate a biological origin.

Gaurdak Deposit   An epigenetic mode of sulfur formation somewhat similar to that in the salt domes in the United States took place in the Gaurdak Deposit of the eastern Turkmen S.S.R. (Ivanov, 1968). This deposit resides in rock of the Upper Jurassic and was probably emplaced in the Quaternary as plutonic waters entered the formation as a result of tectonic activity. The plutonic waters picked up organic carbon from the Kugitang Suite containing bituminous lime-stone, and sulfate from the anhydrite-carbonate rocks of the Gaurdak Suite. Sulfate-reducing bacteria then attacked the sulfate and reduced it to H_2S with the help of the reduced carbon derived from the bituminous material. *T. thio-parus* oxidized the H_2S to S^0 after it reached the plutonic water/surface water interface. Where oxygenated groundwater presently reaches the sulfur deposit, intense biooxidation of the surfur to sulfuric acid has been noted, causing dis-appearance of secondary calcite and deposition of secondary gypsum. The bacteria *T. thioparus* and *T. thiooxidans* have been found in significant numbers respectively in sulfuretted waters in the sulfur deposits with paragenetic calcite, and in acidic sulfur deposits with secondary gypsum.

Shor-Su Deposit   Another example of epigenetic, microbial sulfur accumulation is illustrated by the Shor-Su Deposit in the northern foothills of the Altai Range in the U.S.S.R. Here an extensive, folded sedimentary formation of lagoonal origin and mainly of Paleocene and Cretaceous age contains major sulfuretted regions in lower Paleocene strata (Bukhara and Sazuk) of the second anticline and to a lesser extent in Quaternary conglomerates (Fig. 14.5) (see Ivanov, 1968, pp. 33-34). The sulfur of the main deposit occurs in heavily broken rock, is surrounded by gypsified rock and contains some relict gypsum lenses. It is enclosed in a variety of cavernous rock and associated with calcite and celestite in the Bukhara stratum and in cavities and slitlike caves in the Suzak stratum. Petroleum and natural gas deposits are associated with the fourth anticline of the Shor-Su sedimentary formation. This structure is hydraulically connected with the second anticline, which contains most of the native sulfur. One basis for the claim of hydraulic connections is that the waters of the two anticlines are chemically very similar in composition. Sulfate-reducing bacteria occur in these plutonic waters, flowing through pervious strata from the fourth to the second anticline. It is believed that these bacteria have been reducing the sulfate in the plutonic water derived by progressive solubilization of gypsum and anhydrite in surrounding rock. They are presumed to have been using petroleum hydrocarbons as a source of reducing power and carbon for this process. The

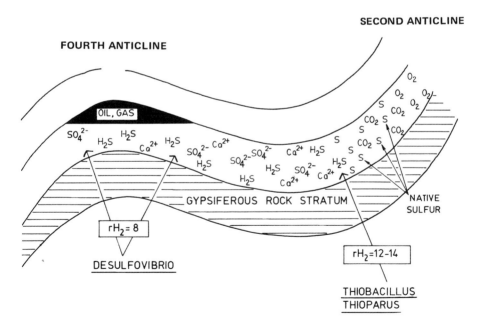

Figure 14.5   Diagrammatic representation of essential features of the Shor-Su Deposit. (Based on a description by Ivanov, 1968.)

presence of sulfate-reducing bacteria has also been reported in the waters of the second anticline and in any rock in which sulfur occurs (Ivanov, 1968). These bacteria were demonstrated to be able to reduce sulfate to H_2S under in situ condituions at a measurable rate (0.009-0.179 mg of H_2S liter^{-1} day^{-1}). Sulfur has been forming where rising plutonic water has been mixing with downward-seeping oxygenated surface water. In this zone of mixing of the two waters, T. thioparus was detected and shown to oxidize H_2S of the plutonic water to native sulfur. Measurements have shown that sulfate reduction predominates where plutonic waters carry sulfate which largely derives from gypsiferous rock, and organic matter which derives from the petroleum deposit, and where these waters have a sufficiently low redox potential (often below rH_2 8) to support growth and activity of sulfate-reducing bacteria (Ivanov, 1968). The moving plutonic water has been carrying the H_2S to a region in the second anticline where it has been encountering aerated surface water from above, raising the redox potential to about rH_2 12-14 (16.5 maximum) and making for an environment that favors the growth and activity of T. thioparus with the resultant precipitation of native sulfur. Where the redox potential exceeds rH_2 16.5, owing to extensive exposure to surface water, for instance in the sulfur of the outcroppings in the western conglomerate of the Shor-Su, the sulfur is undergoing extensive oxidation to sulfuric acid by T. thiooxidans. The pH is found to drop from neutrality to less than 1 where the bacteria are most active. The sulfur deposition in the main deposit (strata) began to be laid down in the Quarternary, according to Ivanov (1968), and continues to the present day. Thus, interpretation of events in the geologic past is supported by the observations of bacterial distribution and activity in contemporary sulfur deposition in Quaternary strata of the Shor-Su, which follow the pattern ascribed to the main sulfur deposit (Ivanov, 1968).

Kara Kum Deposit   Spatially, a somewhat different mechanism of epigenetic sulfur deposition has been recognized in the Kara Kum Deposit north of Ashkhabad (Ivanov, 1968), even though sulfate-reducing bacteria and H_2S-oxidizing bacteria were involved in the sulfur genesis. Here sulfate reduction took place in a different stratum from that involving sulfur deposition. Hence, paragenetic secondary calcite is not found associated with the sulfur in the deposit, and, furthermore, sulfuric acid from the oxidation of sulfur by T. thiooxidans on more direct exposure to air is not neutralized by $CaCO_3$ and converted to gypsum but reacts with the sandstone, liberating aluminum and iron, which are then leached and precipitated as oxides in a more neutral environment.

14.6 SUMMARY

Sulfur wich occurs in organic and inorganic form in nature, is essential to life. Different organisms may assimilate it in organic or inorganic form. However, plants normally take it up as sulfate. Microbes are thus important in mineralizing organic sulfur compounds in soil and aqueous environments. The biochemistry of organic sulfur mineralization, as well as the synthesis of organic sulfur compounds, has been studied in some detail.

Inorganic sulfur may exist in various oxidation states in nature, chiefly as sulfide, sulfur, and sulfate, although thiosulfate may also occur in significant amounts in some environments. Microbes in soil and water play an important role in the interconversion of these oxidation states. Certain bacteria, including under special circumstances cyanobacteria, play important roles in these transformations. Chemolithotrophic, mixotrophic, and photolithotrophic bacteria, and some cyanobacteria can oxidize H_2S to elemental sulfur (S^0). The chemolithotrophs and mixotrophs use oxygen as oxidant under partially reduced conditions. In the absence of oxygen, some chemolithotrophs may substitute nitrate as oxidant. The photolithotrophic bacteria (purple and green bacteria) and select cyanobacteria use fixed CO_2 as an oxidant under anaerobic conditions. Elemental sulfur may be oxidized to sulfuric acid aerobically by chemolithotrophic or mixotrophic bacteria in neutral or acid environments. It may also be oxidized to sulfuric acid anaerobically by some photosynthetic bacteria. Sulfur oxidation has been noted in mesophilic as well as in thermophilic environments. Thiosulfate has been found to serve as energy source for some marine mixotrophic pseudomonads, being oxidized to tetrathionate.

Oxidized forms of inorganic sulfur may be reduced by certain microorganisms. Elemental sulfur, serving as a terminal electron acceptor, has been reported to be reduced to H_2S by *Desulfuromonas acetoxidans, Desulfovibrio gigas,* and a few other sulfate-reducing bacteria. Two types of fungi, *Rhodotorula* and *Trichosporon,* have also been found to be able to reduce elemental sulfur to H_2S.

Sulfate serving as a terminal electron acceptor in the absence of oxygen has been found to be reduced by *Desulfovibrio* spp., *Desulfotomaculum* spp., and *Desulfomonas pigra* (dissimilatory sulfate reduction). This microbial activity is of major importance geologically because under natural conditions on the earth's surface, sulfate cannot be reduced by purely chemical means because of the high activation-energy requirement of the process. Sulfate is reduced aerobically by various microbes and plants but only in small amounts without any extracellular accumulation of H_2S (assimilatory sulfate reduction). Biochemically, the mechanisms of dissimilatory and assimilatory sulfate reduction differ.

Some reducers and oxidizers of sulfur can distinguish between ^{32}S and ^{34}S and can bring about isotope fractionation. Geologically, this is useful in determining whether ancient sulfur deposits were biogenically or abiogenically formed.

Contemporary biogenic sulfur deposition involving sulfate-reducing and aerobic and anaerobic sulfide-oxidizing bacteria have been identified in several lacustrine environments. These represent syngenetic deposits. Bacterial oxidation of elemental sulfur to sulfuric acid in certain hot springs has also been reported.

Ancient epigenetic sulfur deposits of microbial origin have been identified in salt domes and other geologic formations associated with hydrocarbon (petroleum) deposits in various parts of the world. The sulfur in these instances arose from bacterial reduction of sulfate derived from anhydrite or gypsum followed by bacterial oxidation under partially reduced conditions to elemental sulfur. On full exposure to air, some of the sulfur is presently being oxidized by bacteria to sulfuric acid.

Less spectacular oxidative and reductive transformations of sulfur go on in in soil, where they play an important role in maintenance of soil fertility.

15

Biogenesis and Biodegradation of Sulfide Minerals on the Earth's Surface

A number of metal sulfides are of geomicrobial interest, either because they are formed as a result of microbial activity or because they are degraded by microbial activity. Examples of metal sulfide minerals of geomicrobial interest are listed in Table 15.1.

15.1 METAL SULFIDE BIOGENESIS

Origin of Igneous Metal Sulfide Deposits   Most metal sulfides of commercial interest are of igneous origin. Current theory of their formation invokes plate tectonics as being responsible (Strahler, 1977; Bonatti, 1978; see also Hammond, 1975a). In this process, metal-ion species, especially Fe, Cu, and sometimes Zn, are seen as rising to the earth's surface during hydrothermal and/or magmatic activity at the mid-ocean ridges and becoming fixed as metal sulfates in magmatic eruptives (e.g., pillow lavas) (see Strahler, 1977; Bonatti, 1978; also Hammond, 1975b). A somewhat modified process is deducible from visual observations and direct sampling at hydrothermally active regions at seafloor-spreading centers at 2,500-2,600 m in the eastern Pacific Ocean in the Galápagos Rift and the East Pacific Rise (Ballard and Grassle, 1979; Corliss et al., 1979). In these areas, hot, metal-laden and H_2S-charged hydrothermal solutions spew forth from vents or chimneys. Sulfides such as sphalerite and chalcopyrite precipitate around the vents as the solutions mix with cold seawater. The metal-laded hydrothermal solutions appear to arise when seawater penetrates hot volcanic rock in the upper oceanic crust to depths as great as 10 km (Bonatti, 1978). In its heated and pressurized state, the seawater leaches the metals from the rock while its sulfate is reduced to H_2S by metal species such as Fe(II) (Mottl et al., 1979). The hot, metal-enriched solution is forced up through the hydrothermal vents. Not all the H_2S is used up in the metal precipitation. So much is left over, in fact, that it serves as a primary energy source to bacteria, which become the food for tube and polychaete worms,

Table 15.1 Metal Sulfides of Geomicrobial Interest

Mineral or synthetic compound	Formula	References
Antimony trisulfide	Sb_2S_3	Silver and Torma (1974); Torma and Gabra (1977)
Argentite	Ag_2S	Baas Becking and Moore (1961)
Arsenopyrite	$FeAsS$	Ehrlich (1964a)
Bornite	Cu_5FeS_4	Cuthbert (1962); Bryner et al. (1954)
Chalcocite	Cu_2S	Bryner et al. (1954); Ivanov (1962); Razzell and Trussell (1963); Sutton and Corrick (1963, 1964); Fox (1967); Nielsen and Beck (1972)
Chalcopyrite	$CuFeS_2$	Bryner and Anderson (1957)
Cobalt sulfide	CoS	Torma (1971)
Covellite	CuS	Bryner et al. (1954); Razzell and Trussell (1963)
Digenite	Cu_9S_5	Baas Becking and Moore (1961); Nielsen and Beck (1972)
Enargite	$3Cu_2S \cdot As_2S_5$	Ehrlich (1964a)
Galena	PbS	Silver and Torma (1974)
Galium sulfide	Ga_2S_3	Torma (1978)
Marcasite, pyrite	FeS_2	Leathen et al. (1953); Silverman et al. (1961)
Millerite	NiS	Razzell and Trussell (1963)
Molybdenite	MoS_2	Bryner and Anderson (1957); Bryner and Jameson (1958); Brierley and Murr (1973)
Orpiment	As_2S_3	Ehrlich (1963a)
Nickel sulfide	NiS	Torma (1971)
Pyrrhotite	Fe_4S_5	Freke and Tate (1961)
Sphalerite	ZnS	Ivanov et al. (1961); Ivanov (1962); Malouf and Prater (1961)
Tetrahedrite	$Cu_8Sb_2S_7$	Bryner et al. (1954)

clams, and mussels, which in turn attract brachyuran crabs, and so on. The area around the hydrothermal vents becomes a special ecologic niche in which the primary producers are not photosynthetic but chemosynthetic (Jannasch and Wirsen, 1979).

Some of the minerals deposited at the spreading centers may be carried toward continental margins during seafloor spreading. Once at the continental margin, they are carried below the continental crust during subduction and may move into the continental crust during magma formation in the subduction process. **Porphyry ore** (small crystals of metal sulfides richly dispersed in host rock) results, especially ore of copper and molybdenum sulfides. Mineral sulfides may also accumulate in island arcs during subduction and associated volcanic activity.

Sedimentary Metal Sulfides   Among metal sulfides of sedimentary origin, iron sulfides are the most common. They are usually associated with reducing zones in sedimentary deposits. Their presence there is frequently the consequence of an interaction of microbiologically generated H_2S with iron compounds, leading to the formation of hydrotroilite ($FeS \cdot nH_2O$) and pyrite (FeS_2). Rapid and extensive microbial pyrite formation has been observed in salt marsh peat on Cape Cod (Howarth, 1979). Here the bacterial sulfate reduction that led to the pyrite production represents the major form of respiration in the salt marsh.

Nonferrous metal sulfide deposits of sedimentary origin are relatively rare. They have been thought to arise syngenetically; that is, the metals in question are precipitated by H_2S from a hydrothermal or microbiological source and then buried in contemporaneously formed sediment. The limiting conditions for syngenetically formed sedimentary metal sulfide deposition have been examined (Rickard, 1973). For bacteriogenic metal sulfide formation, calculations indicate that a minimum of 0.1% organic carbon (dry weight) is required together with an enriched source of metals such as a hydrothermal solution if more than 1% metal is to be deposited.

Examples of sedimentary sulfide deposits include the Permian Kupferschiefer of Mansfeld in Germany (Love, 1962; Stanton, 1972, p. 1139), Black Sea sediments (Bonatti, 1972, p. 51), the Roan Antelope Deposit in Zambia and Katanga (Cuthbert, 1962; Stanton, 1972, p. 1139), the Zechstein Deposit (Serkies et al., 1968), and deposits in the Pernatty Lagoon (Lambert et al., 1971).

Principles of Metal Sulfide Formation   Whether formed biogenically or abiogenically, the metal sulfides result from an interaction between the metal ion and sulfide ion:

$$M^{2+} + S^{2-} \rightarrow MS \qquad\qquad\qquad (15.1)$$

Table 15.2 Solubility Products for Some Metal Sulfides

CdS	1.4×10^{-28}	FeS	1×10^{-19}	NiS	3×10^{-21}
Bi_2S_3	1.6×10^{-72}	PbS	1×10^{-29}	Ag_2S	1×10^{-51}
CoS_2	7×10^{-23}	MnS	5.6×10^{-16}	SnS	8×10^{-29}
Cu_2S	2.5×10^{-50}	Hg_2S	1×10^{-45}	ZnS	4.5×10^{-24}
CuS	4×10^{-38}	HgS	3×10^{-53}	H_2S	1.1×10^{-7}
				HS^-	1×10^{-15}

It is the source of the sulfide that determines whether a biological agent is implicated in metal sulfide formation. If the sulfide results from bacterial sulfate reduction (see Chapter 14) or from bacterial mineralization of organic compounds (Dévigne, 1968a,b; 1973), it is obvioulsy of biogenic origin. If it is derived from volcanic activity, it is generally of abiogenic origin. The metal sulfides, because of their relative insolubility, form readily at ordinary temperatures and pressures by interaction of metal ions and sulfide ions. Table 15.2 lists solubility products for some common simple sulfides.

The following calculation will show that relatively low concentrations of metal ions are needed to form metal sulfides by reacting with H_2S at typical concentrations in some lakes.* Let us examine, for instance, the case of iron. The dissociation constant for iron sulfide (FeS) is

$$[Fe^{2+}][S^{2-}] = 1 \times 10^{-19} \tag{15.2}$$

The dissociation constant for H_2S is

$$[S^{2-}] = 1.1 \times 10^{-22} \frac{[H_2S]}{[H^+]^2} \tag{15.3}$$

since

$$\frac{[HS^-][H^+]}{[H_2S]} = 1.1 \times 10^{-7} \tag{15.4}$$

and

$$\frac{[S^{2-}][H^+]}{[HS^-]} = 1 \times 10^{-15} \tag{15.5}$$

---

*Activities are taken as approximately equal to concentration here because of the low concentrations involved.

Therefore,

$$[Fe^{2+}] = \frac{[H^+]^2}{[H_2S]} \times \frac{1 \times 10^{-19}}{1.1 \times 10^{-22}} = \frac{[H^+]^2}{[H_2S]} (9.1 \times 10^2) \qquad (15.6)$$

Assuming that the bottom water of a lake contains about 34 mg of H_2S per liter (0.001 M) at pH 7, about 5.08×10^{-3} mg of Fe^{2+} per liter (9.1×10^{-8} M) will be precipitated by 3.4 mg of hydrogen sulfide per liter (10^{-4} M). The unused H_2S will ensure reducing conditions, which will keep the iron in the ferrous state. Since ferrous sulfide is one of the more-soluble sulfides, it can be seen that metals whose sulfides have even smaller solubility products will form even more readily.

Biogenesis of Metal Sulfides Metal sulfides have been generated in laboratory experiments utilizing H_2S from bacterial sulfate reduction. Miller (1950) reported that sulfides of Sb, Bi, Co, Cd, Fe, Pb, Ni, and Zn were formed in a lactate broth culture of *Desulfovibrio desulfuricans* to which insoluble salts of selected metals had been added. For instance, he found bismuth sulfide to be formed on addition of $(BiO_2)_2CO_3 \cdot H_2O$; cobalt sulfide on addition of $2CoCO_3 \cdot 3Co(OH)_2$; lead sulfide on addition of $2PbCO_3 \cdot Pb(OH)_2$ or $PbSO_4$; nickel sulfide on addition of $NiCO_3$ or $Ni(OH)_2$; and zinc sulfide on the addition of $2ZnCO_3 \cdot 3Zn(OH)_2$ as starting compounds. The reason for adding insoluble metal salts as starting compounds was to minimize metal toxicity for *D. desulfuricans*. Metal toxicity depends in part on the solubility of the metal compound in question. Obviously, for the corresponding metal sulfide to be formed, the metal sulfide must be even more insoluble than the starting compound of the metal. Miller was not able to demonstrate copper sulfide formation from malachite $[CuCO_3 \cdot Cu(OH)_2]$, probably because malachite is too insoluble relative to copper sulfides in his medium. In the case of Cd and Zn, Miller (1950) showed that these caused an increase in total sulfide generated from sulfate in batch culture when compared to the sulfide yield in the absence of these added metals.

Baas Becking and Moore (1961) also undertook a study of biogenesis of sulfide minerals. Like Miller, they worked with batch cultures of *Desulfovibrio desulfuricans* or *Desulfotomaculum* sp. (called by them *Clostridium desulfuricans*) growing in lactate or acetate/steel wool-containing medium, where the steel wool was presumably meant to serve as a source of hydrogen in the bacterial reduction of sulfate.

$$Fe^0 + 2H_2O \longrightarrow H_2 + Fe(OH)_2 \qquad (15.7)$$

$$4H_2 + SO_4^{2-} \xrightarrow{\text{bacteria}} S^{2-} + 4H_2O \qquad (15.8)$$

These investigators formed ferrous sulfide from strengite ($FePO_4$) and from hematite (Fe_2O_3). Furthermore, they formed covellite (CuS) from malachite ($CuCO_3 \cdot Cu(OH)_2$) and chrysocolla ($CuSiO_3 \cdot 2H_2O$); digenite (Cu_9S_5) from cuprous oxide (Cu_2O); argentite (Ag_2S) from silver chloride (Ag_2Cl_2) or silver carbonate (Ag_2CO_3); galena (PbS) from lead carbonate ($PbCO_3$) or lead hydroxycarbonate [$(PbCO_3)_2 \cdot Pb(OH)_2$]; and sphalerite (ZnS) from zinc wire and from smithsonite ($ZnCO_3$). All mineral products were identified by X-ray powder diffraction studies. Baas Becking and Moore were unable to form cinnabar from mercuric carbonate, probably owing to the toxicity of Hg^{2+} ion, nor were they able to form alabandite (MnS) from rhodochrosite, or bornite or chalcopyrite from a mixture of cuprous oxide or malachite and hematite and lepidochrosite. They probably succeeded in forming covellite from malachite where Miller (1950) failed because they performed their experiment in the presence of 3% NaCl in the medium. The starting materials that were the source of metal are seen to have all been insoluble, as in Miller's experiments. Baas Becking and Moore found that in the formation of covellite and argentite, native copper and silver were respective intermediates that disappeared with continued bacterial H_2S production.

A Model of Sedimentary Metal Sulfide Biogenesis   The relatively high toxicity of many of the heavy metals for sulfate-reducing bacteria has been used as an argument that these organisms could not have been responsible for metal sulfide precipitation in nature (e.g., Davison, 1962a,b). However, in a sedimentary environment, metal ions will be mostly adsorbed to components of the solid phase of the sediment such as clays or complexed by organic matter (Hallberg, 1978), and their toxicity for bacteria will thus be greatly lessened. Such adsorbed or complexed ions are still capable of reacting with sulfide and pre-cipitating as metal sulfides, as has been shown experimentally (Temple and LeRoux, 1964). Separating a clay or ferric hydroxide slurry carrying adsorbed Cu^{2+}, Pb^{2+}, and Zn^{2+} ions, and in the case of clay also carrying Fe^{3+} ions, by an agar plug from an actively growing saline culture of sulfate-reducing bac-teria, a development of banded precipitates of the metal sulfides in the agar plug may be noted with time (Fig. 15.1). The bands form as the upward-diffusing sulfide species and the downward-diffusing, desorbed metal ions encounter each other in the agar. Differential desorption of metal ions from the absorbent and their differential diffusion in the agar accounts for the observed banding by the various sulfides. These results demonstrate that biogenesis of metal sulfides in a sedimentary environment is possible, even in the presence of relatively large amounts of metal ions, provided that the metal ions are in a nontoxic form such as in insoluble mineral oxides, carbonates, or sulfates, or adsorbed. As Temple (1964) has pointed out, syngenetic microbial metal sulfide production in nature is possible. Restrictions on the process, according

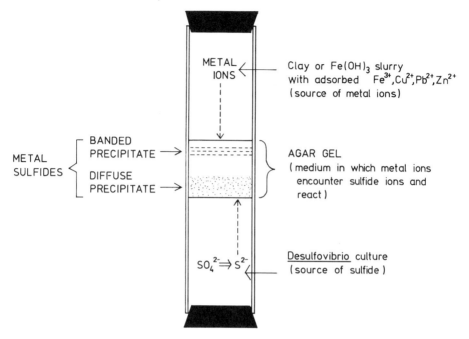

Figure 15.1 Temple and LeRoux column showing how sulfate reducers can precipitate metal sulfides by reaction of biogenic sulfide with metal ions. The adsorbents, clay or $Fe(OH)_3$ slurry, control the concentration of metal ions in solution, and the agar plug prevents physical contact of the sulfate reducer with metal ions. In nature, sediments can act as adsorbents of metal ions. They hold the metal-ion concentration in the interstitial water at such a level that sulfate reducers are not poisoned.

to him, are not metal toxicity, but movement of the bacterially generated sulfide and a need for metal-enriched zones in the sedimentary environment. Temple believes that on biochemical grounds, microbial sulfate reduction had evolved by the early Precambrian.

Metal Sulfide Biogenesis Associated with Mineralization of Organic Sulfur Compounds   Although metal sulfide genesis in nature is usually associated with the H_2S-generating activity of dissimilatory sulfate reducers *Desulfovibrio* spp. and *Desulfotomaculum* spp., at least one case of biogenesis of galena may be attributed to the aerobic mineralizing activity of *Sarcina flava* Bary on sulfur-containing organic compounds (Dévigne, 1968a,b; 1973). The *Sarcina* was isolated from earthy concretions between crystals of galena in an accumulation in a karstic pocket located in the lead-zinc deposit of Djebel Azered, Tunisia.

In laboratory experiments, the organism was shown to produce PbS from Pb^{2+} bound to sulfhydryl groups of peptone.

15.2 METAL SULFIDE BIODEGRADATION

Metal sulfides, whether of abiogenic or biogenic origin, may be subject to microbial oxidation. This may take the form of indirect or direct interaction. In indirect interaction, the microbes are responsible for generating a reagent that causes solubilization of the metal sulfide, most commonly through oxidation, but, under some conditions, through complexation. In direct interaction, the microbes attack the metal sulfide directly by oxidizing it in its insoluble form, rendering it into the soluble sulfate salt.

Biodegradation of Metal Sulfides by Indirect Action   In indirect attack by oxidation, the microbially generated oxidizing agent involved is ferric iron (Fe^{3+}). It may be generated from dissolved ferrous iron (Fe^{2+}) at pH values from 3.5 to 5.0 by *Metallogenium* in a mesophilic temperature range (Walsh and Mitchell, 1972a). At pH values below 3 5, ferric iron may be generated from ferrous iron by *Thiobacillus ferrooxidans* in a mesophilic temperature range, and by *Sulfolobus* sp. and other as-yet-unidentified bacteria in a thermophilic temperature range (Brierley and Lockwood, 1977). Ferric iron may also be generated from iron pyrites (e.g., FeS_2) by *T. ferrooxidans.* The ferric iron, in acid solution, then acts as oxidant on the metal sulfides (Ehrlich and Fox, 1967b):

$$MS + 2Fe^{3+} \rightarrow M^{2+} + S^0 + 2Fe^{2+} \tag{15.9}$$

where M may be any metal in an appropriate oxidation state, which does not always have to be divalent. It should be noted that in this type of chemical reaction the sulfide is only oxidized to sulfur (S^0). Further chemical oxidation to sulfuric acid by oxygen is slow, but is likely to be greatly accelerated by *T. thiooxidans, T. ferrooxidans,* and/or *Sulfolobus* sp. The chemical oxidation of the metal sulfide must occur in acid solution below pH 5.0 to keep enough ferric iron in solution. In nature, this acid may be formed chemically through autoxidation of sulfur, but more likely biologically through bacterial oxidation of sulfur or iron, either as ferrous iron in solution or as pyrite. In ferrous iron oxidation, the acid forms as follows:

$$2Fe^{2+} + \tfrac{1}{2}O_2 + 2H^+ \xrightarrow{\text{bacteria}} 2Fe^{3+} + H_2O \tag{15.10}$$

$$Fe^{3+} + 3H_2O \longrightarrow Fe(OH)_3 + 3H^+ \tag{15.11}$$

If this reation takes place in the presence of sulfate, the ferric hydroxide may convert to the more insoluble jarosite,

$$A^+ + 3Fe(OH)_3 + 2SO_4^{2-} \rightarrow AFe_3(SO_4)_2(OH)_6 + 3OH^- \tag{15.12}$$

where A^+ may represent Na^+, K^+, NH_4^+, or H_3O^+ (Duncan and Walden, 1972). The formation of jarosite decreases the ratio of protons produced per iron oxidized from 2:1 to 1:1. In pyrite oxidation, the acid forms as a result of the following reactions:

$$FeS_2 + \tfrac{7}{2}O_2 + H_2O \rightarrow Fe^{2+} + 2H^+ + 2SO_4^{2-} \tag{15.13}$$

$$FeS_2 + 14Fe^{3+} + 8H_2O \rightarrow 15Fe^{2+} + 2SO_4^{2-} + 16H^+ \tag{15.14}$$

The ferric iron in equation 15.14 will come largely from bacterial oxidation of the ferrous iron in equation 15.13.

It should be noted that in the case of pyrite oxidation, the sulfur is oxidized to sulfate while the iron remains in the ferrous state. The ferrous iron will subsequently be further oxidized to ferric iron by biological or abiological means, as already explained, with the production of additional acid. Since iron pyrites usually accompany other metal sulfides in nature, iron pyrite is an important source of acid for the oxidizing reactions of other metal sulfides. In some cases, the host rock in which metal sulfides, including pyrites, are contained, may itself react with acid and thus raise the pH enough to cause complete precipitation of ferric iron and thereby prevent any oxidation of metal sulfide by it.

If complexation is involved in the degradation of metal sulfide, the role of microbes in this process is in the genesis of the complexing agent. Wenberg et al. (1971) reported, for instance, the isolation of *Penicillium* spp. from mine-tailings ponds of the White Pine Copper Co. in Michigan, which produced unidentified metabolites in Czapek's broth containing sucrose, $NaNO_3$, and cysteine, methionine, or glutamic acid, which could leach copper from the sedimentary ores of the White Pine Deposit. *T. ferrooxidans* could not be employed for leaching of this ore because of the presence of significant quantities of calcium carbonate that would neutralize the required acid. Similar findings were reported by Hartmannova and Kuhr (1974), who found not only *Penicillium* sp. but also *Aspergillus* (e.g., *A. niger*) active in producing complexing compounds. Wenberg et al. (1971) reported that if their fungus was grown in the presence of copper ore (sulfide or native copper minerals with basic gangue constituents), the addition of some citrate lowered the toxicity of the extracted copper for the fungus. Better results were obtained if the fungus was grown in the absence of ore, and the ore was then treated with spent medium. The principle of action of the fungi in these observations is similar to that in the observations of Kee and Bloomfield (1961) in regard to the dissolution of the oxides of several trace elements (e.g., $ZnO, PbO_2, MnO_2, CoO \cdot Co_2O_3$) with anaerobically fermented plant material (Lucerne and Cocksfoot). It is also similar to the principle of action in the observations of Parès (1964a,b,c), according to which *S. marcescens, B. subtilis, B. sphaericus,* and *B. firmus* solubilize copper and some other metals associated with laterites and clays by generating appropriate chelates in special culture media. The chelates of the metal ions are stabler than the original insoluble form of the metal, which forces their solubilization, as illustrated by the following equation:

$$MA + HCh \rightarrow MCh + H^+ + A^- \tag{15.15}$$

where MA is a metal salt, HCh a chelating agent (ligand), MCh the resultant metal chelate, and A^- the anion of the original insoluble metal salt, S^{2-} in the case of the metal sulfides. The S^{2-} may then undergo chemical or bacterial oxidation. The use of carboxylic acids in leaching ores has been proposed as a general process (*Chemical Processing,* 1965).

Biodegradation of Metal Sulfides by Direct Action   In direct microbial action on metal sulfides, the crystal lattice of susceptible metal sulfides is attacked through enzymatic oxidation. For this purpose, the microbes have to be in intimate contact with the mineral they metabolize. Evidence for growth of *T. ferrooxidans* on mineral surfaces of chalcopyrite has been presented by McGoran et al. (1969), and on galena and iron pyrite crystals by Tributsch (1976) and Bennett and Tributsch (1978), respectively. Direct evidence for enzymatic attack of synthetic covellite (CuS) by *T. ferrooxidans* has been obtained through measurement of oxygen consumption and Cu^{2+} and SO_4^{2-} ion production in the presence and absence of the enzyme inhibitor trichloracetate (8 mM) (Rickard and Vanselow, 1978). In this case only the sulfide moiety of the mineral is attacked, the metal moiety being already as oxidized as possible. The oxidation of the mineral probably proceeds in two steps (Fox, 1967):

$$CuS + \tfrac{1}{2}O_2 + 2H^+ \xrightarrow{\text{bacteria}} Cu^{2+} + S^0 + H_2O \tag{15.16}$$

$$S^0 + \tfrac{3}{2}O_2 + H_2O \xrightarrow{\text{bacteria}} H_2SO_4 \tag{15.17}$$

By contrast, *T. thioparus* oxidizes covellite only after its autoxidation to $CuSO_4$ and S^0 (see reaction 15.16 above; no bacterial catalysis is involved in this case) (Rickard and Vanselow, 1978).

In some cases both the oxidizable metal moiety and the sulfide moiety may be simultaneously attacked by separate enzymes, as, for example, in the case of chalcopyrite (Duncan et al., 1967). The overall reaction may be written as follows:

$$4CuFeS_2 + 17O_2 + 4H^+ \xrightarrow{\text{bacteria}} 4Cu^{2+} + 4Fe^{3+} + 8SO_4^{2-} + 2H_2O \tag{15.18}$$

$$4Fe^{3+} + 12H_2O \longrightarrow 4Fe(OH)_3 + 12H^+ \tag{15.19}$$

$$4CuFeS_2 + 17O_2 + 10H_2O \xrightarrow{\text{bacteria}} 4Cu^{2+} + 4Fe(OH)_3 + 8SO_4^{2-} + 8H^+ \tag{15.20}$$

In other cases the oxidizable metal moiety may be attacked before the sulfide, as in the case of chalcocite (Cu_2S) oxidation (Fox, 1967; Nielsen and Beck, 1972):

$$Cu_2S + \tfrac{1}{2}O_2 + 2H^+ \xrightarrow{\text{bacteria}} Cu^{2+} + CuS + H_2O \tag{15.21}$$

$$CuS + \tfrac{1}{2}O_2 + 2H^+ \xrightarrow{\text{bacteria}} Cu^{2+} + S^0 + H_2O \tag{15.22}$$

$$S^0 + \tfrac{3}{2}O_2 + H_2O \xrightarrow{\text{bacteria}} H_2SO_4 \tag{15.23}$$

Digenite (Cu_9S_5) may be an intermediate in the formation of CuS from Cu_2S (Nielsen and Beck, 1972).

The direct microbial oxidaion of iron pyrite (FeS_2) probably proceeds according to the reaction

$$FeS_2 + \tfrac{7}{2}O_2 + H_2O \xrightarrow{\text{bacteria}} Fe^{2+} + 2H^+ + 2SO_4^{2-} \tag{15.24}$$

The ferrous iron generated in this reaction is further oxidized by the bacteria according to reaction 15.10. The resultant ferric iron then can cause chemical oxidation of the pyrite according to the previously described reaction 15.14, whereby Fe^{2+} is regenerated. According to Singer and Stumm (1970), bacterially catalyzed reaction 15.10 is the rate-controlling reaction in pyrite oxidation.

At least two kinds of bacteria have been associated with metal sulfide oxidation. The most extensively studied are classified as *Thiobacillus ferrooxidans*. A more recent discovery has been *Sulfolobus* sp. The organisms, *T. ferrooxidans* and *Sulfolobus* sp., are both strongly acidophilic, growing best in a pH range from about 1.5 to 2.5. *T. ferrooxidans* is mesophilic and *Sulfolobus* is thermophilic.

A third kind of bacterium has been recently detected in leach dumps of mining operations. Although not as yet fully characterized, it is an acidophilic thermophile capable of oxidizing iron. The organism requires small amounts of yeast extract, cysteine, or glutathione for growth. It has been shown to use iron as an energy source at 50 and 55°C. Growth has also been obtained on pyrite (Brierley and Lockwood, 1977; J. A. Brierley, 1978; Brierley et al., 1978).

Nutritionally, *T. ferrooxidans* and *Sulfolobus* are capable of autotrophic growth. However, autotrophic growth of *Sulfolobus* sp. is stimulated by the addition of a trace of yeast extract. Appropriate metal sulfides may serve as energy sources for these organisms. Depending on the oxidation state of the metal in a metal sulfide, both it and the sulfide may serve as energy sources. Thus, in the case of chalcocite (Cu_2S) oxidation by *T. ferrooxidans*, as already described, the initial oxidation step involves the cuprous copper [Cu(I)] of the compound. The organism is able to derive energy from this oxidation, which it can use in CO_2 fixation (Nielsen and Beck, 1972). Cell extracts of *T. ferrooxidans* have been prepared which catalyze the oxidation of cuprous copper in Cu_2S but not of elemental sulfur (Imai et al., 1973). The oxidation is not inhibited by quinacrin (atabrine). It needs the addition of a trace of iron for proper activity. The effect of traces of iron on metal sulfide oxidation had been previously noted in experiments in which the addition of 9 mg of ferrous iron per liter of medium stimulated metal sulfide oxidation by whole cells of *T. ferrooxidans* (Ehrlich and Fox, 1967b).

T. ferrooxidans can use NH_4^+ and some amino acids as a source of nitrogen, but not NO_3^-. At least some strains of *T. ferrooxidans* appear capable of fixing nitrogen (Mackintosh, 1978). The nitrogen requirements of *Sulfolobus* may be as for *T. ferrooxidans*, but this has yet to be thoroughly established. The organism is a strict aerobe.

T. ferrooxidans is very versatile in attacking metal sulfides. It has been reported to oxidize arsenopyrite (FeS_2FeAs_2), bornite (Cu_5FeS_4), chalcocite (Cu_2S), chalcopyrite ($CuFeS_2$), covellite (CuS), enargite ($3Cu_2S \cdot As_2S_5$), galena (PbS), millerite (NiS), orpiment (As_2S_3), pyrite (FeS_2), marcasite (FeS_2), sphalerite (ZnS), stibnite (Sb_2S_3), and tetrahedrite ($Cu_8Sb_2S_7$) (Silverman and Ehrlich, 1964). In addition, the oxidation of gallium sulfide (Ga_2S_3) and of synthetic preparations of CoS, NiS, and ZnS has been reported (Torma, 1971, 1978).

The bacterial oxidation of PbS by *T. ferrooxidans* presents a special problem because the oxidation product, $PbSO_4$, is relatively insoluble. Thus, as oxidation of PbS proceeds, $PbSO_4$ is likely to accumulate on the crystal surface and block further access to PbS by bacteria and oxygen. In the laboratory, agitation of the culture largely eliminates this problem (Silver and Torma, 1974).

The only metal sulfides so far reported putatively to serve as energy sources for *Sulfolobus* sp. include molybdenite and chalcopyrite (Brierley and Murr, 1973). The ability to oxidize molybdenite is a unique property of *Sulfolobus* sp., since *T. ferrooxidans* is strongly inhibited by the oxidation product of molybdate ion (Tuovinen et al., 1971).

The metal-sulfide oxidizing bacteria are naturally associated with metal sulfide-containing deposits, including bituminous coal seams, ore bodies, and the like. Under natural conditions, their growth and activity may be quite limited owing to one or more restricting environmental factors, such as limited access of air and/or moisture (Brock, 1975), limited access to the energy source (the metal sulfide crystals), limited nitrogen source, or unfavorable temperature (Ehrlich and Fox, 1967b).

Acid Mine Drainage; Leaching   When bituminous coal seams are exposed to air and moisture during mining, associated iron pyrites start to oxidize, leading to the production of acid mine drainage. At the same time, iron-oxidizing thiobacilli become readily detectable (Leathen, 1953). Similar processes take place when metal sulfide ore bodies become exposed during mining activities. While production of acid mine drainage from coal mines is a nuisance and detrimental to the environment, the production of acid drainage from metal sulfide ore deposits may be beneficial, representing a means of extracting metal from the ore. Its genesis is often encouraged by mining companies when dealing with low-grade ore, such as ore with a metal content below about 0.5% (wt/wt) whose extraction is uneconomic by the conventional means of milling followed by flotation separation because of the unfavorable waste/metal ratio. Also, waste rock and mine tailings, by-products of mining and conventional metal recovery, may be placed in piles of 100 or more feet in height and watered. In time, as a result of the oxidation of the residual pyrite and metal sulfide in these materials, which is at least in part dependent on the activity or iron-oxidizing bacteria, the effluent from these ore dumps will take on an acidic character and become ferruginous. As this effluent is recirculated through the ore dumps, metal sulfides other than simple iron will begin to be oxidized, and the oxidation product can then be recovered in the effluent. Such metal-containing effluents are known in the trade as **pregnant solutions**. When the concentration of the valuable metal has become sufficient, it can be recovered. In the case of copper, this may be accomplished by precipitation with scrap iron in a basin called a **launder** (Fig. 15.2). The precipitation is the result of the reaction

$$Cu^{2+} + Fe^0 \rightarrow Cu^0 + Fe^{2+} \tag{15.25}$$

Alternatively, the copper may also be recovered electrolytically.

The solution remaining after removal of copper is known as **barren solution** and is rich in ferrous iron. This solution is conducted into shallow oxidation ponds, where *T. ferrooxidans* rapidly oxidizes the Fe^{2+} to Fe^{3+} and in the process generates a little more acid. A significant portion of the ferric iron formed by the bacteria precipitates. This is a very desirable reaction since an

Figure 15.2 Launder used in recovering copper from pregnant solutions from leach dumps. (Courtesy of Duval Corporation.)

excess of iron in the barren solution can interfere with metal extraction when the barren solution is recirculated as leach solution into the ore dumps. In the dumps the ferric iron may precipitate from solution and coat metal sulfide crystals, thereby preventing their oxidation; or the ferric iron may clog drainage channels in the ore dumps, preventing proper circulation of the leach solution. The barren solution from the oxidation ponds is thus an acidic ferric sulfate solution. It is pumped to ore dumps as leach solution and applied to them in a fine spray. The ferric iron in the solution helps in the chemical oxidation of susceptible metal sulfides. The acid in the solution plays at least three roles. It helps to keep the ferric iron in solution. It furnishes protons for acid-consuming reactions, as, for example, reactions 15.16, 15.21, and 15.22, and not least, it helps in weathering the host rock (gangue) (Zimmerley et al., 1958; Moshuyakova et al., 1971), which encloses the metal sulfide crystals in porphyry ore, making them more accessible to oxidation. High acidity may also play a fourth role, preventing adsorption of metal ions, such as those of copper, by the host rock (Ehrlich, 1977; Ehrlich and Fox, 1967b). A typical leach cycle is diagrammed in Figure 15.3 (see C. L. Brierley, 1978, for a further discussion of bacterial leaching).

Iron-oxidizing bacilli developing in an ore dump help to regenerate ferric iron from ferrous iron that results from the chemical interaction of ferric iron with metal sulfides, and they also play a direct role in metal sulfide oxidation, as explained previously. Although it is usually not possible to assess to what extent bacteria solubilize a metal sulfide in an ore dump by direct oxidation and to what extent they solubilize it indirectly by regeneration of ferric iron, it is readily shown in controlled laboratory experiments that bacteria such as iron-oxidizing thiobacilli are able to accelerate the oxidation process significantly. Owing to the fact that metal sulfide oxidation is an exothermic process, the interior temperature of some ore dumps can rise to levels (70-80°C) which are unfavorable for the growth of *T. ferrooxidans*. It has been argued that in these instances the only roles of *T. ferrooxidans* in ore leaching are to initiate self-heating (Lyalikova, 1960) and to regenerate ferric iron in the oxidation ponds. However, the recent discovery of thermophilic *Sulfolobus* sp. with a capacity to oxidize ferrous iron and some sulfides such as molybdenite and

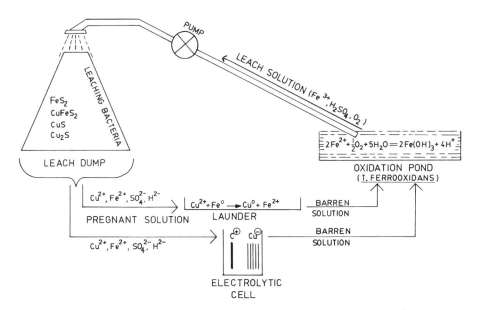

Figure 15.3   Diagrammatic representation of copper sulfide ore dump leaching.

chalcopyrite and the discovery of other as yet unidentified thermophilic bacteria which have at least a capacity to oxidize ferrous iron raise the distinct possibility that in self-heated dumps these latter organisms play an important role in metal sulfide oxidation. Indeed, it has been suggested that the thermophiles may be responsible for regulating the internal temperature of active leach dumps (Murr and Brierley, 1978).

The primary copper mineral in ore bodies of magmatic-hydrothermal origin is chalcopyrite ($CuFeS_2$). This mineral tends to be somewhat refractory to chemical leaching when compared to the secondary copper sulfide minerals chalcocite (Cu_2S) and covellite (CuS). The oxidation of chalcopyrite in nature can thus be significantly enhanced by *T. ferrooxidans* or *Sulfolobus* sp. (Razzell and Trussell, 1963; Brierley, 1974). Ferric iron, when present in excess of, for example, 1,000 ppm, has been found to inhibit chalcopyrite oxidation (Ehrlich, 1977; Duncan and Walden, 1972), probably because it precipitates as jarosite or adsorbs to the surface of residual chalcopyrite crystals and prevents further oxidation. Bacteria themselves may interfere by generating excess ferric iron from ferrous iron which precipitates or is adsorbed by the residual chalcopyrite.

Not all metal sulfide deposits in nature are readily oxidizable by acidophilic, iron-oxidizing bactiera. Some are associated with host rock containing significant amounts of limestone ($CaCO_3$). Limestone, being very reactive with acid, will tend to neutralize it and thus produce pH conditions unfavorable to the development of the acidophilic bacteria. An example of such an ore body is the White Pine Deposit in upper Michigan. Copper from this ore body can be extracted biologically only by an indirect process involving chelation, as mentioned earlier.

Although the acid mine drainage from bituminous coal mines or from exposed metal sulfide ore bodies would seem to be an unfavorable environment for any organism except the acidophilic iron bacteria, other bacteria, as well as certain fungi, algae, and protozoans, are found there. Zoogloeal streamers of bacteria, a few fungi, occasional *Gallionella* (but see Walsh and Mitchell, 1972a), cyanobacteria above pH 3.7, the green algae *Ulothrix zonata* and *Stigioclonium*, the diatom Tabellaria, *Euglena*, and several protozonas, including flagellates, rhizopods, and ciliates, have been found in such acid mine drainage (Lackey, 1938). Yeasts such as *Rhodotorula* and *Trichosporon* and other fungi besides the flagellate *Eutrepia* and amoebae and zoogloeal bacteria have also been found in acid mine drainage from copper mines (Ehrlich, 1964b; Marchlewitz and Schwartz, 1961; Beck, 1960; Corrick and Sutton, 1961; Razzell and Trussell, 1963).

Of the fungi, two strains, one of *Rhodotorula* and one of *Trichosporon*, were isolated which in the presence of glucose were able to reduce elemental sulfur to H_2S which precipitated dissolved copper but not iron from the acid solution generated (Ehrlich and Fox, 1967a). In the presence of added iron, lowered oxygen tension increased the amount of copper precipitated, presumably by favoring S^0 reduction to H_2S, although the lack of autoxidation of the precipitated copper sulfide may also have contributed to the greater copper sulfide recovery.

Uranium Leaching   Although the foregoing leaching processes were described with the extraction of copper mainly in mind, they have also been practically applied to the extraction of residual low-grade uranium ores. In this process, the bacteria belonging to *T. ferrooxidans* play an exclusively indirect role: they do not attack the uranium ore directly, but generate the oxidant, ferric iron. Thus, the bacteria catalyze the following reaction:

$$2Fe^{2+} + \tfrac{1}{2}O_2 + 2H^+ \xrightarrow{\text{bacteria}} 2Fe^{3+} + H_2O \qquad (15.26)$$

and some of the resultant ferric iron hydrolyzes:

$$Fe^{3+} + 3H_2O \rightarrow Fe(OH)_3 + 3H^+ \qquad (15.27)$$

The dissolved ferric iron then reacts with the uraninite to form uranyl ions:

$$UO_2 + 2Fe^{3+} \rightarrow 2Fe^{2+} + UO_2^{2+}$$

The dissolved uranium may be recovered from solution through concentration by ion exchange. The growth of the bacteria, which are naturally present in the mine, is stimulated by intermittent spraying of nutrient-enriched solution on the floors, walls, and mud of mine stopes (Zajic, 1969). If the leachate should become anoxic and rise in pH, and sulfate-reducing bacteria should develop in it, these bacteria can reprecipitate UO_2 as a result of UO_2^{2+} reaction with the biogenic H_2S:

$$UO_2^{2+} + H_2S \rightarrow UO_2 + 2H^+ + S^0 \qquad (15.28)$$

Uranium leaching is another example of an artifically stimulated process. The reactions encouraged by leaching undoubtedly occur only on a very limited scale under natural conditions and thus cause only slow mobilization of uranium.

T. ferrooxidans is not the only organism capable of uranium mobilization. Heterotrophic organisms such as soil microflora and bacteria from granites or mine waters (*P. fluorescens, P. putida, Achromobacter*) can mobilize uranium in granite rocks, ore, and sand by weathering of the rock through interaction with organic acids and chelators produced by them (Zajic, 1969; Magne et al. . 1973, 1974). Magne et al. found experimentally that the addition of thymol to percolation columns of uraniferous material fed with glucose solution selected for a more efficient extractive flora in terms of greater production of oxalic acid. The authors suggested that in nature, phenolic and quinoid compounds of plant origin may serve the role of thymol. The authors also reported that microbes can precipitate uranium by digestion of soluble uranium complexes (Magne et al., 1974), that is, by microbial destruction of the organic moiety that complexes the uranium. The observations may explain how in nature uranium in granitic rock may be mobilized by bacteria and reprecipitated and concentrated elsewhere under the influence of microbial activity.

The foregoing processes in industrial bacterial leaching of metal sulfides are proceeding under optimized conditions because human intervention adds moisture and oxygen into sulfide ore bodies or materials derived from them. In the absence of such human intervention, the foregoing processes can be assumed to occur only in highly localized situations, contributing to a very slow, gradual change from reduced to oxidized ore.

15.3 SUMMARY

Metal sulfides may occur in locally high concentrations, in which case they constitute ores. Although most nonferrous sulfides are abiogenically formed through magmatic and hydrothermal processes, a few sedimentary deposits of nonferrrous sulfides and many more of ferrous sulfides are of biogenic origin. The microbial role in metal sulfide formation is the genesis of H_2S, usually from the reduction of sulfate, but in some special instances, possibly from the mineralization of organic sulfur compounds. Since metal sulfides are highly insoluble, the spontaneous reaction of metal ions with the biogenic sulfide proceeds readily. Biogenesis of specific metal sulfide minerals has been demonstrated in the laboratory. These experiments require relatively insoluble metal compounds as starting materials to limit the toxicity of the metal ions to the sulfate-reducing bacteria. In nature, adsorption of the metal ions by sediment components serves a similar function in lowering the concentration of metal ions below their toxic level for sulfate reducers.

Metal sulfides are also subject to oxidation by bacteria, especially *Thiobacillus ferrooxidans* and *Sulfolobus* spp. The action may be by direct attack of the metal sulfide crystal or by indirect attack. In the latter process, the bacteria are involved in the genesis of an acid ferric sulfate solution which acts then chemically as an oxidant of the metal sulfide. The indirect mechanism is of primary importance also in the solubilization of uraninite (UO_2). This microbial oxidizing activity may be practically applied in the leaching of metal sulfide ore deposits. It is an inexpensive method for the recovery of metals from low-grade ores.

16

Geomicrobiology of Selenium and Tellurium

The elements selenium and tellurium, like sulfur, belong to group VI of the periodic table. They have some properties in common, but selenium and tellurium, especially the latter, have some metallic attributes, unlike sulfur. Selenium and tellurium are much less abundant in the earth's crust than is sulfur. Selenium amounts to only 0.14-0.05 ppm (Rapp, 1972, p. 1080) and tellurium to 10^{-2}-10^{-5} ppm. Both are associated with metal sulfides in nature and occur in distinct minerals [e.g., ferroselite ($FeSe_2$), challomenite ($CuSeO_3 \cdot 2H_2O$), hirsite (Ag_2Te), and tetradymite (Bi_2Te_2S)].

16.1 SELENIUM

Selenium occurs in small amounts in various soils. Selenium concentrations in the range 0.01-100 ppm have been measured, the high concentrations being associated with arid, alkaline soils that contain some free $CaCO_3$ (Rosenfeld and Beath, 1964). Some plants, such as *Astragalus* spp. and *Stanleya* spp., can accumulate large amounts of selenium in the form of organic selenium compounds, but not all forms of selenium in soil are available for assimilation by plants. Although toxic at high concentrations, selenium is required as a trace element by at least some bacteria, plants, and animals in their nutrition (Stadtman, 1974). It is probably of benefit in human nutrition. Selenium has been found as an essential component in the enzyme glutathione peroxidase in mammalian red blood corpuscles (Rotruck et al. 1973). The enzyme catalyzes the reaction

$$2GSH + H_2O_2 \rightarrow GSSG + 2H_2O \tag{16.1}$$

Selenium has also been found essential together with molybdenum in the structure of formate dehydrogenase in the bacteria *E. coli, Clostridium thermoaceticum, C. sticklandii,* and *Methanococcus vannielii* (Pinsent, 1954, Lester and

299

DeMoss, 1971; Shum and Murphy, 1972; Andreesen and Ljungdahl, 1973; Enoch and Lester, 1972; Stadtman, 1974). The enzyme catalyzes the reaction

$$HCOOH + NAD^+ \rightarrow CO_2 + NADH + H^+ \tag{16.2}$$

Finally, selenium has been found to be essential in protein A of glycine reductase in clostridia (Stradtman, 1974), an enzyme that catalyzes the reaction

$$CH_2COOH + R(SH)_2 + P_i + ADP \rightarrow CH_3COOH + NH_3 + R\begin{matrix} S \\ / | \\ \\ \backslash | \\ S \end{matrix} + ATP \tag{16.3}$$
$$| \atop NH_2$$

No biological requirement has yet been reported for tellurium.

Some inorganic selenium compounds have been reported to be oxidizable by microorganisms. *Micrococcus selenicus* isolated from mud (*Bergey's Manual,* 1957), a rod-shaped bacterium isolated from soil (Lipman and Waksman, 1923), and a purple bacterium (Sapozhnikov, 1937) have been reported to oxidize Se^0 to SeO_4^{2-}. *Thiobacillus ferrooxidans* has been reported to oxidize copper selenide (CuSe) to cupric copper (Cu^{2+}) and elemental selenium (Se^0) (Torma and Habashi, 1972). The reaction may be written

$$CuSe + 2H^+ + \tfrac{1}{2}O_2 \rightarrow Cu^{2+} + Se^0 + H_2O \tag{16.4}$$

Other inorganic selenium compounds have been found to be reducible by microorganisms. Hydrogenase from *M. lactilyticus* has been shown to reduce Se^0 to HSe^- with hydrogen (Woolfolk and Whiteley, 1962). A variety of bacteria, actinomycetes, and fungi have been shown to reduce selenite to Se^0 (Bautista and Alexander, 1972; Zalokar, 1953). A significant portion of selenite, when reduced by *E. coli,* is deposited as Se^0 on its cell membranes but not in its cytoplasm (Gerrard et al., 1974), while another portion is incorporated as selenide into organic compounds such as selenomethionine (Ahluwalia et al., 1968). Some soil microbes reduce selenate or selenite to dimethylselenide [$(CH_3)_2 Se$] at high selenium concentrations (Kovalskii et al., 1958; Fleming and Alexander, 1972; Alexander, 1977).

Candida albicans, a fungus, contains the enzyme selenite reductase (Falcone and Nickerson, 1963; Nickerson and Falcone, 1963), which reduces selenite to Se^0. A characterization of the enzyme has shown that it requires a quinone, a thiol compound (e.g., glutathione), a pyridine nucleotide (NADP), and an electron donor (e.g., glucose 6-phosphate) for activity. Electron transfer between NADP and quinone is probably mediated by flavin mononucleotide in this system.

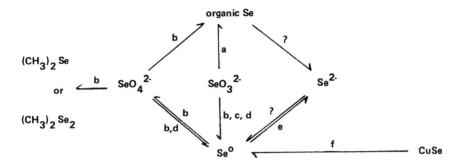

Figure 16.1   The selenium cycle: a, *E. coli;* b, bacteria; c, actinomycetes; d, fungi; e, *M. lactilyticus;* f, *Thiobacillus ferrooxidans*. (See also Doran and Alexander, 1977.)

A selenium cycle in nature has been suggested by Shrift (1964). A modified form is presented in Figure 16.1.

16.2   TELLURIUM

Tellurium metabolism by microbes has so far been mainly observed in the form of bacterial reduction of tellurates and tellurites to Te^0 or $(CH_3)_2Te$ (Silverman and Ehrlich, 1964; Nagai, 1965; Woolfolk and Whiteley, 1962). However, the fungus *Penicillium* sp. has been reported to produce $(CH_3)_2Te$ from several inorganic tellurium compounds, provided only that reducible selenium compounds were also present (Fleming and Alexander, 1972). The amount of dialkyl telluride formed was related to the relative Se and Te concentration in the medium. Microbial oxidation of tellurides has so far not been reported. In view of the low tellurium concentration in nature, this element is probably not important geomicrobiologically.

16.3   SUMMARY

Selenium, although a very toxic element, is nutritionally required by some bacteria, plants, and animals. Microorganisms have been described that can oxidize reduced selenium compounds. At least one, *Thiobacillus ferrooxidans,* can use selenide in the form of CuSe as a sole source of energy, oxidizing the compound to elemental selenium (Se^0) and Cu^{2+}. Oxidized forms of inorganic selenium compounds can be reduced by microorganisms, including bacteria and fungi. Selenate and selenite may be reduced to Se^0 or dimethylselenide $[(CH_3)_2Se]$. The reductions are enzymatic. These microbial reactions contribute to a selenium cycle in nature.

Tellurium occurs in such low concentration in nature that it does not seem geomicrobiologically important. Nevertheless, microbial tellurate and tellurite reduction to elemental tellurium (Te^0) and dimethyltelluride $[(CH_3)_2Te]$ have been observed. Microbial oxidation of tellurides has not been reported.

17

The Fossil Fuels

Although most organic carbon in the biosphere is continually recycled, a significant amount has been and is being trapped in special sedimentary formations, leaving it inaccessible to mineralization by microbes until it becomes reexposed to air and water through natural causes or human intervention (Table 17.1). Much of the trapped organic carbon is in a fairly reduced form, depending on the length of time it has been trapped and depending on any secondary chemical changes that it has undergone. Its reduced nature makes it of value as a fuel. Because of the organic origin and the great age of much of this material, it is often referred to as **fossil fuel**. Among such fossil fuels we include methane gas, natural gas (which is largely methane), petroleum, oil shale, peat, and coal.

17.1 METHANE

Methanogenesis Although methane (CH_4) may be formed separately in some freshwater and marine sedimentary deposits and in water bodies (Davis, 1967, p. 193), it is also found with petroleum deposits and coal deposits (Nesterov et al., 1971). It is a gas at ordinary temperatures and pressures. The methane that is found at the earth's surface at the present time is mostly of biogenic origin; but some may be of volcanic origin. At the time of formation of the primitive earth, methane may have been a constituent of the earth's atmosphere which was later converted to graphite and organic matter when enough hydrogen had escaped from the atmosphere into space (Miller and Orgel, 1974). According to these authors, the graphite then reacted with water to yield CO_2 and H_2,

$$C + 2H_2O \rightarrow CO_2 + 2H_2 \tag{17.1}$$

the CO_2 being trapped as $CaCO_3$. These reactions are estimated to have been completed in the first billion years of the earth's existence.

Table 17.1 Comparison of the Estimated Quantities of Living, Dead Decaying, and Dead Trapped Organic Matter on the Earth's Surface

Category	Quantity ($\times 10^9$ metric tons)	References
Living matter, land	300	Wehmiller (1972)
Decaying matter, land	2,600	Wehmiller (1972)
Living matter, oceans	30	Wehmiller (1972)
Decaying matter, oceans	10,000	Wehmiller (1972)
Trapped organic matter,		
Total	3,800,000	Degens (1972)
In shales	3,600,000	Degens (1972)
Coal	6,000	Degens (1972)
Petroleum	200	Degens (1972)

The methane in sedimentary deposits can be assumed to have been formed primarily from such compounds as acetate with or without H_2, formate plus H_2, or CO_2 plus H_2, which resulted from anaerobic decomposition of organic matter. Methane-forming bacteria are known that are able to transform these substances anaerobically to methane. The organic matter from which the methane precursors are formed anaerobically today and from which they were probably formed anaerobically in the geologic past is usually mostly of vegetable origin, representing buried algal and plant debris. During burial, various kinds of anaerobic, heterotrophic bacteria are known to attack it, with ultimate production of methane precursors. The initial step in the decomposition of the organic debris can actually be aerobic and involve fungi, until sedimentation has progressed to the point where further supply of air to the decomposing matter is blocked. Methane precursors need not always be of biogenic origin, however. One case of biogenic methane formation has recently been described that utilizes CO_2 and H_2 of abiogenic origin (i.e., volcanic origin). This is found to take place in Lake Kivu, an African Rift lake (Deuser et al., 1973). In marine environments, microbial methane formation is generally detected in strata below those where active bacterial sulfate reduction occurs (Martens and Berner, 1974). Similar findings have been made in a freshwater lake (Cappenberg, 1974a,b; Cappenberg and Prins, 1974). A probable explanation for why methane is not formed in those environments where significant sulfate concentrations occur, is not inhibition of methanogenesis by sulfate but rather the inability of methane-forming bacteria to compete for H_2 with sulfate-reducing bacteria on the one hand and possible utilization of methane by sulfate-reducing bacteria on the other (Davis and Yarbrough, 1966; Martens, 1976; Oremland and Taylor, 1978;

Panganiban and Hanson, 1976; Abram and Nedwell, 1978). (See Chapter 14 for a discussion of the physiology of sulfate reduction.)

A typical reaction whereby methanogenic bacteria form methane is by reducing CO_2 with H_2 (Zeikus, 1977):

$$CO_2 + 4H_2 \rightarrow CH_4 + 2H_2O \tag{17.2}$$

This reaction is exothermic ($\Delta F, -33.21$ kcal) and yields energy whereby the organism can assimilate CO_2 and convert it to organic carbon. All methanogenic bacteria are able to use this mechanism and are, therefore, autotrophs.* These organisms may also convert formate to methane in the presence of H_2 (Zeikus, 1977):

$$HCOOH + 3H_2 \rightarrow CH_4 + 2H_2O \tag{17.3}$$

Whether they first cleave the formate into H_2 and CO_2 before generating methane has not been clearly established. *Methanosarcina barkeri* and the thermophilic *Methanobacterium thermoautotrophicum* can also convert acetate to methane in the presence of hydrogen (Zeikus et al., 1975). The overall reaction may be written

$$CH_3COOH + 4H_2 \rightarrow 2CH_4 + 2H_2O \tag{17.4}$$

The hydrogen concentration in the reaction mixture for these organisms is critical. The addition of CO_2/HCO_3^- buffer stimulates acetate utilization. Both methyl and carboxyl carbon are found to be converted to methane, although more methyl than carboxyl carbon ends up as the gas. Equation 17.4, therefore, is not a fully correct description of methane formation from acetate. In practice, less than 4 mol of H_2 will be used for every mole of acetate that is reduced to methane. The unreduced carboxyl carbon may be expected to be assimilated into cell carbon. In the presence of trypticase, yeast extract, and hydrogen, *M. barkeri* converts acetate to methane of which 63-82% comes from the methyl carbon of acetate (Weimer and Zeikus, 1978).

Methanosarcina strain 227, isolated from anaerobic sewage sludge, can convert acetate to methane in the absence of H_2 (Mah et al., 1978; Smith and Mah, 1978),

$$CH_3COOH \rightarrow CH_4 + CO_2 \tag{17.5}$$

---

*An exception has recently been reported by Zehnder et al. (1980). *Methanobacterium soehngenii* or "fat rod" from sewage is unable to form methane from $H_2 + CO_2$ because it is unable to oxidize H_2. It forms methane instead from acetate by decarboxylation.

The mechanism by which this reaction occurs remains unknown, nor is it understood how the organism can derive useful energy from this reaction. *Methanosarcina barkeri* can also use methanol as energy source for growth (Zeikus, 1977):

$$4CH_3OH \rightarrow 3CH_4 + CO_2 + 2H_2O \tag{17.6}$$

Methanobacterium thermoautotrophicum has been shown to grow, although poorly, on carbon monoxide (CO) as a sole source of carbon and energy, converting CO to methane for energy by the overall reaction (Daniels et al., 1977)

$$4CO + 2H_2O \rightarrow 3CO_2 + CH_4 \tag{17.7}$$

In the presence of H_2, CO can be substituted for CO_2 as the terminal electron acceptor.

Methane-generating bacteria (**methanogens**) are presently assigned to the genera *Methanobacterium, Methanobrevibacter, Methanococcus, Methanomicrobium, Methanogenium, Methanospirillum,* and *Methanosarcina* (Fig. 17.1) (Balch et al., 1979). All of these have been found in anaerobic aquatic environments, especially muds, and in some cases in soil. One species, *Methanobacterium thermoautotrophicum,* is thermophilic (Zeikus and Wolfe, 1972). Its optimum temperature range for growth is 65-70°C, its maximum is 75°C, and its minimum 40°C. Some methane bacteria are also found in the digestive tract of animals. They include both gram-positive and gram-negative forms. Almost all are capable of autotrophic growth on a H_2 and CO_2 mixture in the complete absence of oxygen. They thus constitute the only known obligately anaerobic chemoautotrophs. Like *Sulfolobus,* they lack a typical prokaryotic cell wall (Zeikus, 1977); that is, their cell envelope contains no typical murein, although it may contain pseudomurein (Balch et al., 1979).

The biochemical mechanism of **methanogenesis** (methane formation) and CO_2 assimilation is still very incompletely understood. Methanogenesis is known to involve a unique low-molecular-weight compound, coenzyme F_{420}, and special coenzymes F_{342} and F_{430}, as electron carriers, and coenzyme M (HSCoM), identified as 2-mercaptoethane sulfonic acid, as a methyl carrier (Zeikus, 1977; Gunsalus and Wolfe, 1978):

$$CH_3-S-CoM \xrightarrow[\text{methyl reductase}]{H_2, Mg^{2+}, ATP,} CH_4 + HSCoM \tag{17.8}$$

CO_2 assimilation does not seem to involve the typical reductive pentose pathway (Calvin cycle), the serine pathway, or the hexulose-phosphate pathway (Daniels and Zeikus, 1978). After brief exposure of intact methanogenic cells (2-45 sec) to $^{14}CO_2$, radioactive carbon was found in coenzyme M derivative, alanine,

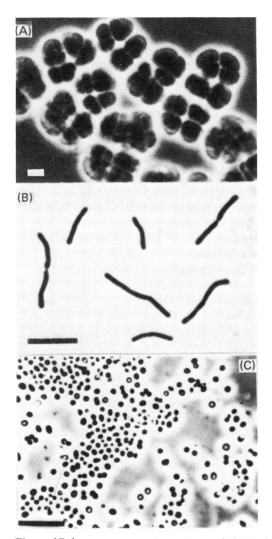

Figure 17.1   Methanogenic bacteria. (A) *Methanosarcina barkeri* strain PS.
(B) *Methanobacterium thermoautotrophicum* strain AO. (C) *Methanococcus*
sp. Bar indicates 5 μm. (From Zeikus, 1977; by permission.)

aspartate, glutamate, and several unidentified compounds. Organic phosphates
contained no label until after 1 min of exposure to $^{14}CO_2$. Pyruvate appeared
to be an early intermediate in CO_2 fixation (Daniels and Zeikus, 1978; see also
Zeikus et al., 1977). Fumarate has been found to be transformed via succinate
to α-ketoglutarate, a precursor of glutamate, arginine, and proline (Fuchs et al.,

1978). Acetate has been found to be assimilated by *M. barkeri* into alanine and aspartate, with pyruvate and oxalacetate as intermediates, and into glutamate with citrate, isocitrate, and α-ketoglutarate as intermediates (Weimer and Zeikus, 1979). ATP is synthesized by a proton motive force across the cell membrane associated with the H_2 metabolism (Doddema et al., 1979). A typical Mg^{2+}-requiring adenosine triphosphatase (ATPase) has been detected in *M. thermoautotrophicum*, which is similar to that of other aerobic and anaerobic bacteria (Doddema et al., 1978).

Methanogens, because they possess a special kind of 16S ribosomal RNA, may represent a special form of life that should be classified with archaebacteria rather than eubacteria, to which most bacteria belong (Woese and Fox, 1977). Fox et al. (1977) have proposed that methanogens represent offshoots of the earliest forms of life on earth that may have been specially adapted to use H_2 and CO_2 for growth in the premordial, reducing atomosphere.

In nature, the methane in coal and petroleum deposits can be assumed to have arisen in the early stage of their formation by organisms similar to present-day types and from substrates similar to those employed in present-day methanogenesis. The organic raw material for methane in coal formation was mostly terrestrial plant matter, whereas that for methane in petroleum deposits was mostly marine phytoplankton. After initial aerobic attack by microbial heterotrophs, anaerobic heterotrophs must have converted the organic residues to CO_2, H_2, acetate, and formate, after which the methanogens must have done their work. The biogenic origin of methane in coal and petroleum deposits can be inferred from its $\delta^{13}C$ value, which shows a decrease relative to a universal standard (Davis, 1967, pp. 85-86). The methane associated with coal seams, known as **coal damp**, is a major hazard in coal mining. Because it is so flammable, accumulations of the gas in mine shafts can lead to serious explosions, often with extensive loss of human life. The methane associated with petroleum deposits is contained in the **natural gas** and frequently represents a major component of it.

Methane Oxidation   Under appropriate environmental conditions, methane can be oxidized to CO_2 and H_2O by gram-negative, obligately methylotrophic bacteria such as *Methylomonas, Methylococcus, Methylobacter, Methylosinus*, and *Methylocystis*, and by some facultative **methylotrophs** (Fig. 17.2). On the basis of the internal membrane organization, the first three genera are grouped in type I and the last two genera, plus the facultative methylotrophs, are groups in type II (Davies and Whittenbury, 1970). Type I contains internal, stacked membranes, whereas type II contains paired membranes concentric with the plasma membrane and forming vesiclelike or tubular structures (Fig. 17.3). Thus, organisms belonging to type I are all obligate methylotrophs, whereas organisms belonging to type II may be either obligate or facultative methylotrophs.

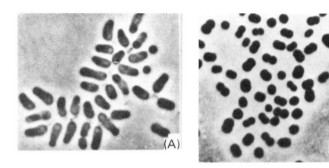

Figure 17.2 Methane-oxidizing bacteria (methylotrophs) (19,000X). (A) *Methylosinus trichosporium* in rosette arrangement. Organisms are anchored by visible holdfast material. (B) *Methylococcus capsulatus*. [From R. Whittenbury, K. C. Phillips, and J. F. Wilkinson, Enrichment, isolation and some properties of methane-utilizing bacteria, *J. Gen. Microbial.* 61:205-218 (1970). By permission Cambridge University Press.]

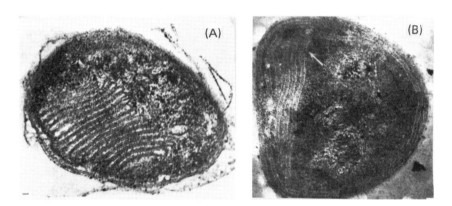

Figure 17.3 Fine structure of methane-oxidizing bacteria (80,000X). (A) Section of *Methylococcus* (subgroup minimus) showing type I membrane system. (B) Peripheral arrangement of membranes in *Methylosinus* (subgroup *sporium*) characteristic of type II membrane systems. [From S. L. Davies and R. Whittenbury, Find structure of methane and other hydrocarbon-utilizing bacteria, *J. Gen. Microbiol.* 61:227-232 (1970). By permission Cambridge University Press.]

Methane and methanol seem to be the only compounds that can serve as carbon and energy source for the obligate methylotrophs. For purposes of energy production, the bacteria oxidize methane via the following steps:

$$CH_4 \xrightarrow{\frac{1}{2}O_2} CH_3OH \xrightarrow{-\frac{1}{2}O_2} HCHO \xrightarrow{\frac{1}{2}O_2} HCOOH \xrightarrow{\frac{1}{2}O_2} CO_2 + H_2O$$

(17.9)

The first step involves catalysis by an oxygenase (Doelle, 1975) and is therefore not an energy-producing step. The subsequent steps involve dehydrogenases, at least some of which can couple to oxidative phosphorylation pathways (e.g., Patel et al., 1972). For purposes of carbon assimilation, the bacteria oxidize methane to the formaldehyde level, as shown in equation 17.9, and then incorporate the formaldehyde carbon either by the hexulose monophosphate (HuMP) pathway or by the serine pathway (Fig. 17.4) (Strom et al., 1974; Wagner and Levitch, 1975; Quayle and Ferenci, 1978). The HuMP pathway is taken by the type I methylotrophs, whereas the serine pathway is taken by the type II methylotrophs (Quayle, 1972). Quayle and Ferenci (1978) have speculated that type I methylotrophs may be the offshoots of the forerunners of the true

HuMP PATHWAY OF HCHO-CARBON ASSIMILATION:

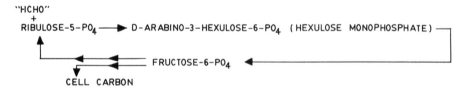

SERINE PATHWAY OF HCHO-CARBON ASSIMILATION:

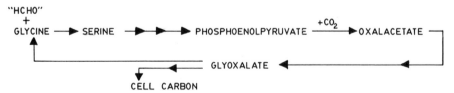

Figure 17.4   Alternative pathways of formaldehyde-carbon assimilation in methane oxidizing bacteria. The hexulose monophosphate (HuMP) pathway is used by type I methylotorphs and the serine pathway is used by type II methylotrophs.

autotrophs. Recently, facultative methylotrophs have been isolated which can use methane or C_1 compounds (Patt et al., 1974; Patel et al., 1978). When utilizing C_1 compounds, they act like type II methylotrophs and feature peripheral intracytoplasmic membranes. The C_1-utilizing enzyme system seems to be inducible (Patel et al., 1978).

Methane-oxidizing organisms are active in the aerated surface strata overlying methanogenic strata of sediments, thus preventing much methane from entering the atmosphere.

17.2   PEAT AND COAL

Peat   Although peat and coal are two different substances, their modes of origin have included common initial steps. Indeed, peat formation may have been an intermediate step in coal formation. Peat is mostly derived from plant remains that have accumulated in marshes and bogs (Fig. 17.5). According to Francis (1954), these remains may have derived from (1) sphagnum as well as grasses and heather (higher moor peat); (2) reeds, grasses, sedges, shrubs, and bushes (low moor peat); (3) trees, branches, and debris of large forests in low-lying wet ground (forest peat); or (4) plant debris accumulated in swamps (sedimentary or lake peat). In all these instances, plant growth outstripped the decay of the plant remains, assuring a continual supply of new raw material for peat formation.

The plant remains may have first undergone attack by some of their own enzymes but soon were attacked by fungi, which degraded the relatively resistant polymers such as cellulose, hemicellulose, and lignin. Bacteria degraded the more easily oxidizable substances and the breakdown products of fungal activity which were not consumed by the fungi themselves. Fungal activity continued for as long as the organisms had access to air, but as they and their remaining substrate became buried and conditions became anaerobic, bacterial fermentation and anaerobic respiration set in and continued until arrested by accumulation of inhibitory (toxic) wastes, lack of sufficient moisture, or other factors (Francis, 1954; Kuznetsov et al., 1963; Rogoff et al., 1962). Uppermost aerobic layers of peat may harbor a viable microbial flora in places even today, indicating that peat formation may be occurring at the present time (Kuznetsov et al., 1963). Rogoff et al. (1962) claim that viable anaerobic bacteria and actinomycetes are present even in deep layers of peat.

The peat in its formation is ultimately enriched in lingnin, ulmins, and humic acids. The first of these compounds is a relatively resistant polymer of woody tissue and the second and third are complex material from the incomplete breakdown of plant matter, including lignin. Peat also contains other compounds relatively resistant to microbial attack, such as resins and waxes from cuticles,

Figure 17.5 Peat. (A) Section of ditch near Vestaburg, showing light sphagnum peat over dark peat: a, living sphagnum; b, sphagnum peat; c, shrub remains; d, sedge rootstocks; e, pond lily rootstocks; f, laminated peat. (B) View of surface near ditch, showing corresponding vegetative zones: a, shrub zone; b, grass zone; c, sedge zone; d, pond lily zone. (From Davis, 1907. Photograph is plate XIII from Annual Report, Geological Surevy of Michigan; reproduced by permision.)

stems, and spore exines of the peat-forming plants. When compared to the C, H, O, N, and S content of the original undecomposed plant material, peat is slightly enriched in carbon, nitrogen, and sulfur, but depleted in oxygen and sometimes hydrogen (Francis, 1954). This enrichment in C, N, and S over oxygen may be explained in part in terms of volatilization of the less-resistant components by microbial attack and the buildup of residues (resins, waxes, lignin) that are relatively oxygen-deficient owing to hydrocarbonlike and armoatic properties. The plant origin of peat is still clearly recognizable by examination of its structure.

Coal   Coal has been defined by Francis (1954) as "a compact, stratified mass of mummified plants which have been modified chemically in varying degrees, interspersed with smaller amounts of inorganic matter." Peat may be distinguished from coal in chemical terms by its much lower carbon content (51-59% dry wt) and higher hydrogen content (5.6-6.1% dry wt) than coal (carbon, 75-95% dry wt; hydrogen, 2.0-5.8% dry wt) (Francis, 1954, p. 295). The average carbon content of typical wood has been given as 49.2% (dry wt), and its average hydrogen content has been given as 6.1% (dry wt) (Francis, 1954). Coalification can thus be seen to have resulted in an enrichment in carbon and a slight depletion of hydrogen of the substance that gave rise to coal. Coal is generally found buried below layers of sedimentary strata. Its geologic age is generally advanced. Significant deposits formed in the Upper Paleozoic between 300×10^6 and 210×10^6 years ago, in the Mesozoic between 180×10^6 and 100×10^6 years ago, and in the Tertiary between 60×10^6 and 2.5×10^6 years ago. Peats generally have developed from about 1×10^6 years ago up to the present.

Coal is classifed by rank. According to the ASTM classification system, four major classes are recognized (*Mineral Facts and Problems,* 1965). They are, starting with the least developed coal, lignitic coal, subbituminous coal, bituminous coal, and anthracite coal. Lignitic coal is the least-developed coal, structurally resembling peat, having the highest moisture and lowest carbon content (59.5-72.3% dry wt; Francis, 1954, p. 335) as well as the lowest heat value of any of the coals. This coal formed in Tertiary times. Subbituminous coal has a somewhat lower moisture content but a higher carbon content (72.3-80.4% dry wt) and higher heat value than lignitic coal. Bituminous coal (Fig. 17.6) has a carbon content ranging from 80.4 to 90.9% (dry wt) and a high heat value. Both types of bituminous coal are mostly of late Paleozoic and Mesozoic age. Anthracite coals have very low moisture content, few volatiles, and a high carbon content (92.9-94.7% dry wt). They are of late Paleozoic age. Cannel coal is a special variety of bituminous coal which was derived mainly from wind-blown spores and pollen rather than from woody plant tissue.

As mentioned before, coal deposits developed at special periods in geologic time. In these periods the climate, landscape features, and biological activity

Figure 17.6   Block of bituminous coal showing pyrite along the bedding. The white circle is the outline of an English coin. (From Williamson, 1967; by permission.)

were suitable. Plant debris accumulated in swamps or shallow lakes because, owing to climatic conditions, plant growth was so profuse that there was a continual fresh supply of plant debris. At times this would become buried under water and be covered up by sediment (clays and sands) before a new layer of plant debris accumulated. Subsidence followed and is believed to have been an integral part in the formation of coal deposits (Francis, 1954).

Bacteria and fungi are generally believed to have played an important role in coalification only in the initial stages, as in peat formation. They destroyed the easily metabolizable substances, such as sugars, amino acids, and volatile acids, in a short time, and degraded more slowly the more stable polymers, such as cellulose, hemicellulose, lignins, waxes, and resins. Many of the latter were degraded only very incompletely before microbial activity is believed to have ceased. Hyphal remains, sclerotia, and fungal spores have been identified in some coal remains. Initial microbial attack is believed to have been aerobic and mainly of a fungal nature. Later attack occurred under progressively more anaerobic conditions and mostly by bacteria. Tauson believed, however, that in coal formation the anaerobic phase was abiogenic (Kuznetsov et al., 1963). Conversion of the residue of microbial activity (peat?) to coal is presently believed to have been due to physical and chemical agencies of unidentified nature, but may have involved heat and pressure which resulted in loss of volatile components.

Different reports on the isolation of microorganisms from different kinds of coal have claimed the detection of such fungi as *Aspergillis minor, Penicillium, Verticillium, Trichoderma, Fusarium,* and *Mucor.* Similarly, claims for the detection of bacteria such as *Sarcina, Clostridium welchii, Bacillus megaterium, B. subtilis, Pseudomonas fluorescens, Galionella,* and *Thiothrix* exist (see the review by Rogoff et al., 1962). However, because most of these organisms are common in soil and water, it is not clear if they represent an indigenous flora of coal. The most frequent claims for microbial association with coal were in connection with the lower-rank coals (lignites), whose flora includes *Pseudomonas,* molds, and actinomycetes (see Rogoff et al., 1962). Despite a claim to the contrary (Lipman, 1931), anthracite coal probably does not harbor an indigenous flora (Farrell and Turner, 1932; Burke and Wiley, 1937).

Iron-oxidizing thiobacilli have been found associated with bituminous coal. They grow at the expense of FeS_2 (pyrite or marcasite) associated with the coal seams. The FeS_2 was undoubtedly formed during the anaerobic phase of coal formation and must have been at least in part, if not wholly, the result of microbial activity. The eoclogy of iron-oxidizing bacteria associated with coal has been studied (Belly and Brock, 1974).

Microbial Development in a Coal Spoil   Microbial succession in an artificial coal spoil has been studied in the laboratory (Harrison, 1978). The coal spoil consisted of a homogeneous mixture of crushed and sifted coal (1 part) and, from the overburden of the coal deposit, shale (2 parts) and subsoil (8 parts), heaped in a deep plastic tray into a conical mound 50 cm in diameter at its base and 25 cm high. The mound was inoculated with 20 liters of an emulsion of acid soil and drainage water, with mud from an old coal strip mine spoil poured into the bottom of the plastic tray. The inoculum suspension was taken into the mound and migrated upward to the summit, presumably by capillary action and evaporation. Evaporation losses were made up periodically by additions of distilled water to the free liquid in the tray.

Sampling at the base of the mount initially yielded evidence of the presence of heterotrophic bacteria, which dominated first. They reached a population density of about 10^7 cells per gram within 2 weeks. After 8 weeks, heterotrophs were still dominant, although the pH had dropped from 7 to 5. Between 12 and

20 weeks the population decreased by about an order of magnitude, coinciding with a slight increase in acidity to just below pH 5 caused by a burst in sulfur-oxidizing bacteria, which died progressively thereafter. At that time the hetero-trophic population increased again to just below 10^7 cells per gram.

Sampling near the summit of the mound, heterotrophs also predominated for the first 16 weeks but then decreased dramatically from 10^6 to less than 10^2 cells per gram, concomitant with a drop in pH to 2.6. The pH drop was corre-lated with a marked rise in the population density of sulfur- and iron-oxidizing autotrophic bacteria, the former dominating briefly over the latter in the initial weeks.

Protozoans, algae, an arthropod, and a moss were also noted, mostly at the higher pH values. The *Metallogenium* of the type of Walsh and Mitchell (1972a) was not seen.

The sulfur-oxidizing bacteria were assumed to be utilizing elemental sulfur resulting from the chemical oxidation of the pyrite (FeS_2) in the coal by ferric iron:

$$FeS_2 + Fe_2(SO_4)_3 = 3FeSO_3 + 3S^0. \tag{17.10}$$

However, pyrite or marcasite are not generally oxidized chemically to form elemental sulfur (S_8^0) (Sato, 1960). It is possible that sulfur arose indirectly from microbial sulfate reduction in anaerobic zones in the coal spoil, yielding H_2S that then became the energy source for the sulfur-oxidizing autotrophs. Anaerobic bacteria were not sought in this study.

After 7 weeks of incubation of the mound, a mineral efflorescence developed on the surface consisting mainly of sulfates of Mg, Ca, Na, Al, and Fe. The magnesium sulfate was in the form of hexahydrite rather than epsomite. The metals were leached from the coal, but magnesium was also leached from the overburden material.

A thermophilic, acidophilic *Thermoplasma acidophilus* has been isolated from a coal refuse pile which had become self-heated (Darland et al., 1970). This organism lacks a true prokaryotic cell wall and has been found related to myco-plasmas. Its growth temperature optimum was $59°C$ (range 45-$62°C$), and its optimum pH for growth was between 1 and 2 (range, pH 0.96-3.5). The organ-ism is a heterotroph, growing readily in a medium of 0.02% $(NH_4)_2SO_4$, 0.05% $MgSO_4$, 0.025% $CaCl_2 \cdot 2H_2O$, 0.3% KH_2PO_4, 0.1% yeast extract, and 1.0% glucose at pH 3.0. Its relation to coal needs to be clarified.

Coal is not a very suitable substrate for support of microbial growth because it contains inhibitory substances ("antibiotics") that may suppress growth. These "antibiotics" seem to be associated with the waxy or resinous part of coal, extractable with methanol (Rogoff et al., 1962). In experiments with coal slurries, only marginal bacterial growth was obtainable, the limiting factors being the presence of inhibitory substances and lack of assimilable nutrients (Koburger, 1964). Growth of *Escherichia freundii* and *Pseudomonas rathonis* in such slurries improved if the coal had first been treated with H_2O_2.

17.3   PETROLEUM AND NATURAL GAS

Origin and Development of Deposits   Whereas coal derived chiefly from terrestrial plant remains that were deposited in shallow bodies of water or bogs and periodically buried, petroleum and natural gas derived chiefly from planktonic remains that were deposited in marine sediments in depressions of shallow seas and ultimately buried under heavy layers of sediment, deposited perhaps by turbidity currents. Over geologic time, the organic matter became converted to petroleum and natural gas, the former a liquid or a solid, and the latter a gas consisting chiefly of compounds of carbon and hydrogen with a minor content of oxygen, nitrogen, and sulfur. Petroleum is found in rock formations that include sands, sandstone, limestone, and conglomerates but rarely fissured shale or igneous rock. The enclosing rocks are usually of marine origin. The age of such host rocks may range from the Late Cambrian (500×10^6 years) to the Pliocene (1×10^6 to 13×10^6 years). Very extensive petroleum reservoirs are found in rock of Tertiary age (70×10^6 years).

Many theories have been advanced to explain the origin of petroleum and associated natural gas (see Beerstecher, 1954). None of these have been fully accepted. Some theories invoke heat or pressure or both as agents promoting abiological conversion of planktonic residues to the hydrocarbons of petroleum and natural gas. The source of the heat is viewed as the radioactivity in the earth's interior. This heat diffuses outward. Other theories have invoked inorganic catalysis with or without the influence of heat and pressure and with or without a prior acid or alkaline hydrolysis. Still other theories have proposed that petroleum constitutes a residue of naturally occurring hydrocarbons in the planktonic remains after all the other components have been biologically destroyed. One theory has been that biological agents caused aerobic or anaerobic reduction of fatty acids, proteins, and amino acids, carbohydrates, carotenoids, sterols, glycerol, chlorophyll, and lignin-humus complexes, together with appropriate decarboxylations and deaminations. Finally, a theory has it that biogenically formed methane from the planktonic debris became polymerized under high temperature and pressure and in the possible presence of catalysts, or, as an alternative, that bacteria modified the planktonic materials to substances closely resembling petroleum components, which were then converted to petroleum and natural gas constituents by heat and pressure.

At present, it is generally thought that bacteria have played a role in the initial stage of oil formation, but what this role was remains obscure, except in the case of methane formation. ZoBell (1952, 1963) has suggested that the planktonic debris may have been fermented, leading to compounds enriched in hydrogen and depleted in oxygen, sulfur, and phosphorus. Davis (1967, p. 23) has visualized microbial processes not unlike those in peat formation, involving initially aerobic attack of the sedimented planktonic debris followed by anaerobic activity after initial burial, including hydrolytic, decarboxylating, deaminating, and sulfate-reducing reactions, causing accumulation of marine humus. Progressively deeper burial resulted in compaction and in cessation of microbial activity, accompanied by the evolution of small amounts of hydrocarbon substance plus petroleum precursors. Such hydrocarbon could have resulted in part from hydrogenation reactions whose source of hydrogen might have been as follows (Davis, 1967):

$$FeS \cdot nH_2O + H_2S \rightarrow FeS_2 + H_2 + nH_2O \qquad (17.11)$$

Both hydrotroilite ($FeS \cdot nH_2O$) and H_2S in this reaction are considered to have been of biogenic origin. The catalytic action of clays, which furthers reducing reactions, is believed to have caused additional transformation of petroleum precursors into hydrocarbons. As hydrocarbon built up, the more volatile compounds generated a gas pressure that helped to force the more liquid components through porous rock (limestone, sandstone, etc.) to anticlinal folds where the petroleum could become trapped below a stratum of impervious rock and thus accumulate. Such accumulations constitute the oil reservoirs that are worked today. The migration of petroleum from the source rock to the reservoir rock was probably also helped by groundwater movement and by the action of natural detergents such as fatty acid soaps and other surface-active compounds of microbial origin. ZoBell (1952) has suggested that bacteria themselves may help to liberate oil from sediment particles and thereby promote its migration by dissolving carbonate and sulfate minerals to which oil may adhere and by generating CO_2, whose gas pressure may help to force migration of petroleum. Bacterially produced methane may lower the viscosity of petroleum liquid by dissolving in it and thus help its migration (ZoBell, 1952).

Oil deposits exist in reducing environments. Hence, any presently active bacteria associated with oil deposits would be expected to be anaerobes. Three specific types have been detected in petroleum-associated brines. The first type includes *Desulfovibrio desulfuricans* and *Desulfotomaculum nigrificans* (Kuznetsov et al., 1963; Nazina and Rozanova, 1978). To the second type belong *Methanococcus mazei, Sarcina methanica,* and *Methanobacterium omelianskii* (Kuznetsov et al., 1963). The third type is represented by *Rhodopseudomonas palustris* (Rozanova, 1971). The first type consists of sulfate reducers, the second of methanogenic bacteria, and the third of photoheterotrophs, such as

R. capsulata, which can grow heterotrophically in the dark under aerobic conditions but may also be capable of anaerobic respiration (Madigan and Gest, 1978; Yen and Marrs, 1977). The petroleum-associated brines may be connate seawaters whose mineral content has been somewhat altered through contact with enclosing rock strata (see Chapter 4). Such brines may be low in sulfate but high in chlorides and not very conducive to microbial growth. But sulfate-containing groundwaters and alkaline carbonate waters, on mixing with the brines, provide a milieu suitable for the activity of sulfate-reducing bacteria (Table 17.2). These waters furnish needed moisture, and in the case of sulfate-reducing bacteria, the needed terminal electron acceptor sulfate. By current understanding of them (see the discussion of methanogenesis in Section 17.1), methane-forming bacteria probably rely chiefly on CO_2 and H_2 or on acetate for methane formation during petroleum genesis. Sulfate-reducing bacteria have been claimed to obtain their carbon and energy requirements from petroleum components, but it seems more likely that other, as yet unidentified bacteria convert petroleum components to compounds that can be used by sulfate reducers (Ivanov, 1967). Jobson et al. (1979), on the basis of some laboratory experiments, have concluded that the petroleum degraders that furnish utilizable carbon to the sulfate reducers must be aerobes. Another study, however, recently revealed a sulfate-reducing bacterium in a lake sediment that could use methane as a source of reducing power, with acetate as the carbon source (Panganiban and Hanson, 1976).

The sulfate-reducing bacteria may be responsible for the sealing of oil deposits. They form a layer of secondary calcite at the interface between the oil-bearing stratum and the stratal waters (Ashirov and Sazanova, 1962; Davis, 1967; Kuznetsov et al., 1963). The calcite formation results from the reduction of sulfate:

$$CaSO_4 + 2(CH_2O) \rightarrow H_2S + CO_2 + H_2O + CaCO_3 \qquad (17.12)$$

where (CH_2O) represents an organic energy source directly or indirectly derived from petroleum. Davis (1967) has suggested the reaction

Table 17.2   Composition of Petroleum-Associated Brines (Percent Equivalents)

Cl^-	7.4-49.90	Ca^{2+}	0.33-11.02
SO_4^{2-}	0.03-10.06	Mg^{2+}	0.04-4.70
CO_3^{2-}	0.03-42.2	K^+ and Na^+	34.28-49.34

Source: After Kuznetsov et al. (1963), p. 17.

$$CaSO_4 + 8(H) + CO_2 \rightarrow CaCO_3 + 3H_2O + 3H_2O + H_2S \qquad (17.13)$$

where H is derived from the dehydrogenation of petroleum compounds or represents molecular hydrogen (e.g., from reaction 17.11). However, CO_2 cannot serve as an exclusive source of carbon for sulfate-reducing bacteria (see Chapter 14).

Liquid and gaseous petroleum may seep to the surface through cracks and fissures, resulting from folding, faulting, or erosion (Davis, 1967, p. 97). The amount of petroleum seepage may be large or miniscule (Wilson et al., 1974). An example of a large natural oil seep is Coal Oil Point, Santa Barbara, California (Allen et al., 1970).

Microbial Degradation of Petroleum   When the petroleum enters an aerobic environment, it becomes subject to microbial attack. An extensive literature has built up around this subject, largely because of its application to the management of oil pollution. Only the major principles of microbial petroleum degradation will be discussed here.

A variety of bacteria and fungi are able to metabolize hydrocarbons. Some examples are listed in Table 17.3. The mode of attack of hydrocarbons by microorganisms depends on the kind of microorganisms. Alkanes may be attacked monoterminally to form an alcohol, the first step involving an oxygenase (Doelle, 1975).

$$RCH_2CH_3 \xrightarrow{\frac{1}{2}O_2} RCH_2CH_2OH \xrightarrow{-2H} RCH_2CHO \xrightarrow{+H_2O, -2H} RCH_2COOH$$
$$(17.14)$$

They may also be monoterminally attacked to form a ketone (Fredericks, 1967) or a hydroperoxide (Stewart et al., 1959). Alkanes may, furthermore, be attacked diterminally (Doelle, 1975). For instance, *Pseudomonas aeruginosa* can attack 2-methylhexane at either end of the chain, forming a mixutre of 5-methylhexanoic and 2-methylhexanoic acid (Foster, 1962). Furthermore, alkanes may be desaturated terminally or subterminally, forming alkenes (Chouteau et al., 1962; Abbott and Casida, 1968). Subterminal desaturation may proceed as follows:

$$hexadecane \xrightarrow{n(2H)} \begin{array}{l} 7\text{-hexadecane,} \\ 8\text{-hexadecane,} \\ 6\text{-hexadecane} \end{array} \qquad (17.15)$$

Alkenes may be attacked by formation of epoxides, which may then be further metabolized (Abbott and Hon, 1973); diols may be formed in the process.

Aromatic compounds are attacked through oxygenation of the benezene ring either between the two adjacent oxygenated carbon atoms or adjacent to one of them (Dagley, 1975).

Table 17.3   Microorganisms Capable of Hydrocarbon Metabolism

Organisms	Substrates	Mode of attack	References
Pseudomonas oleovorans	Octane	Desaturation	Abbott and Hon (1973)
P. fluorescens, P. aeruginosa	Aromatic hydrocarbons	Oxidation	Van der Linden and Thijsse (1965)
Nocardia salmonicolor	Hexadecane	Desaturation	Abbott and Casida (1968)
Yeasts			Ahearn et al. (1971)
Trichosporon	n-Paraffins		Barna et al. (1970)
Arthrobacter	n-Alkane Aromatics		Klein et al. (1968); Stevenson (1967)
Mycobacterium	Butane		Nette et al. (1965); Phillips and Perry (1974)
Brevibacterium erythrogenes	Alkane	Oxidation	Pirnik et al. (1974)
Nocardia	Mono- and dicyclic hydrocarbons	Oxidation	Raymond et al. (1967)
Cladosporium	n-Alkane		Teh and Lee (1973); Walker and Cooney (1973)
Graphium	Ethane	Cooxidation	Volesky and Zajic (1970)

The ability to attack hydrocarbons does not necessarily mean that an organism can use such a compound as a sole source of carbon and energy. Many cases are known in which hydrocarbons are oxidized in a process known as cooxidation, wherein another compound, which may be quite unrelated, is the carbon and energy source but which somehow permits the simultaneous oxidation of the hydrocarbon. Examples are the oxidation of ethane to acetic acid, the oxidation of propane to propionic acid and acetone, and the oxidation of butane to butanoic acid and methyl ethyl ketone by *Pseudomonas methanica* growing on methane, as first shown by Leadbetter and Foster (1959). Methane is the only hydrocarbon on which this organism can grow. Another example is the oxidation of alkyl benzenes by a strain of *Micrococcus cerificans* growing on n-paraffins (Donos and Frankenfeld, 1968). Still other examples have been summarized by Horvath (1972).

Chain length and branching of aliphatic hydrocarbons affect microbial attack. For instance, some bacteria that attack alkanes of chain lengths C_8 -C_{20} may not be able to attack chain lengths of C_1 -C_6, whereas others cannot grow on alkanes of chain length greater than C_{10} (Johnson, 1964). Lack of growth on short-chain alkanes may have to do with toxic effects on membrane lipids, whereas lack of growth on long-chain alkanes may have to do with the relative insolubility on these compounds (Johnson, 1964). Fungi are, however, known that can grow on alkanes of chain lengths up to C_{34}. It has also been noted that certain placements of methyl and propyl groups in the alkane carbon chain lessens or prevents utilization of these compounds (McKenna and Kallio, 1965).

Hydrocarbon oxidation is generally considered a strictly aerobic process because the initial attack is usually an oxygenation. This notion has been the chief basis for explaining why petroleum has accumulated in nature in the geologic past. However, a few claims for anaerobic utilization do exist in the literature. Sulfate-reducing bacteria in oil-well brines have been claimed to be able to derive their energy and/or carbon from petroleum constituents, especially methane (Davis, 1967, p. 243; Davis and Yarbrough, 1966; Panganiban and Hanson, 1976; Panganiban et al., 1979). Even if the sulfate reducers are not able to attack the hydrocarbon themselves, satellite organisms have been thought to convert such compounds to products utilizable by sulfate reducers (Dutova, 1962). Kuznetsova and Gorlenko (1965) claim to have observed anaerobic hydrocarbon attack by a strain of *Pseudomonas* in a mineral salts medium with "oil" as the only carbon source. The bacterial population in these experiments increased a maximum of a millionfold, and the redox potential dropped from +40 to -110 mV. Similarly, Kvasnikov et al. (1973) have obtained growth of *Clostridium (Bacillus) polymyxa* anaerobically with n-alkanes as the sole source of carbon. Interestingly, Simakova et al. (1968) found that methane and high paraffinaceous oil were more intensely attacked aerobically than anaerobically by mixed cultures from oil deposits. They found very similar products, including fatty acids of high and low molecular weight, amino acids, alcohols, and aldehydes under either condition. However, hydroxy acids were only formed aerobically. This work showed that petroleum degradation is much slower anaerobically than aerobically. Ward and Brock (1978) recently reported the finding of very slow anaerobic conversion of [1-^{14}C]hexadecane added to reducing sediments and bottom water from Lake Mendota in Madison, Wisconsin. They reported that 13.7% of the [1-^{14}C]hexadecane was converted by the sediment to $^{14}CO_2$ and ^{14}C-cell carbon in 375 hr of incubation. Aerobically, the hexadecane was degraded much more rapidly by the same sediment samples. Methanogens are able to oxidize small amounts of methane they form anaerobically (Zehnder and Brock, 1979). The methane oxidation mechanism appears to differ from the methane-forming mechanism. The slow rate of anaerobic hydrocarbon degradation, when it occurs, helps to explain in part why petroleum has remained preserved over the eons of time. However, prolonged periods of an absence of degradative activity of any kind must have been a more important factor in preserving petroleum deposits. A case of in situ microbial conversion of petroleum into bitumin has recently been attributed to the Alberta (Canada) oil sands, based on laboratory simulation (Rubinstein et al., 1977).

Microbiological Prospecting for Petroleum   Prospecting for petroleum by the use of hydrocarbon-utilizing microorganisms has been proposed. The basis for this method is the detection of microseepage of petroleum or some of its constituents, especially the more volatile components, in the ground overlying a deposit by looking for microbes that can metabolize gaseous hydrocarbons (Davis, 1967). Methane-oxidizing bacteria are poor indicators for this purpose, because methane can occur in the absence of petroleum deposits, and, moreover, some methane-oxidizing bacteria are unable to oxidize other aliphatic hydrocarbons. Since ethane and propane produced in anaerobic fermentation are evolved in only small quantities, detection of bacteria that can oxidize these compounds should be presumptive indication of a petroleum reservoir (Davis, 1967). Only bacteria capable of oxidizing ethane or longer-chain, volatile hydrocarbons are useful indicator organisms (see Davis, 1967, p. 225). Detection of such organisms in soil or water samples depends on an enrichment in mineral salts solution with added volatile hydrocarbon (ethane, propane, butane, or isobutane) and measurement of hydrocarbon consumption. Likely organisms active in such enrichments may include *Mycobacterium paraffinicum* and *Streptomyces* spp. Recently, an enrichment culture method using ^{14}C-labeled hydrocarbon was developed, which quantifies the activity of hydrocarbon-oxidizing bacteria in water and sediments (Caparello and LaRock, 1975). With this method it was found that the hydrocarbon-oxidizing potential of samples reflects the hydrocarbon burden of the environment from which the sample was taken.

Ozokerite   Petroleum when oozing to the surface from its reservoir along fissures in disrupted overlying structures may deposit paraffinic material, called ozokerite, forming strings or veins (Davis, 1967). Ozokerite has been a commercial source of paraffin. The material may be modified by microbes, including bacteria and fungi, examples of these organisms being mycobacteria, proactinomycetes, and penicillia (Rozanova and Shturm, 1965), which apparently oxidize some paraffinic constituents and contribute cell material to it (i.e., humic material).

17.4 SUMMARY

Not all organic carbon in the biosphere is continually being recycled. Some is trapped in special sedimentary formations, where it is inaccessible to microbial attack. Some of this carbon is in the form of methane, some is in the form of peat and coal, and some of it is in the form of petroleum.

Most methane in sedimentary formations is of biogenic origin. It may occur by itself or in association with coal or petroleum deposits. Its biogenic formation is a strictly anaerobic process involving methanogenic bacteria that may reduce CO_2, formate, or acetate with H_2, or transform acetate to methane and CO_2 in the absence of H_2. The detailed biochemistry of methanogenesis is still poorly understood. The process can occur mesophilically or thermophilically.

Methane may be oxidized and assimilated by a special group of microorganisms called methylotrophs. This process is generally aerobic, although a report does exist of anaerobic methane oxidation by a sulfate reducer. For assimilation, methane is oxidized to the formaldehyde level and then assimilated either by the hexulose monophosphate or the serine pathway.

Peat is the result of partial degradation of plant remains accumulating in marshes and bogs. Aerobic attack by enzymes in the plant debris and by fungi and some bacteria initiates the process. It is followed by anaerobic attack by bacteria during burial through continual sedimentation until inhibited by accumulating wastes, lack of sufficient moisture, and so on. A viable microbial flora can usually be detected at the present time, even though peat formed over a geologically extended period. Coal is thought to have formed like peat, except that in the advanced stages, as a result of deeper burial, it was subject to physical and chemical influences that converted peat to coal. Different ranks of coal exist which differ from each other largely in carbon content and heat value. It is questionable whether coal itself harbors an indigenous flora. Bituminous coal has pyrite or marcasite associated with it. Upon exposure to air during mining, this iron disulfide becomes subject to attack by acidophilic, iron-oxidizing thiobacilli and is the source of acid mine drainage.

Whereas peat and coal are derived from terrestrial plant matter, petroleum and associated natural gas (mostly methane) is derived from phytoplankton remains that accumulated in depressions of shallow seas. Microbial attack altered these remains biochemically until complete burial by accumulating sediment stopped the organisms. Further burial caused a buildup of heat and pressure which caused further chemical alterations, possibly catalyzed by agents such as clay minerals, until petroleum hydrocarbons resulted. At least some of the natural gas may represent biogenic methane formed in the initial stages of plankton debris fermentation. At its site of formation, petroleum is highly dispersed. As a result of gas (natural gas, CO_2) and hydrostatic pressure and lubrication by bacteria, petroleum may be forced to migrate through pervious sedimentary strata until caught in a trap such as an anticlinal fold. It is such traps that constitute commercially exploitable petroleum reservoirs. Sulfate-reducing bacteria may assist in the trapping of petroleum by laying down impervious calcite layers. This calcite may, however, also interfere with petroleum recovery.

At least some petroleum hydrocarbons are oxidizable in air by certain bacteria and fungi. Isolated reports of anaerobic attack by bacteria exist, but the process described in the reports is very slow. Hydrocarbon-utilizing microorganisms may be used in prospecting for petroleum.

References

Abbott, B. J., and L. E. Casida, Jr. (1968). *J. Bacteriol.* 96:925-930.

Abbott, B. J., and C. T. Hon (1973). *Appl. Microbiol.* 26:86-91.

Abd-el-Malek, Y., and S. G. Rizk (1963a). *J. Appl. Bacteriol.* 26:14-19.

Abd-el-Malek, Y., and S. G. Rizk (1963b). *J. Appl. Bacteriol.* 26-20-26.

Abram, J. W., and D. B. Nedwell (1978). *Arch. Microbiol.* 117:89-92.

Adams, F., and J. P. Conrad (1953). *Soil Sci.* 75:361-371.

Adams, J. K., and B. Burkhart (1967). Diagenetic phosphates from the northern Atlantic coastal plain. *Abstr. Ann. Geophys. Soc. Am. and Assoc. Soc. Joint Meet., New Orleans, La., Nov. 20-22*, Program, p. 2.

Adeny, W. E. (1894). *Sci. Proc. R. Dublin Soc.,* Chap. 27, pp. 247-251.

Agate, A. D., and W. Vishniac (1970). *Bacteriol. Proc.,* p. 50.

Agnihotri, V. P. (1970). *Can. J. Microbiol.* 16:877-880.

Ahearn, D. G., S. P. Meyers, and P. G. Standard (1971). *Dev. Ind. Microbiol.* 12:126-134.

Ahluwalia, G. S., Y. R. Saxena, and H. H. Williams (1968). *Arch. Biochem. Biophys.* 124:79-84.

Ahmadjian, V. (1967). *The Lichen Symbiosis.* Blaisdell (Xerox), Greenwich, Conn.

Akagi, J. M., M. Chan, and V. Adams (1974). *J. Bacteriol.* 120:240-244.

Alberts, J. J., J. E. Schindler, R. W. Miller, and D. E. Nutter, Jr. (1974). *Science* 184:895-897.

Aleem, M. I. H. (1966a). *J. Bacteriol.* 91:729-736.

Aleem, M. I. H. (1966b). *Biochim. Biophys. Acta* 128:1-12.

Aleem, M. I. H. (1969). *Antonie van Leeuwenhoek J. Microbiol. Serol.* 85:379-391.

Aleem, M. I. H., H. Lees, and D. J. D. Nicholas (1963). *Nature (Lond)* 200: 759-761.

Alexander, M. (1977). *Introduction to Soil Microbiology* 2nd ed. Wiley, New York.

Alexandrov, E. A. (1972). Manganese: element and geochemistry. In *Encyclopedia of Geochemistry and Environmental Sciences*. Encyclopedia of Earth Science Series, Vol. IVA. R. W. Fairbridge, ed. Van Nostrand Reinhold, New York, pp. 607-671.

Ali, S. H., and J. L. Stokes (1971). *Antonie van Leeuwenhoek J. Microbiol. Serol.* 37:519-528.

Allen, A. A., R. S. Schlueter, and P. J. Mikolaj (1970). *Science* 170:974-977.

Altmann, H. J. (1972). *Z. Anal. Chem.* 262:97-99.

Altschuler, S., R. S. Clarke, Jr., and E. J. Young (1958). Geochemistry of uranium in apatite and phosphorite. *U. S. Geol. Surv. Prof. Pap. 314D.*

Aminuddin, M., and D. J. D. Nicholas (1973). *Biochim. Biophys. Acta* 325:81-93.

Anderson, A. (1967). *Grundfoerbaettring* 20:95-105.

Anderson, D. Q. (1940). *J. Mar. Res.* 2:225-235.

Anderson, G. M. (1972). Silica solubility. In *The Encyclopedia of Geochemistry and Environmental Sciences*. Encyclopedia of Earth Sciences Series, Vol. IVA. R. W. Fairbridge, ed. Van Nostrand Reinhold, New York, pp. 1085-1088.

Andreesen, J. R., and L. G. Ljungdahl (1973). *J. Bacteriol.* 116:869-873.

Arcuri, E. J. (1978). Identification of the cytochrome complements of several strains of marine manganese oxidizing bacteria and their involvement in manganese oxidation. Ph.D. thesis. Rensselaer Polytechnic Institute, Troy, N.Y.

Arcuri, E. J., and H. L. Ehrlich (1979). *Appl. Environ. Microbiol.* 37:916-923.

Aristovskaya, T. V. (1961). *Dokl. Akad. Nauk SSSR* 136:954-957.

Aristovskaya, T. V. (1963). *Pochvovedenie,* no. 1, pp. 30-43.

Aristovskaya, T. V., and R. S. Kutuzova (1968). *Pochvovedenie,* no. 12, pp. 59-66.

Aristovskaya, T. V., and O. M. Parinkina (1963). *Izv. Akad. Nauk SSSR Ser. Biol.* 218:49-56.

Aristovskaya, T. V., and G. A. Zavarzin (1971). Biochemistry of iron in soil. In *Soil Biochemistry,* Vol. 2. A. D. McLaren and J. Skujins, eds. Marcel Dekker, New York, pp. 385-408.

Arnon, D. I., M. Losada, M. Nogaki, and K. Takagawa (1961). *Nature (Lond.)* 190:601-610.

Ashirov, K. B., and I. V. Sazanova (1962). *Mikrobiologiya* 31:680-683.

Avakyan, A. A., and G. I. Karavaiko (1970). *Mikrobiologiya* 39:855-861.

Ayyakkamin, K., and D. Chandramotean (1971). *Mar. Biol.* 11:201-205.

Azam, F. (1974). *Planta (Berl.)* 121:205-212.

Azam, F., and B. E. Volcani (1974). *Arch. Microbiol.* 101:1-8.

Azam, F., B. B. Hemmingsen, and B. E. Volcani (1973). *Arch. Mikrobiol.* 92: 11-20.

Azam, F., B. B. Hemmingsen, and B. E. Volcani (1974). *Arch. Mikrobiol.* 97:103-114.

Baalsrud, K., and K. S. Baalsrud (1954). *Arch. Mikrobiol.* 20:34-62.

Baas Becking, L. G. M., and I. R. Kaplan (1956). *Trans. R. Soc. S. Austr.* 29: 52-65.

Baas Becking, L. G. M., and D. Moore (1961). *Econ. Geol.* 56:259-272.

Baas Becking, L. G. M., and G. S. Parks (1927). *Physiol. Rev.* 7:85-106.

Baas Becking, L. G. M., I. R. Kaplan, and D. Moore (1960). *J. Geol.* 68:243-284.

Badziong, W., and R. K. Thauer (1978). *Arch. Microbiol.* 117:209-214.

Badziong, W., R. K. Thauer, and J. G. Zeikus (1978). *Arch. Microbiol.* 116: 41-49.

Balashova, V. V., and G. A. Zavarzin (1972). *Mikrobiologiya* 41:909-911.

Balashova, V. V., I. Ya. Vedenima, G. E. Markosyan, and G. A. Zavarzin (1974). *Mikrobiologiya* 43:581-588.

Balch, W. E., G. E. Fox, L. J. Magrum, C. R. Woese, and R. S. Wolfe (1979). *Microbiol. Rev.* 43:260-296.

Ballard, R. D., and J. F. Grassle (1979). *Natl. Geogr. Mag.* 156:680-703.

Bambach, R. K., C. R. Scotese, and A. M. Ziegler (1980). *Am. Sci.* 68:26-38.

Barghoorn, E. S., and J. W. Schopf (1965). *Science* 150:337-339.

Barghoorn, E. S., and J. W. Schopf (1966). *Science* 152:785-763.

Barna, R. K., S. D. Bhagat, K. R. Pillai, H. D. Singh, J. N. Barnah, and M. S. Iyengar (1970). *Appl. Microbiol.* 20:657-661.

Barrenscheen, H. K., and H. A. Beckh-Widmanstetter (1923). *Biochem. Z.* 140: 279-283.

Baturin, G. N. (1972). *Dokl. Akad. Nauk SSSR Earth Sci. Sect.* 189:1359-1362.

Baturin, G. N., K. I. Merkhulova, and P. I. Chalov (1969). *Mar. Geol.* 13:M37-M41.

Bauld, J., and T. D. Brock (1973). *Arch. Mikrobiol.* 92:267-284.

Bautista, E. M., and M. Alexander (1972). *Soil Sci. Soc. Am. Proc.* 36:918-920.

Bavendamm, W. (1932). *Arch. Mikrobiol.* 3:205-276.

Bechamp, E. (1868). *Ann. Chim. Phys.* 13:103.

Beck, J. V. (1960). *J. Bacteriol.* 79:502-509.

Beck, J. V., and D. G. Brown (1968). *J. Bacteriol.* 96:1433-1434.

Beck, J. V., and S. R. Elsden (1958). *J. Gen. Microbiol.* 19:i.

Beerstecher, E. (1954). *Petroleum Microbiology.* Elsevier, New York.

Beijerinck, M. W. (1895). *Zentralbl. Bakteriol. Parasitenkd. Infektionskr. Hyg. Abt. I Orig.* 1:1-9; 49-59; 104-114.

Beijerinck, M. W. (1913). *Folia Microbiol. (Delft)* 2:123-134.

Belly, R. T., and T. D. Brock (1974). *J. Bacteriol.* 117:726-732.

Bennett, J. C., and H. Tributsch (1978). *J. Bacteriol.* 134:310-317.

Bennett, R. L., and M. H. Malamy (1970). *Biochem. Biophys. Res. Commun.* 40:496-503.

Bergey's Manual of Determinative Bacteriology (1957). 6th ed. R. S. Breed, E. G. D. Murray, and A. P. Hitchens, eds. Williams & Wilkins, Baltimore, Md.

Bergey's Manual of Determinative Bacteriology (1974). 8th ed. R. E. Buchanan, and N. F. Gibbons, eds. Williams & Wilkins, Baltimore, Md.

Berner, R. A. (1962). Experimental studies of the formation of sedimentary iron sulfides. In *Biogeochemistry of Sulfur Isotopes.* M. L. Jensen, ed. Yale University Press, New Haven, Conn., pp. 107-120.

Berner, R. A. (1968). *Science* 159:195-197.

Berthelin, J. (1977). Queleques aspects des mécanismes de transformation des minéraux des sols par les micro-organismes hétérotrophes. *Sci. Sol, Bull. A.F.E.S.,* no. 1. pp. 13-24.

Berthelin, J., and D. Boymond (1977). Some aspects of the role of heterotrophic microorganisms in the degradation of minerals in waterlogged acid soils. In *Environmental Biogeochemistry and Geomicrobiology,* Vol. 2: *The Terrestrial Environment.* W. E. Krumbein, ed. Ann Arbor Science Publishers, Ann Arbor, Mich., pp. 659-673.

Berthelin, J., and Y. Dommergues (1972). *Rev. Ecol. Biol. Sol* 9:397-406.

Berthelin, J., and A. Kogblevi (1972). *Rev. Ecol. Biol. Sol* 9:407-419.

Berthelin, J., and A. Kogblevi (1974). *Rev. Ecol. Biol. Sol* 11:499-509.

Berthelin, J., A. Kogblevi, and Y. Dommergues (1974). *Soil Biol. Biochem.* 6: 393-399.

Bertilsson, L., and H. Y. Neujahr (1971). *Biochemistry* 10:2805-2808.

Biebl, H., and N. Pfennig (1977). *Arch. Microbiol.* 112:115-117.

Bignelli, P. (1901). *Gazz. Chim. Ital.* 31:58.

Bignot, G., and L. Dangeard (1976). *C. R. Somm. Soc. Geol. Fr.,* Fasc. 3, pp. 96-99.

Blank, G. S., and C. W. Sullivan (1979). *Arch. Microbiol.* 123:157-164.

Blasco, F., C. Gaudin, and R. Jeanjean (1971). *C.R. Acad. Sci. (Paris), Ser. D.* 273:812-815.

Blaylock, B. A., and A. Nason (1963). *J. Biol. Chem.* 238:3453-3462.

Bloomfield, C. (1950). *J. Soil Sci.* 1:205-211.

Bloomfield, C. (1951). *J. Soil Sci.* 2:196-211.

Bloomfield, C. (1953a). *J. Soil Sci.* 4:5-16.

Bloomfield, C. (1953b). *J. Soil Sci.* 5:17-23.

Bonatti, E. (1966). *Science* 153:534-357.

Bonatti, E. (1972). Authigenesis of minerals—marine. In *The Encyclopedia of Geochemistry and Environmental Sciences.* Encyclopedia of Earth Sciences Series, Vol. IVA. R. W. Fairbridge, ed. Van Nostrand Reinhold, New York, pp. 48-56.

Bonatti, E. (1978). *Sci. Am.* 238:54-61.

Bonatti, E., and P. R. Hamlyn (1978). *Science* 201:249-251.

Bowen, H. J. M. (1966). *Trace Elements in Biochemistry.* Academic Press, New York.

Brierley, C. L. (1974). Leaching. Use of a high-temperature microbe. *Solution Min. Symp., Proc. 103rd AIME Annu. Meet., Dallas, Tex., Feb. 25-27*, pp. 461-469.

Brierley, C. L. (1978). *CRC Crit. Rev.* 6:207-262.

Brierley, C. L., and J. A. Brierley (1973). *Can. J. Microbiol.* 19:183-188.

Brierley, C. L., and L. E. Murr (1973). *Science* 179:488-499.

Brierley, J. A. (1966). Contribution of chemoautotrophic bacteria to the acid thermal waters of Geyser Spring Group in Yellowstone National Park. Ph.D. thesis. Montana State University, Bozeman, Mont.

Brierley, J. A. (1978). *Appl. Environ. Microbiol.* 36:523-525.

Brierley, J. A., and N. W. LeRoux (1977). A facultative thermophilic Thiobacillus-like bacterium: oxidation of iron and pyrite. In *Conference—Bacterial Leaching.* Gesellschaft fuer Biotechnologische Forschung mbH, pp. 55-66. Braunschweig-Stoeckheim. W. Schwartz, ed. Verlag Chemie, Weinheim, West Germany.

Brierley, J. A., and S. J. Lockwood (1977). *FEMS Microbiol. Lett.* 2:163-165.

Brierley, J. A., P. R. Norris, D. P. Kelly, and N. W. LeRoux (1978). *Eur. J. Appl. Microbiol. Biotechnol.* 5:291-299.

Brock, T. D. (1967). *Science* 158:1012-1019.

Brock, T. D. (1974). *Biology of Microorganisms,* 2nd ed. Prentice-Hall, Englewood Cliffs, N.J.

Brock, T. D. (1975). *Appl. Microbiol.* 29:495-501.

Brock, T. D. (1978). *Thermophilic Microorganisms and Life at High Temperatures.* Springer-Verlag, New York.

Brock, T. D., and M. L. Brock (1968). *J. Bacteriol.* 95:811-815.

Brock, T. D., and J. Gustafson (1976). *Appl. Environ. Microbiol.* 32:567-571.

Brock, T. D., M. L. Brock, T. L. Bott, and M. R. Edwards (1971). *J. Bacteriol.* 107:303-314.

Brock, T. D., K. M. Brock, R. T. Belly, and R. L. Weiss (1972). *Arch. Mikrobiol.* 84:54-68.

Brock, T. D., S. Cook, S. Petersen, and J. L. Messer (1976). *Geochim. Cosmochim. Acta* 40:493-500.

Broeze, R. J. (1978). The effects of low temperature on growth, survival, and metabolism of bacteria. Ph.D. thesis. Rensselaer Polytechnic Institute, Troy, N.Y.

Bromfield, S. M. (1953). *J. Gen. Microbiol.* 8:378-390.

Bromfield, S. M. (1954a). *J. Soil Sci.* 5:129-139.

Bromfield, S. M. (1954b). *J. Gen. Microbiol.* 11:1-6.

Bromfield, S. M. (1956). *Austr. J. Biol. Sci.* 9:238-252.

Bromfield, S. M. (1958a). *Plant Soil* 9:325-337.

Bromfield, S. M. (1958b). *Plant Soil* 10:147-160.

Bromfield, S. M. (1974). *Soil Biol. Biochem.* 6:383-392.

Bromfield, S. M. (1978). *Austr. J. Soil Res.* 16:91-100.

Bromfield, S. M. (1979). *Soil Biol. Biochem.* 11:115-118.

Bromfield, S. M., and D. J. David (1976). *Soil Biol. Biochem.* 8:37-43.

Bromfield, S. M., and D. J. David (1978). *Austr. J. Soil Sci.* 16:79-89.

Bromfield, S. M., and V. B. D. Skerman (1950). *Soil Sci.* 69:337-348.

Brooks, R. R., and I. R. Kaplan (1972). Biogeochemistry In *The Encyclopedia of Geochemistry and Environmental Sciences.* Encyclopedia of Earth Sciences Series, Vol. IVA. R. W. Fairbridge, ed. Van Nostrand Reinhold, New York, pp. 74-82.

Brown, K. A., and C. Ratledge (1975). *FEBS Lett.* 53:262-266.

Brunker, R. L., and T. L. Bott (1974). *Appl. Microbiol.* 27:870-873.

Bryner, L. C., and R. Anderson (1957). *Ind. Eng. Chem.* 49:1721-1724.

Bryner, L. C., and A. K. Jameson (1958). *Appl. Microbiol.* 6:281-287.

Bryner, L. C., J. V. Beck, D. B. Davis, and D. G. Wilson (1954). *Ind. Eng. Chem.* 46:2587-2592.

Buckman, H. O., and N. C. Brady (1969). *The Nature and Properties of Soild,* 7th ed. Revised by N. C. Brady. MacMillan, New York.

Bunt, J. S., and A. D. Rovira (1955). *J. Soil Sci.* 6:121-128.

Bunting, B. T. (1967). *Geography of Soil.* Hutchinson University Library, London.

Burke, V., and J. Wiley (1937). *J. Bacteriol.* 34:475-481.

Burns, R. G., V. M. Burns, and W. Sung (1974). Ferromanganese nodule mineralogy: suggested terminology of the principal manganese oxide phases. *Abstr. Annu. Meet., Geol. Soc. Am.*

Butkevitch, V. S. (1928). *Wiss. Meeresinst. Ber.* 3:7-80.

Butlin, K. (1953). *Research* 6:184-191.

Butlin, K. R., and J. R. Postgate (1952). The microbiological formation of sulfur in the Cyrenaikan Lakes. In *Biology of Deserts.* Institute of Biology, London.

Button, D. K., S. S. Dunker, and M. L. Moore (1973). *J. Bacteriol.* 113:599-611.

Caldwell, D. E., and P. Hirsch (1973). *Can. J. Microbiol.* 19:53-58.

Calvin, M. (1975). *Am. Sci.* 63:170-177.

Campbell, L. L., M. A. Kasprzycki, and J. R. Postgate (1966). *J. Bacteriol.* 92: 1122-1127.

Campbell, N. E. R., and H. Lees (1967). The nitrogen cycle. In *Soil Biochemistry,* Vol. 1. A. D. McLaren and G. H. Petersen, eds. Marcel Dekker, New York, pp. 194-215.

Caparello, D. M., and P. A. LaRock (1975). *Microb. Ecol.* 2:28-42.

Cappenberg, T. E. (1974a). *Antonie van Leeuwenhoek J. Microbiol. Serol.* 40: 285-295.

Cappenberg, T. E. (1974b). *Antonie van Leeuwenhoek J. Microbiol. Serol.* 40: 297-306.

Cappenberg, T. E., and R. A. Prins (1974). *Antonie van Leeuwenhoek J. Microbiol. Serol.* 40:457-469.

Carapella, S. C., Jr. (1972). Arsenic: element and geochemistry. In *The Encyclopedia of Geochemistry and Environmental Sciences.* Encyclopedia of Earth Sciences Series, Vol. IVA. R. W. Fairbridge, ed. Van Nostrand, Reinhold, New York, pp. 41-42.

Casida, L. E., Jr. (1960). *J. Bacteriol.* 80:237-241.

Casida, L. E., Jr. (1962). *Can. J. Microbiol.* 8:115-119.

Casida, L. E., Jr. (1965). *Appl. Microbiol.* 13:327-334.

Casida, L. E., Jr. (1971). *Appl. Microbiol.* 21:1040-1045.

Castenholz, R. W. (1976). *J. Phycol.* 12:54-68.

Castenholz, R. W. (1977). *Microb. Ecol.* 3:79-105.

Challenger, F. (1951). *Adv. Enzymol.* 12:429-491.

Challenger, F., C. Higginbottom, and L. Ellis (1933). *J. Chem. Soc. (Lond.),* pp. 95-101.

Chambers, L. A., and P. A. Trudinger (1975). *J. Bacteriol.* 123:36-40.

Charles, A. M., and I. Suzuki (1966). *Biochim. Biophys. Acta* 128:522-534.

Chatterjee, A. K. and P. Nandi (1964). *Trans. Bose Res. Inst. (Calcutta)* 27:115-120.

Chemical Processing (1965). 11:24-25.

Cheng, C.-N., and D. D. Focht (1979). *Appl. Environ. Microbiol.* 38:494-498.

Cholodny, N. (1924). *Ber. Dtsch. Bot. Ges.* 42:35-44.

Cholodny, N. (1926). *Die Eisenbakterien: Beitraege zu einer Monographie.* Gustav Fischer, Jena, East Germany.

Chouteau, J., E. Azoulay, and J. C. Zenez (1962). *Bull. Soc. Chim. Biol.* 44: 1670-1672.

Cloud, P. (1965). *Science* 148:27-35.

Cloud, P. (1973). *Econ. Geol.* 68:1135-1143.

Cloud, P. (1974). *Am. Sci.* 62:54-66.

Cloud, P. E., Jr., and G. R. Licari (1968). *Proc. Natl. Acad. Sci. U.S.A.* 61: 779-786.

Cohen, Y., E. Padan, and M. Shilo (1975). *J. Bacteriol.* 123:855-861.

Cohen, Y., W. E. Krumbein, M. Goldberg, and M. Shilo (1977a). *Limnol. Oceanogr.* 22:597-608.

Cohen, Y., W. E. Krumbein, and M. Shilo (1977b). *Limnol. Oceanogr.* 22:609-620.

Cohen, Y., W. E. Krumbein, and M. Shilo (1977c). *Limnol. Oceanogr.* 22:635-656.

Cohn, F. (1870). *Beitr. Biol. Pflanz.* 1:108-131.

Collins, J. F., and S. W. Buol (1970). *Soil Sci.* 110:111-116.

Colmer, A. R., K. L. Temple, and M. E. Hinkle (1950). *J. Bacteriol.* 59:317-328.

Coombs, J., and B. E. Volcani (1968). *Planta* 80:264-279.

Corliss, J. B., J. Dymond, L. I. Gordon, J. M. Edmont, R. P. von Herzen, R. D. Ballard, K. Green, D. Williams, A. Bainbridge, K. Crane, and T. H. van Andel (1979). *Science* 203:1073-1083.

Corrick, J. D., and J. A. Sutton (1961). Three chemosynthetic autotrophic bacteria important to leaching operations at Arizona copper mines. *U. S. Bur. Mines Rep. Invest. 5718.*

Cosgrove, D. J. (1967). Metabolism of organic phosphates in soil. In *Soil Biochemistry*, Vol. 1. A. D. McLaren and G. H. Petersen, eds. Marcel Dekker, New York, pp. 216-228.

Cosgrove, D. J. (1977). *Adv. Microb. Ecol.* 1:95-134.

Cox, C. D. (1980). *J. Bacteriol.* 141:199-204.

Cox, D. P., and M. Alexander (1973). *Appl. Microbiol.* 25:408-413.

Cox, J. C., and D. H. Boxer (1978). *Biochem. J.* 174:497-502.

Craig, P. J., and P. D. Bartlett (1978). *Nature (Lond.)* 275:635-637.

Crerar, D. A., and H. L. Barnes (1974). *Geochim. Cosmochim. Acta* 38:279-300.

Crerar, D. A., G. W. Knox, and J. L. Means (1979). *Chem. Geol.* 24:111-135.

Cromack, K., Jr., P. Solkins, W. C. Graustein, K. Speidel, A. W. Todd, G. Spycher, C. Y. Li, and R. L. Todd (1979). *Soil Biol. Biochem.* 11:463-468.

Cronan, D. S., and R. L. Thomas (1970). *Can. J. Earth Sci.* 7:1346-1349.

Cuthbert, M. E. (1962). *Econ. Geol.* 57:38-41.

Da Costa, E. W. B. (1971). *Nature (Lond.) New Biol.* 231:32.

Da Costa, E. W. B. (1972). *Appl. Microbiol.* 23:46-53.

Dagley, S. (1975). *Am. Sci.* 63:681-689.

d'Anglejan, B. F. (1967). *Mar. Geol.* 5:15-44.

d'Anglejan, B. F. (1968). *Can. J. Earth Sci.* 5:81-87.

Daniels, L., and J. G. Zeikus (1978). *J. Bacteriol.* 136:75-84.

Daniels, L., G. Fuchs, R. K. Thauer, and J. G. Zeikus (1977). *J. Bacteriol.* 132:118-126.

Danon, A., and W. Stoeckenius (1974). *Proc. Natl. Acad. Sci. U.S.A.* 71:1234-1238.

Darland, G., T. D. Brock, W. Samsonoff, and S. F. Conti (1970). *Science* 170:1416-1418.

Darley, W. M. (1969). Silicon requirement for growth and macromolecular synthesis in synchronized cultures of diatoms. *Navicula pelliculosa* (Brebisson) Hilse and *Cylindrotheca fusiformis* Reimann and Lewin. Ph.D. thesis. University of California, San Diego, Calif.

Darley, W. M., and B. E. Volcani (1969). *Exp. Cell Res.* 58:334-343.

Davies, S. L., and R. Whittenbury (1970). *J. Gen. Microbiol.* 61:227-232.

Davis, C. A. (1907). *Peat: Essays on the Origin and Distribution in Michigan.* Wynkoop Hallenbeck Crawford, State Printers, Lansing, Mich.

Davis, E. A., and E. J. Johnson (1967). *Can. J. Microbiol.* 13:873-884.

Davis, J. B. (1967). *Petroleum Microbiology.* Elsevier, New York.

Davis, J. B., and H. F. Yarbrough (1966). *Chem. Geol.* 1:137-144.

Davison, C. F. (1962a). *Econ. Geol.* 57:265-274.

Davison, C. F. (1962b). *Econ. Geol.* 57:1134-1137.

Dean, W. E. (1970). Iron-manganese oxidate crusts in Oneida Lake, New York. *13th Proc. Conf. Great Lakes Res.* 1:217-222.

Dean, W. E., and S. K. Ghosh (1978). *J. Res. Geol. Surv.* 6:231-240.

De Castro, A. F., and H. L. Ehrlich (1970). *Antonie van Leeuwenhoek J. Microbiol. Serol.* 36:317-327.

Degens, E. T. (1969). Biogeochemistry of stable isotopes. In *Organic Geochemistry: Methods and Results.* G. Eglinton and M. T. J. Murphy, eds. Springer-Verlag, New York, pp. 304-329.

Degens, E. T. (1972). Geochemistry of sediment: ancient. In *The Encyclopedia of Geochemistry and Environmental Sciences.* Encyclopedia of Earth Sciences Series, Vol. IVA. R. W. Fairbridge, ed. Van Nostrand Reinhold, New York, pp. 417-428.

Degens, E. T. (1979). *Chem. Geol.* 25:257-269.

Delfino, J. J., and G. F. Lee (1968). *Environ. Sci. Technol.* 12:1094-1100.

De Simone, R. E., M. W. Penley, L. Charbonneau, S. G. Smith, J. M. Wood, H. A. O. Hill, J. M. Pratt, S. Ridsdale, and J. P. Williams (1973). *Biochim. Biophys. Acta* 304:851-863.

De Toni, J. B., and V. Trevisan (1889). Schizomycetaceae Naeg. In *Sylloge Fungorum*. P. A. Saccardo, ed. pp. 923-1073.

Deuser, W. G., E. T. Degens, G. R. Harvey, and M. Rubin (1973). *Science* 181: 51-54.

Dévigne, J.-P. (1968a). *C.R. Acad. Sci. (Paris)* 267:935-937.

Dévigne, J.-P. (1968b). *Arch. Inst. Pasteur (Tunis)* 45:341-358.

Dévigne, J.-P. (1973). *Cah. Geol.* no. 89, pp. 35-37.

Dewey, J. F., and J. M. Bird (1970). *J. Geophys. Res.* 75:2625-2647.

Dietrich, G., and K. Kalle (1965). *Allgemeine Meereskunde: Eine Einfuehrung in die Ozeanographie.* Gebrueder Borntraeger, Berlin.

Dietz, R. S., and J. C. Holden (1970). *J. Geophys. Res.* 75:4939-4956.

Dimmick, R. L., H. Wolochow, and M. A. Chatigny (1979). *Appl. Environ. Microbiol.* 37:924-927.

Din, G. A., I. Suzuki, and H. Lees (1967a). *Can. J. Biochem.* 45:1523-1546.

Din, G. A., I. Suzuki, and H. Lees (1967b). *Can. J. Microbiol.* 13:1413-1419.

Doddema, H. J., T. J. Hutton, C. van der Drift, and G. D. Vogels (1978). *J. Bacteriol.* 136:19-23.

Doddema, H. J., C. van der Drift, G. D. Vogels, and M. Veenhuis (1979). *J. Bacteriol.* 140:1081-1089.

Dodson, E. O. (1979). *Can. J. Microbiol.* 25:651-674.

Doelle, H. W. (1975). *Bacterial Metabolism*, 2nd ed. Academic Press, New York.

Doemel, W. N., and T. D. Brock (1974). *Science* 184:1083-1085.

Doemel, W. N., and T. D. Brock (1976). *Limnol. Oceanogr.* 21:237-244.

Doemel, W. N., and T. D. Brock (1977). *Appl. Environ. Microbiol.* 34:433-452.

Doetsch, R. N., and T. M. Cook (1973). *Introduction to Bacteria and Their Ecobiology.* University Park Press, Baltimore, Md.

Doherty, M. W. (1898). *Australasian Assoc. Ad. Sci. Rpt.*, p. 339.

Dommergues, Y., and F. Mangenot (1970). *Ecologie microbienne du sol.* Masson et Cie, Paris.

Donos, J. R., Jr., and J. W. Frankenfeld (1968). *Appl. Microbiol.* 16:532-533.

Doran, J. W., and M. Alexander (1977). *Appl. Environ. Microbiol.* 33:31-37.

Dorff, P. (1934). *Die Eisenorganismen: Systematik und Morphologie. Pflanzenforschung.* R. Kolwitz, ed. G. Fischer, Jena, G.D.R.

Douka, C. E. (1977). *Soil Biol. Biochem.* 9:89-97.

Douka, C. E. (1980). *Appl. Environ. Microbiol.* 39:74-80.

Drake, H. L., and J. M. Akagi (1978). *J. Bacteriol.* 136:916-923.

Drew, G. H. (1911). *J. Mar. Biol. Assoc.* 9:142-155.

Drew, G. H. (1914). On the precipitation of the calcium carbonate in the sea by marine bacteria, and on the action of denitrifying bacteria in tropical and temperate seas. *Carnegie Inst. Wash., Pap. Tort. Lab., Publ. 182*, 5:7-45.

Drosdoff, M., and C. C. Nikiforoff (1940). *Soil Sci.* 49:333-345.

Dubinina, G. A. (1970). *Z. Allg. Mikrobiol.* 10:309-320.

Dubinina, G. A. (1978a). *Mikrobiologiya* 47:591-599.

Dubinina, G. A. (1978b). *Mikrobiologiya* 47:783-789.

Dubinina, G. A., and Z. P. Deryugina (1971). *Dokl. Akad. Nauk SSSR* 201:714-716.

Dubinina, G. A., and A. V. Zhadnov (1975). *Int. J. Syst. Bacteriol.* 25:340-350.

Duff, R. B., and D. M. Webley (1959). *Chem. Ind.,* pp. 1376-1377.

Duff, R. B., D. M. Webley, and R. O. Scott (1963). *Soil Sci.* 95:105-114.

Dugan, P. R., and D. G. Lundgren (1964). *Anal. Biochem.* 8:312-318.

Dugan, P. R., and D. G. Lundgren (1965). *J. Bacteriol.* 89:825-834.

Dugolinsky, B. K., S. V. Margolis, and W. C. Dudley (1977). *J. Sediment. Petrol.* 47:428-445.

Duncan, D. W., and C. C. Walden (1972). *Dev. Ind. Microbiol.* 13:66-75.

Duncan, D. W., J. Landesman, and C. C. Walden (1967). *Can. J. Microbiol.* 13: 397-403.

Dutova, E. N. (1962). The significance of sulfate-reducing bacteria in prospecting for oil as exemplified in the study of ground water in central Asia. In *Geologic Activity of Microorganisms.* S. I. Kuznetsov, ed. Consultants Bureau, New York, pp. 76-78.

Dyke, K. G. H., M. T. Parker, and M. H. Richmond (1970). *J. Med. Microbiol.* 3:125-136.

Egorova, A. A., and Z. P. Deryugina (1963). *Mikrobiologiya* 32:439-446.

Ehrenberg, C. G. (1836). *Poggendorfs Ann.* 38:213-227.

Ehrenberg, C. G. (1838). *Die Infusionsthierchen als vollkommene Organismen.* L. Voss, Leipzig, East Germany.

Ehrlich, G. G., and R. Schoen (1967). Possible role of sulfur-oxidizing bacteria in surficial acid alteration near hot springs. *U. S. Geol. Surv. Prof. Pap. 575C,* pp. C110-C112.

Ehrlich, H. L. (1963a). *Econ. Geol.* 58:991-994.

Ehrlich, H. L. (1963b). *Appl. Environ. Microbiol.* 11:15-19.

Ehrlich, H. L. (1964a). *Econ. Geol.* 59:1306-1312.

Ehrlich, H. L. (1964b). *J. Bacteriol.* 86:350-352.

Ehrlich, H. L. (1966). *Dev. Ind. Microbiol.* 7:279-286.

Ehrlich, H. L. (1968). *Appl. Microbiol.* 16:197-202.

Ehrlich, H. L. (1970). *Microbiology of Manganese Nodules*. AD 1970, no. 716, 508. U.S. Clearing House, Fed. Sci. Tech. Inf., Springfield, Va.

Ehrlich, H. L. (1973). The biology of ferromanganese nodules. Determination of the effect of storage by freezing on the viable nodule flora, and a check on the reliability of the results from a test to identify MnO_2-reducing cultures. In *Interuniversity Program of Research of Ferromanganese Deposits on the Ocean Floor*. Phase I Report, Seabed Assessment Program, International Decade of Ocean Exploration. National Science Foundation, Washington, D.C., pp. 217-219.

Ehrlich, H. L. (1975). *Soil Sci.* 119:36-41.

Ehrlich, H. L. (1976). Manganese as an energy source for bacteria. In *Environmental Biogeochemistry*, Vol. 2. J. O. Nriagu, ed. Ann Arbor Science Publishers, Ann Arbor, Mich., pp. 633-644.

Ehrlich, H. L. (1977). Bacterial leaching of a low-grade chalcopyrite ore with different lixiviants. In *Conference – Bacterial Leaching*. Gesellschaft fuer Biotechnologische Forschung mbH, Braunschweig-Stoeckheim. W. Schwartz, ed. Verlag Chemie, Weinstein, West Germany. pp. 145-155.

Ehrlich, H. L. (1978a). *Geomicrobiol. J.* 1:65-83.

Ehrlich, H. L. (1978b). Conditions for bacterial participation in the initiation of manganese deposition around marine sediment particles. In *Environmental Biogeochemistry and Geomicrobiology*, Vol. 3: *Methods, Metals, and Assessment*. W. E. Krumbein, ed. Ann Arbor Science Publishers, Ann Arbor, Mich., pp. 839-845.

Ehrlich, H. L., and S. I. Fox (1967a). *Appl. Microbiol.* 15:135-139.

Ehrlich, H. L., and S. I. Fox (1967b). *Biotech. Bioeng.* 9:471-485.

Ehrlich, H. L., W. C. Ghiorse, and G. L. Johnson II (1972). *Dev. Ind. Microbiol.* 13:57-65.

Ehrlich, H. L., S. H. Yang, and J. D. Mainwaring, Jr. (1973). *Z. Allg. Mikrobiol.* 13:39-48.

Eleftheriadis, D. K. (1976). Mangan- und Eisenoxydation in Mineral- und Terman-Quellen-Mikrobiologie, Chemie und Geochemie. Ph.D. thesis. Universitaet des Saarlandes, Saarbruecken, West Germany.

Emiliani, C., J. Hudson, E. A. Shinn, and R. Y. George (1978). *Science* 202: 627-629.

Enoch, H. G., and R. L. Lester (1972). *J. Bacteriol.* 110:1032-1040.

Eren, J., and D. Pramer (1966). *Soil Sci.* 101:39-49.

Ernst, J. F., and G. Winkelmann (1977). *Biochim. Biophys. Acta* 500:27-41.

Fagerstroem, T., and A. Jerneloev (1971). *Water Res.* 5:121-122.

Falcone, G., and W. J. Nickerson (1963). *J. Bacteriol.* 85:754-762.

Farrell, M. A., and H. G. Turner (1932). *J. Bacteriol.* 23:155-162.

Feeley, H. W., and J. L. Kulp (1957). *Bull. Am. Assoc. Petrol. Geol.* 41:1802-1853.

Fleming, R. W., and M. Alexander (1972). *Appl. Microbiol.* 24:424-429.

Fliermans, C. B., and T. D. Brock (1972). *J. Bacteriol.* 111:343-350.

Fooden, J. (1972). *Science* 175:894-898.

Foster, J. W. (1962). *Antonie van Leeuwenhoek J. Microbiol. Serol.* 28:241-274.

Foster, T. L., L. Winans, Jr., and J. S. Helms (1978). *Appl. Environ. Microbiol.* 35:937-944.

Fox, G. E., L. J. Magrum, W. E. Balch, R. S. Wolfe, and C. R. Woese (1977). *Proc. Natl. Acad. Sci. U.S.A.* 74:4537-4541.

Fox, S. I. (1967). Bacterial oxidation of simple copper sulfides. Ph.D. thesis. Rensselaer Polytechnic Institute, Troy, N.Y.

Francis, W. (1954). *Coal: Its Formation and Composition.* Edward Arnold, London.

Fredericks, K. M. (1967). *Antonie van Leeuwenhoek J. Microbiol. Serol.* 33:41-48.

Fredericks-Jantzen, C. M., H. Herman, and P. Herley (1975). *Nature (Lond.)* 258:270.

Freke, A. M., and D. Tate (1961). *J. Biochem. Microbiol. Technol. Eng.* 3:29-39.

Freney, J. R. (1976). Sulfur-containing organics. In *Soil Biochemistry*, A. D. McLaren and G. H. Petersen, eds. Marcel Dekker, New York, pp. 229-259.

Fridovich, I. (1977). *BioScience* 27:462-466.

Friedman, E. I., and R. Ocampo (1976). *Science* 193:1247-1249.

Friedman, E. I., W. C. Roth, J. B. Turner, and R. S. McEwen (1972). *Science* 177: 891-893.

Fuchs, G., E. Stupperich, and R. K. Thauer (1978). *Arch. Microbiol.* 119:215-218.

Fuller, W. H. (1974). Desert soils. In *Desert Biology*, Vol. 2. G. W. Brown, Jr., ed. Academic Press, New York, pp. 31-101.

Furukawa, K., and K. Tonomura (1971). *Agric. Biol. Chem.* 35:604-610.

Furukawa, K., and K. Tonomura (1972a). *Agric. Biol. Chem.* 36:217-226.

Furukawa, K., and K. Tonomura (1972b). *Agric. Biol. Chem.* 36, Suppl. 13:2441-2448.

Gale, N. L., and J. V. Beck (1967). *J. Bacteriol.* 94:1052-1059.

Garey, C. L., and S. A. Barber (1952). *Soil Sci. Soc. Am., Proc.* 16:173-179.

Garlick, S., A. Oren, and E. Padan (1977). *J. Bacteriol.* 129:623-629.

Garrels, R. M. (1965). *Science* 148:69.

Geloso, M. (1927). *Ann. Chim.* 7:113-150.

Gerrard, T. L., J. N. Telford, and H. H. Williams (1974). *J. Bacteriol.* 119: 1057-1060.

Gerretsen, F. C. (1937). *Ann. Bot.* 1:207-230.

Gest, H. (1972). *Adv. Microb. Physiol.* 7:243-282.

Ghiorse, W. C., and H. L. Ehrlich (1976). *Appl. Environ. Microbiol.* 31:977-985.

Gibson, F., and D. I. Magrath (1969). *Biochim. Biophys. Acta* 192:175-184.

Gillette, N. J. (1961). *Conservationist,* Apr.-May, p. 19.

Goldberg, E. D. (1954). *J. Geol.* 62:249-265.

Goldberg, E. D., and G. O. S. Arrhenius (1958). *Geochim. Cosmochim. Acta* 13: 153-212.

Goldschmidt, V. M. (1954). *Geochemistry.* Clarendon Press, Oxford, pp. 621-642.

Golovacheva, R. S., and G. I. Karavaiko (1978). *Mikrobiologiya* 47:815-822.

Golubic, S. (1969). *Am. Zool.* 9:747-751.

Golubic, S. (1973). The relationship between blue-green algae and carbonate deposits. In *The Biology of Blue-Green Algae.* H. G. Carr and B. A. Whitton, eds. University of California Press, Berkeley, Calif., pp. 434-472.

Golubic, S., R. D. Perkins, and K. J. Lukas (1975). Boring microorganisms and microborings in carbonate substrates. In *The Study of True Fossils.* R. W. Frey, ed. Springer-Verlag, New York, pp. 229-259.

Gornitz, V. (1972). Antimony: element and geochemistry. In *The Encyclopedia of Geochemistry and Environmental Sciences.* Encyclopedia of Earth Sciences Series, Vol. IVA. R. W. Fairbridge, ed. Van Nostrand Reinhold, New York, pp. 33-36.

Gosio, B. (1897). *Ber. (Dtsch. Chem. Ges.)* 30:1024.

Graham, J. W. (1959). *Science* 129:1428-1429.

Graham, J. W., and S. C. Cooper (1959). *Nature (Lond.)* 183:1050-1051.

Graustein, W. C., K. Cromack, Jr., and P. Sollins (1977). *Science* 198:1252-1254.

Green, H. H. (1918). *S. Afr. J. Sci.* 14:465-467.

Greenfield, L. J. (1963). *Ann. N.Y. Acad. Sci.* 109:23-45.

Greenslate, J. (1974a). *Science* 186:529-531.

Greenslate, J. (1974b). *Nature (Lond.)* 249:181-183.

Gross, M. G. (1972). *Oceanography: A View of the Earth.* Prentice-Hall, Englewood Cliffs, N.J.

Gruner, J. W. (1922). *Econ. Geol.* 17:407-460.

Guay, R., and M. Silver (1975). *Can. J. Microbiol.* 21:281-288.

Guittoneau, G. (1927). *C.R. Acad. Sci. (Paris)* 184:45-46.

Guittoneau, G., and J. Keilling (1927). *C.R. Acad. Sci. (Paris)* 184:898-901.

Gulbrandsen, R. A. (1969). *Econ. Geol.* 64:365-382.

Gunsalus, R. P., and R. S. Wolfe (1978). *FEMS Microbiol. Lett.* 3:191-193.

Guyard, A. (1864). *Bull. Soc. Chim.* 1:89.

Haight, R. D., and R. Y. Morita (1962). *J. Bacteriol.* 83:112-120.

Hajj, H., and J. Makemson (1976). *Appl. Environ. Microbiol.* 32:699-702.

Hale, M. E., Jr. (1967). *The Biology of Lichens.* Edward Arnold, London.

Hall, F. R. (1972). Silica cycle. In *The Encyclopedia of Geochemistry and Environmental Sciences.* Encyclopedia of Earth Sciences Series, Vol. IVA. R. W. Fairbridge, ed. Van Nostrand Reinhold, New York, pp. 1082-1085.

Hallberg, R. O. (1978). Metal-organic interaction at the redoxcline. In *Environmental Biogeochemistry and Geomicrobiology,* Vol. 3: *Methods, Metals, and Assessment.* W. E. Krumbein, ed. Ann Arbor Science Publishers, Ann Arbor, Mich., pp. 947-953.

Hammann, R., and J. C. G. Ottow (1974). *Z. Pflanzenernaehr. Bodenkd.* 137:108-115.

Hammond, A. L. (1975a). *Science* 189:779-781.

Hammond, A. L. (1975b). *Science* 189:868-869; 915.

Hanert, H. (1968). *Arch. Mikrobiol.* 60:348-376.

Hanert, H. (1973). *Arch. Mikrobiol.* 88:225-243.

Hanert, H. (1974a). *Arch. Mikrobiol.* 96:59-74.

Hanert, H. (1974b). *Z. Kulturtech. Flurbereinig.* 15:80-90.

Hansen, T. A., and H. van Gemerden (1972). *Arch. Mikrobiol.* 86:49-56.

Hansen, T. A., and H. Veldkamp (1973). *Arch. Mikrobiol.* 92:45-58.

Harada, K. (1978). *Mem. Fac. Sci. Kyoto Univ., Ser. Geol. Mineral.* 45:111-132.

Harder, E. C. (1919). Iron depositing bacteria and their geologic relations. *U.S. Geol. Surv. Prof. Pap. 113.* 89 pp.

Hariya, Y., and T. Kikuchi (1964). *Nature (Lond.)* 202:416-417.

Harris, R. F., W. A. Gardner, A. A. Adebayo, and L. E. Sommers (1970). *Appl. Microbiol.* 19:536-537.

Harrison, A. G., and H. G. Thode (1957). *Trans. Faraday Soc.* 53:1-4.

Harrison, A. G., and H. G. Thode (1958). *Trans. Faraday Soc.* 54:84-92.

Harrison, A. P., Jr. (1978). *Appl. Environ. Microbiol.* 36:861-869.

Harriss, R. C. (1972). Silica—biogeochemical cycle. In *The Encyclopedia of Geochemistry and Environmental Sciences.* Encyclopedia of Earth Sciences Series, Vol. IVA. R. W. Fairbridge, ed. Van Nostrand Reinhold, New York, pp. 1080-1082.

Harriss, R. C., and A. G. Troup (1969). *Science* 166:604-606.

Harriss, R. C., and A. G. Troup (1970). *Limnol. Oceanogr.* 15:702-712.

Hartmannova, V., and I. Kuhr (1974). *Rudy* 22:234-238.

Healy, F. P., J. Coombs, and B. E. Volcani (1967). *Arch. Mikrobiol.* 59:131-142.

Hedges, R. W., and S. Baumberg (1973). *J. Bacteriol.* 115:459-460.

Heezen, B. C., and C. Hollister (1971). *The Face of the Deep.* Oxford University Press, New York.

Heinen, W. (1960). *Arch. Mikrobiol.* 37:199-210.

Heinen, W. (1962). *Arch. Mikrobiol.* 41:229-246.

Heinen, W. (1963a). *Arch. Mikrobiol.* 45:145-161.

Heinen, W. (1963b). *Arch. Mikrobiol.* 45:162-171.

Heinen, W. (1963c). *Arch. Mikrobiol.* 45:172-178.

Heinen, W. (1968). *Acta Bot. Neerl.* 17:105-113.

Heinen, W., and A. M. Lauwers (1974). *Arch. Mikrobiol.* 95:267-274.

Helbig, M. (1914). *Naturw. Ztschr. Forst Landw.* 12:385-393.

Helmche, J.-G. (1954). *Naturwissenschaften* 11:254-255.

Hem, J. D. (1963). Chemical equilibria and rate of manganese oxidation. *U.S. Geol. Surv. Water Supply Pap. 1667-A.*

Henderson, M. E. K., and R. B. Duff (1963). *J. Soil Sci.* 14:236-246.

Hendrie, M. S., A. J. Holding, and J. M. Shewan (1974). *Int. J. Syst. Bacteriol.* 24:534-550.

Heye, D., and H. Beiersdorf (1973). *Z. Geophys.* 39:703-726.

Hinkle, P. C., and R. E. McCarty (1978). *Sci. Am.* 238:104-123.

Hirsch, P. (1968). *Arch. Mikrobiol.* 60:201-216.

Hoehnl, G. (1955). *Vom Wasser* 22:176-193.

Hogan, V. C. (1973). Electron transport and manganese oxidation in *Leptothrix discophorus.* Ph.D. thesis. Ohio State University, Columbus, Ohio.

Hohn, H. W., and M. F. Cox (1975). *Appl. Microbiol.* 29:491-494.

Holm-Hansen, O., and C. R. Booth (1966). *Limnol. Oceanogr.* 11:510-519.

Holzapfel, L., and W. Engel (1954a). *Naturwissenschaften* 41:191.

Holzapfel, L., and W. Engel (1954b). *Z. Naturforsch.* 9b:602-606.

Hood, D. W. (1963). *Oceanogr. Mar. Biol. Annu. Rev.* 1:129-155.

Hood, D. W., and J. F. Slowey (1964). *Texas A and M Univ. Prog. Rep. Proj. 276,* AEC Contract no. AT-(40-1)-2799.

Hoppe-Seyler, F. (1886). *Z. Physiol. Chem.* 10:201;401

Horowitz, N. H. (1945). *Proc. Natl. Acad. Sci. U.S.A.* 31:153-157.

Horvath, R. S. (1972). *Bacteriol. Rev.* 36:146-155.

Howard, H. H., and S. W. Chisholm (1975). *Am. Mid. Nat.* 93:188-197.

Howarth, R. W. (1979). *Science* 203:49-51.

Huisingh, J. J., J. McNeil, and G. Matrone (1974). *Appl. Microbiol.* 28:489-497.

Hutchinson, M., K. I. Johnstone, and D. White (1966). *J. Gen. Microbiol.* 44: 373-381.

Imai, K. H., H. Sakaguchi, T. Sugio, and T. Tano (1973). *J. Ferment. Technol.* 51:865-870.

Imshenetsky, S. V., S. V. Lysenko, and G. A Kazakov (1978). *Appl. Environ. Microbiol.* 35:1-5.

Imura, N., E. Sukegawa, S.-K. Pan, K. Nagao, J.-Y. Kim, T. Kwan, and T. Ukita (1971). *Science* 172:1248-1249.

Ingledew, W. J., J. C. Cox, and P. J. Halling (1977). *FEMS Microbiol. Lett.* 2:193-197.

Isenberg, H. D., and L. S. Lavine (1973). Protozoan calcification. In *Biological Mineralization.* I. Zipkin, ed. Wiley, New York, pp. 649-686.

Ivanov, M. V. (1967). *Mikrobiologiya* 36:849-859.

Ivanov, M. V. (1968). *Microbiological Processes in the Formation of Sulfur Deposits.* Israel Program for Scientific Translations. U.S. Department of Agriculture and the National Science Foundation, Washington, D.C.

Ivanov, V. I. (1962). *Mikrobiologiya* 31:795-799.

Ivanov, V. I., and N. N. Lyalikova (1962). *Mikrobiologiya* 31:468-469.

Ivanov, V. I., F. I. Nagirynyak, and B. A. Stepanov (1961). *Mikrobiologiya* 30: 688-692.

Ivarson, K. C. (1973). *Can. J. Soil Sci.* 53:315-323.

Ivarson, K. C., and M. Sojak (1978). *Can. J. Soil Sci.* 58:1-17.

Iverson, W. P. (1968). *Nature (Lond.)* 217:1265.

Izaki, K., Y. Tashiro, and T. Funaba (1974). *J. Biochem. (Tokyo)* 75:591-599.

Jackson, T. A. (1975). *Soil Sci.* 119:56-64.

Jannasch, H. W. (1967). *Limnol. Oceanogr.* 12:264-271.

Jannasch, H. W. (1969). *J. Bacteriol.* 99:156-160.

Jannasch, H. W., and C. O. Wirsen (1973). *Science* 180:641-643.

Jannasch, H. W., and C. O. Wirsen (1979). *BioScience* 29:592-598.

Jannasch, H. W., K. Eimhjellen, C. O. Wirsen, and A. Farmanfarmian (1971). *Science* 171:672-675.

Jensen, M. L. (1962). *Int. Conf. Saline Deposits.* Spec. Pap. 88. Geological Society of America, Boulder, Colo.

Jensen, S., and A. Jerneloev (1969). *Nature (Lond.)* 233:753-754.

Jobson, A. M., F. D. Cook, and D. W. S. Westlake (1979). *Chem. Geol.* 24:355-365.

Johnson, A. H., and J. L. Stokes (1966). *J. Bacteriol.* 91:1543-1547.

Johnson, D. L. (1972). *Nature (Lond.)* 240:44-45.

Johnson, M. J. (1964). *Chem. Ind.* 36:1532-1537.

Jonasson, I. P. (1970). Mercury in the natural environment: a review of recent work. *Geol. Surv. Can. Pap. 1970, 70-57.* 39 pp.

Jones, G. E., and R. L. Starkey (1957). *Appl. Microbiol.* 5:111-118.

Jones, G. E., and R. L. Starkey (1962). Some necessary conditions for fractionation of stable isotopes of sulfur by *Desulfovibrio desulfuricans.* In *Biogeochemistry of Sulfur Isotopes.* N. S. F. Symp. M. L. Jensen, ed. Yale University, New Haven, Conn., pp. 61-79.

Jørgensen, B. B. (1977a). *Limnol. Oceanogr.* 22:814-832.

Jørgensen, B. B. (1977b). *Mar. Biol.* 41:7-18.

Jørgensen, B. B., and T. Fenchel (1974). *Mar. Biol.* 24:189-201.

Jørgensen, B. B., J. G. Kuenen, and Y. Cohen (1979a). *Limnol. Oceanogr.* 24: 799-822.

Jørgensen, B. B., N. P. Revsbech, H. Blackburn, and Y. Cohen (1979b). *Appl. Microbiol.* 38:46-58.

Jørgensen, E. G. (1955). *Physiol. Plant.* 8:846-851.

Jung, W. K., and R. Schweisfurth (1979). *Z. Allg. Mikrobiol.* 19:107-115.

Kalinenko, V. O., O. V. Belokopytova, and G. G. Nikolaeva (1962). *Okeanologiya* 11:1050-1059.

Kamalov, M. R. (1967). *Izv. Akad. Nauk Kaz. SSR, Ser. Biol.* 5:47-50.

Kamalov, M. R., G. I. Karavaiko, A. N. Ilyaledinov, and S. A. Abdrashitova (1973 (1973). *Izv. Akad. Nauk Kaz. SSR, Ser. Biol.* 11:37-44.

Kaplan, I. R., and S. C. Rittenberg (1962). The microbiological fractionation of sulfur isotopes. In *Biogeochemistry of Sulfur Isotopes:* N.S.F. Symp. M. L. Jensen, ed. Yale University, New Haven, Conn.

Kaplan, I. R., and S. C. Rittenberg (1964). *J. Gen. Microbiol.* 34:195-212.

Kaplan, I. R., T. A. Rafter, and J. R. Hulston (1960). *N. Z. J. Sci.* 3:338-361.

Kardos, L. T. (1955). Soil fixation of plant nutrients. In *Chemistry of the Soil.* F. E. Bear, ed. Van Nostrand Reinhold, New York, pp. 177-199.

Karl, D. M. (1978). *Appl. Environ. Microbiol.* 36:344-355.

Karl, D. M., and P. A. LaRock (1975). *J. Fish. Res. Board Can.* 32:599-607.

Kay, M. (1955). Sediments and subsidence through time. In *Crust of the Earth: A Symposium.* Spec. Pap. 62. A. Poldervaart, ed. pp. 665-684. Geological Society of America, Boulder, Colo.

Kazutami, I., and T. Tano (1967). *Hakko Kyokaishi* 25:166-167.

Kee, N. S., and C. Bloomfield (1961). *Geochim. Cosmochim. Acta* 24:206-225.

Kelly, D. P. (1971). *Annu. Rev. Microbiol.* 25:177-210.

Kelly, D. P., and O. H. Tuovinen (1972). *Int. J. Syst. Bacteriol.* 22:170-172.

Kemp. A. L. W., and H. G. Thode (1968). *Geochim. Cosmochim. Acta* 32:71-91.

Kester, D. R. (1975). Dissolved gases other than CO_2. In *Chemical Oceanography*, Vol. 1, 2nd ed. J. P. Riley and G. Skirrow, eds. Academic Press, New York, pp. 497-547.

Khak-mun, Ten. (1967). *Mikrobiologiya* 36:337-344.

Kharkar, D. P., K. K. Turekian, and K. K. Bertine (1968). *Geochim. Cosmochim. Acta* 32:285-298.

Kindle, E. M. (1932). *Am. J. Sci.* 24:496-504.

Kinsel, N. A. (1960). *J. Bacteriol.* 80:628-632.

Klaveness, D. (1977). *Hydrobiologia* 56:23-33.

Klein, D. A., J. A. Divis, and L. E. Casida (1968). *Antonie van Leeuwenhoek J. Microbiol. Serol.* 34:495-503.

Kleinmann, R. L. P., and D. A. Crerar (1979). *Geomicrobiology J.* 1:373-388.

Knoll, A. H., and E. S. Barghoorn (1975). *Science* 190:52-54.

Knoll, A. H., and E. S. Barghoorn (1977). *Science* 198:396-398.

Kobayashi, K., S. Tashibana, and M. Ishimoto (1969). *J. Biochem. (Tokyo)* 65: 155-157.

Kobayashi, K., E. Takahashi, and M. Ishimoto (1972). *J. Biochem. (Tokyo)* 72: 879-887.

Koburger, J. A. (1964). *Proc. W.Va. Acad. Sci.* 36:26-30.

Komura, I., and K. Izaki (1971). *J. Biochem. (Tokyo)* 70:885-893.

Komura, I., K. Izaki, and H. Takahashi (1970). *Agric. Biol. Chem.* 34:480-482.

Kovalskii, V. V., V. V. Ermakov, and S. V. Letunova (1968). *Mikrobiologiya* 37:122-130.

Kowallik, U., and E. G. Pringsheim (1966). *Am. J. Bot.* 53:801-806.

Kertz, R.(1972). Silicon: Element and geochemistry. In *The Encyclopedia of Geochemistry and Environmental Sciences.* Encyclopedia of Earth Sciences Series, Vol. IVA. R. W. Fairbridge, ed. Van Nostrand Reinhold, New York, pp. 1091-1092.

Kriss, A. E. (1970). *Mikrobiologiya* 39:362-371.

Krul, I.M., P. Hirsch, and J. T. Staley (1970). *Antonie van Leeuwenhoek J. Microbiol. Serol.* 36:409-420.

Krumbein, W. E. (1968). *Z. Allg. Mikrobiol.* 8:107-117.

Krumbein, W. E. (1969). *Geol. Rundsh.* 58:333-365.

Krumbein, W. E. (1971). *Naturwissenschaften* 58:56-57.

Krumbein, W. E. (1972). *Rev. Ecol. Biol. Sol* 9:283-319.

Krumbein, W. E. (1973). *Dtsch. Kunst Denkmalpfl.* 31:54-71.

Krumbein, W. E. (1974). *Naturwissenschaften* 61:167.

Krumbein, W. E. (1979). *Geomicrobiology J.* 1:139-203.

Krumbein, W. E., and H. J. Altmann (1973). *Helgolaender Wiss. Meeresunters.* 25:347-356.

Ku, T. L., and W. S. Broecker (1969). *Deep-Sea Res. Oceanogr. Abstr.* 16:625-635.

Kucera, S., and R. S. Wolfe (1957). *J. Bacteriol.* 74:344-349.

Kulibakin, V. G., and S. F. Laptev (1971). *Sb. Tr. Tsentr. Nauchno-Issled Inst. Olonyan Prom.*, no. 1, pp. 75-76.

Kullmann, K.-H., and R. Schweisfurth (1978). *Z. Allg. Mikrobiol.* 18:321-327.

Kutuzova, R. S. (1969). *Mikrobiologiya* 38:714-721.

Kuznetsov, S. I. (1975). *Soil Sci.* 119:81-88.

Kuznetsov, S. I., and V. I. Romanenko (1963). Izddatel'stvo Akad. Nauk SSSR, Moscow.

Kuznetsov, S. I., M. V. Ivanov, and N. N. N. Lyalikova (1963). *Introduction to Geological Microbiology.* English transl. McGraw-Hill, New York.

Kuznetsova, V. A., and V. M. Gorlenko (1965). *Prokl. Biokhim. Mikrobiol.* 1: 623-626.

Kvasnikov, E. I., V. V. Lipshits, and N. V. Zubova (1973). *Mikrobiologiya* 42: 925-930.

LaBerge, G. L. (1967). *Geol. Soc. Am. Bull.* 78:331-342.

Lackey, J. B. (1938). *Public Health Rep.* 53:1499-1507.

Lambert, I. B., J. McAndrew, and H. E. Jones (1971). Geochemical and bacteriological studies of the cupriferous environment at Pernatty Lagoon, South Australia. *Australas. Inst. Min. Met., Proc.*, no. 240, pp. 15-23.

Landesman, J., D. W. Duncan, and C. C. Walden (1966). *Can. J. Micribiol.* 12: 25-33.

Landner, L. (1971). *Nature (Lond.)* 230:452-454.

Lang, A. R. G. (1967). *Austr. J. Chem.* 20:2017-2023.

LaRock, P. A. (1969). The bacterial oxidation of manganese in a fresh water lake. Ph.D. thesis. Rensselaer Polytechnic Institute, Troy, N.Y.

LaRock, P.A., and H. L. Ehrlich (1975). *Microb. Ecol.* 2:84-96.

Lascelles, J., and V. A. Burke (1978). *J. Bacteriol.* 134:585-589.

Latimer, W. M., and J. H. Hildebrand (1940). *Reference Book of Inorganic Chemistry*, Rev. ed. MacMillan, New York.

Lauwers, A. M., and W. Heinen (1974). *Arch. Mikrobiol.* 95:67-78.

Lawton, K. (1955). Chemical composition of soils. In *Chemistry of the Soil.* F. E. Bear, ed. Van Nostrand Reinhold, New York, pp. 53-84.

Lazaroff, N. (1963). *J. Bacteriol.* 85:78-83.

Leadbetter, E. R., and J. W. Foster (1959). *Arch. Biochem. Biophys.* 82:491-492.

Leathen, W. W. (1953). *Proc. Pa. Acad. Sci.* 27:37-44.

Leathen, W. W., S. A. Braley, and L. D. McIntyre (1953). *Appl. Microbiol.* 1: 65-68.

Leathen, W. W., N. A. Kinsel, and S. A. Braley, Sr. (1956). *J. Bacteriol.* 72: 700-704.

Lee, J. P., and H. D. Peck (1971). *Biochem. Biophys. Res. Commun.* 45:583-589.

Lee, J. P., C. S. Yi, J. LeGall, and H. D. Peck (1973). *J. Bacteriol.* 115:453-455.

Leeper, G. W. (1947). *Soil Sci.* 63:79-94.

Leeper, G. W., and R. J. Swaby (1940). *Soil Sci.* 49:163-169.

Lees, H., S. C. Kwok, and I. Suzuki (1969). *Can. J. Microbiol.* 15:43-46.

LeGall, J., and J. R. Postgate (1973). *Adv. Microb. Physiol.* 19:81-133.

Legge, J. W. (1954). *Austr. J. Biol. Sci.* 7:504-514.

Legge, J. W., and A. W. Turner (1954). *Austr. J. Biol. Sci.* 7:496-503.

Lehninger, A. L. (1970). *Biochemistry,* 1st ed. Worth, New York.

Lehninger, A. L. (1975). *Biochemistry,* 2nd ed. Worth, New York.

LeRoux, N., D. S. Wakerley, and S. D. Hunt (1977). *J. Gen. Microbiol.* 100: 197-201.

Lester, R. L., and J. A. DeMoss (1971). *J. Bacteriol.* 105:1006-1014.

Lewin, J. C. (1957). *J. Gen. Physiol.* 39:1-10.

Lewin, J. C. (1965a). Calcification. In *Physiology and Biochemistry of Algae.* R. A. Lewin, ed. Academic Press, New York, pp. 457-465.

Lewin, J. C. (1965b). Silicification. In *Physiology and Biochemistry of Algae.* R. A. Lewin, ed. Academic Press, New York, pp. 447-455.

Licari, G. R., and P. Cloud (1968). *Proc. Natl. Acad. Sci. U.S.A.* 59:1053-1060.

Liebert, F. (1927). *Zentralbl. Bakteriol. Parasitenkd. Infektionskr. Hyg.* 72: 369-374.

Lieske, R. (1919). *Zentralbl. Bakteriol. Parasitenkd. Infektionskr. Hyg. Abt. II* 49:413-425.

Lipman, C.-B. (1929). Further studies on marine bacteria with special reference to the Drew hypotheses on $CaCO_3$ precipitation in the sea. *Carnegie Inst. Wash. Publ. 391, Pap. Tort. Lab.* 26:231-248.

Lipman, C. B. (1931). *J. Bacteriol.* 22:183-198.

Lipman, J. G., and S. A. Waksman (1923). *Science* 57:60.

Lipman, J. G., H. McLean, and H. C. Lint (1916). *Soil Sci.* 1:533-539.

Ljunggren, P. (1960). *Econ. Geol.* 55:531-538.

Loewenstam, H. A., and S. Epstein (1957). *J. Geol.* 65:364-375.

Loganathan, P., and R. G. Burau (1973). *Geochim. Cosmochim. Acta* 37:1277-1293.

London, J. (1963). *Arch. Mikrobiol.* 46:329-337.

London, J., and S. C. Rittenberg (1964). *Proc. Natl. Acad. Sci. U.S.A.* 52:1183-1190.

London, J., and S. C. Rittenberg (1966). *J. Bacteriol.* 91:1062-1069.

London, J., and S. C. Rittenberg (1967). *Arch. Mikrobiol.* 59:218-225.

Loutit, J. S. (1970). *Genet. Res.* 16:179-184.

Love, L. G. (1962). *Econ. Geol.* 57:350-366.

Lowe, D. R. (1980). *Nature (Lond.)* 284:441-443.

Lunde, G. (1973). *Acta Chem. Scand.* 27:1586-1594.

Lundgren, D. G., K. G. Anderson, C. C. Remsen, and R. P. Mahoney (1964). *Dev. Ind. Microbiol.* 6:250-259.

Lundgren, D. G., J. R. Vestal, and F. R. Tabita (1972). The microbiology of mine drainage pollution. In *Water Pollution Microbiology.* R. Mitchell, ed. Wiley-Interscience, New York, pp. 69-88.

Lyalikova, N. N. (1958). *Mikrobiologiya* 27:556-559.

Lyalikova, N. N. (1960). *Mikrobiologiya* 29:382-387.

Lyalikova, N. N. (1961). *Tr. Inst. Mikrobiol. Akad. Nauk SSSR*, no. 9, pp. 134-143.

Lyalikova, N. N. (1972). *Dokl. Akad. Nauk SSSR, Ser. Biol.* 205:1228-1229.

Lyalikova, N. N. (1974). *Mikrobiologiya* 43:941-948.

Lyalikova, N. N., L. B. Shlain, O. G. Unanova, and L. S. Anisimova (1972). *Izv. Akad. Nauk. SSSR, Ser. Biol.* 4:564-567.

Lyalikova, N. N., I. Ya. Vedemina, and A. K. Romanova (1976). *Mikrobiologiya* 45:552-554.

Lyalikova, N. N., L. B. Shlain, and V. G. Trofimov (1974). *Izv. Akad. Nauk SSSR, Ser. Biol.* No. 3, 440-444.

McBride, B. C., and R. S. Wolfe (1971). *Biochemistry* 10:4312-4317.

McConnell, D. (1965). *Econ. Geol.* 60:1059-1062.

McGoran, C. J. M., D. W. Duncan, and C. C. Walden (1969). *Can. J. Microbiol.* 15:135-138.

Maciag, W. J., and D. G. Lundgren (1964). *Biochem. Biophys. Res. Commun.* 17:603-607.

McKenna, E. J., and R. E. Kallio (1964). Hydrocarbon structure: its effect on bacterial utilization of alkanes. In *Principles and Applications in Aquatic Microbiology.* H. Heukelakian and N. C. Dondero, eds. Wiley, New York, pp. 1-14.

McKenna, M. C. (1972). *BioScience* 22:519-525.

Mackenzie, F. T., M. Stoffyn, and R. Wollast (1978). *Science* 199:680-682.

Mackintosh, M. E. (1978). *J. Gen. Microbiol.* 105:215-218.

MacLeod, R. A. (1965). *Bacteriol. Rev.* 29:9-23.

Macnamara, J., and G. H. Thode (1951). *Research* 4:582-583.

MacRae, I. S., and J. S. Celo (1975). *Appl. Microbiol.* 29:837-840.

Madigan, M. T., and T. D. Brock (1975). *J. Bacteriol.* 122:782-784.

Madigan, M. T., and H. Gest (1978). *Arch. Microbiol.* 117:119-122.

Magne, R., J. Berthelin, and Y. Dommergues (1973). *C.R. Acad. Sci. (Paris) Ser. D* 276:2625-2628.

Magne, R., J. R. Berthelin, and Y. Dommergues (1974). Solubilisation et insolubilisation de l'uranium des granites par des bactéries hétérotrophes. In *Formation of Uranium Ore Deposits.* International Atomic Energy Agency, Vienna, pp. 73-88.

Mah, R. A., M. R. Smith, and L. Baresi (1978). *Appl. Environ. Microbiol.* 35:1174-1184.

Mahler, H. R., and E. H. Cordes (1966). *Biological Chemistry.* Harper & Row, New York.

Malacinski, G., and W. A. Konetzka (1966). *J. Bacteriol.* 91:578-582.

Malacinski, G. M., and W. A. Konetzka (1967). *J. Bacteriol.* 93:1906-1910.

Malouf, E. E., and J. D. Prater (1961). *J. Met.* 13:353-356.

Mann, P. J. G., and J. H. Quastel (1946). *Nature (Lond.)* 158:154-156.

Marchlewitz, B., and W. Schwartz (1961). *Z. Allg. Mikrobiol.* 1:100-114.

Margalith, P., M. Silver, and D. G. Lundgren (1966). *J. Bacteriol.* 92:1706-1709.

Margolis, J. V., and R. G. Burns (1976). *Annu. Rev. Earth Planet. Sci.* 4:229-263.

Margulis, L. (1970). *Origin of Eukaryotic Cells.* Yale University Press, New Haven, Conn.

Marine Chemistry (1971). A report of the Marine Chemistry Panel of the Committee of Oceanography. National Academy of Sciences, Washington, D.C.

Markosyan, G. E. (1972). *Biol. Zh. Arm.* 25:26.

Markosyan, G. E. (1973). *Dokl. Akad. Nauk SSSR, Ser. Biol.* 211:1205-1208.

Marshall, K. C. (1971). Sorption interactions between soil particles and microorganisms. In *Soil Biochemistry,* Vol. 2. A. D. McLaren and J. Skujins, eds. Marcel Dekker, New York, pp. 409-445.

Martens, C. S. (1976). *Science* 192:998-1000.

Martens, C. S., and R. A. Berner (1974). *Science* 185:1167-1169.

Martin, J. P., and S. A. Waksman (1940). *Soil Sci.* 50:29-47.

Martin, J. P., and S. A. Waksman (1941). *Soil Sci.* 52:381-394.

Matsumara, F. Y., Gotoh, and G. M. Brush (1971). *Science* 173:49-51.

Matthews, S. W. (1973). *Natl. Geogr. Mag.* 143:1-37.

Mechalas, B. J., and S. C. Rittenberg (1960). *J. Bacteriol.* 80:501-507.

Menzel, D. W., and J. H. Ryther (1970). Distribution and cycling of organic matter in the oceans. In *Organic Matter in Natural Waters*. D. W. Wood, ed. Inst. Mar. Sci. Occas. Publ. no. 1. University of Alaska, Fairbanks, Alaska, pp. 31-54.

Mercy, E. (1972). Mantle geochemistry. In *The Encyclopedia of Geochemistry and Environmental Sciences*. Encyclopedia of Earth Sciences Series, Vol. IVA. R. W. Fairbridge, ed. Van Nostrand Reinhold, New York, pp. 677-683.

Merkle, F. G. (1955). Oxidation-reduction processes in soils. In *Chemistry of the Soil*. F. E. Bear, ed. Van Nostrand Reinhold, New York, pp. 200-218.

Mero, J. L. (1962). *Econ. Geol.* 57:747-767.

Merrill, G. P. (1895). *Geol. Soc. Am. Bull.* 6:321-332.

Milhaud, G., J.-P. Aubert, and J. Millet (1958). *C.R. Acad. Sci. (Paris)* 246: 1766-1769.

Miller, L. P. (1950). *Boyce Thompson Inst. Contr.* 16:85-89.

Miller, S. L., and L. E. Orgel (1974). *The Origin of Life on Earth*. Prentice-Hall, Englewood Cliffs, N.J.

Milliman, J. D., and E. Boyle (1975). *Science* 189:995-997.

Mineral Facts and Problems (1965). U.S. Bur. Mines Bull. 630.

Minshall, G. W. (1978). *BioScience* 78:767-771.

Mitchell, R. L. (1955). Trace elements. In *Chemistry of the Soil*. F. E. Bear, ed. Van Nostrand Reinhold, New York, pp. 253-285.

Moese, J. R., and H. Brantner (1966). *Zentralbl. Bakteriol. Parasitenkd. Infektionskr. Hyg. Abt. II* 120:480-495.

Molisch, H. (1910). *Die Eisenbakterien*. Gustav Fischer, Jena, East Germany.

Monty, C. L. V. (1967). *Ann. Soc. Geol. Belgique* 90:55-99.

Monty, C. L. V. (1972). *Geol. Rundsch.* 61:742-783.

Moore, L. R. (1969). Geomicrobiology and geomicrobial attack on sedimented organic matter. In *Organic Geochemistry: Methods and Results*. pp. 264-303. G. Eglinton and M. T. J. Murphy, eds. Springer-Verlag, New York.

Moore, W. E. C., J. L. Johnson, and L. V. Holdeman (1976). *Int. J. Syst. Bacteriol.* 26:238-252.

Moriarty, D. J. W., and D. J. D. Nicholas (1969). *Biochim. Biophys. Acta* 184: 114-123.

Moriarty, D. J. W., and D. J. D. Nicholas (1970a). *Biochim. Biophys. Acta* 197: 143-151.

Moriarty, D. J. W., and D. J. D. Nicholas (1970b). *Biochim. Bipphys. Acta* 216: 130-138.

Morita, R. Y. (1967). *Oceanogr. Mar. Biol. Annu. Rev.* 5:187-203.

Morita, R. Y. (1975). *Bacteriol. Rev.* 39:144-167.

Moshuyakova, S. A., G. I. Karavaiko, and E. V. Shchetinia (1971). *Mikrobiologiya* 40:1100-1107.

Mosser, J. L., A. G. Mosser, and T. D. Brock (1973). *Science* 179:1323-1324.

Mosser, J. L., B. B. Bohlool, and T. D. Brock (1974). *J. Bacteriol.* 118:1075-1081.

Mottl, M. J., H. D. Holland, and R. F. Corr (1979). *Geochim. Cosmochim Acta* 43:869-884.

Muentz, A. (1890). *C.R. Acad. Sci. (Paris)* 110:1370-1372.

Munch, J. C., and J. C. G. Ottow (1977). *Z. Pflanzenernaehr. Bodenkd.* 140: 549-562.

Murr, L. E., and J. A. Brierley (1978). The use of large-scale test facilities in studies of the role of microorganisms in commercial leaching operations. In *Metallurgical Applications of Bacteria.* Academic Press, New York, pp. 491-520.

Murray, J., and R. Irvine (1894). *Edinb. R. Soc. Trans.* 37:721-742.

Myers, D. J., M. E. Heimbrook, J. Osteryoung, and S. M. Morrison (1973). *Environ. Lett.* 5:53-61.

Nadson, G. A. (1903). Microorganisms as geologic agents. I. *Tr. Komissii Isslect. Min. Vod g. Slavyanska,* St. Petersburg.

Nadson, G. A. (1928). *Arch. Hydrobiol.* 19:154-164.

Nagai, S. (1965). *J. Bacteriol.* 90:220-222.

Nathansohn, A. (1902). *Mitt. Zool. Sta. Neapel* 15: 655-680.

Nazina, T. N., and E. P. Rozanova (1978). *Mikrobiologiya* 47:142-148.

Nealson, K. H., and J. Ford (1980). *Geomicrobiol. J.* 2:21-37.

Neilands, J. B., ed. (1974). *Microbial Iron Metabolism.* Academic Press, New York.

Nelson, J. D., Jr., and R. R. Colwell (1974). Metabolism of mercury compounds by bacteria in Chesapeake Bay. *Proc. 3rd Int. Congr. Mar. Corros. Fouling, 1972.* R. F. Acker, B. F. Brown, and J. R. de Palma, eds. Northwestern University Press, Evanston, Ill., pp. 767-777.

Nelson, J. D., and R. R. Colwell (1975). *Microb. Ecol.* 1:191-218.

Nelson, J. D., W. Blair, F. E. Brinckman, R. R. Colwell, and W. P. Iverson (1973). *Appl. Microbiol.* 26:321-326.

Nesterov, A. I., E. M. Moskalenko, I. A. Molchanov, and E. D. Chermisina (1971). *Prkl. Biokhim. Mikrobiol.* 7:305-309.

Nette, I. T., N. N. Grechushkina, and I. L. Rabotnova (1965). *Prikl. Biokhim. Mikrobiol.* 1:167-174.

Neufeld C. A. (1904). *Unters. Nahrungs. Genussmittel* 7:478.

Nickerson, W. J., and G. Falcone (1963). *J. Bacteriol.* 85:763-771.

Nielsen, A. M., and J. V. Beck (1972). *Science* 175:1124-1126.

Nierenberg, W. A. (1978). *Am. Sci.* 66:20-29.

Novick, R. P. (1967). *Fed. Proc.* 26:29-38.

Omeliansky, W. (1906). *Zentralbl. Bakteriol. Parasitenkd. Infektionskr. Hyg. Abt. II* 15:673.

Oparin, A. I. (1938). *The Origin of Life.* MacMillan, New York.

Oremland, R. S., and B. F. Taylor (1978). *Geochim. Cosmochim. Acta* 42:209-214.

Oren, A., and M. Shilo (1979). *Arch. Microbiol.* 122:77-84.

Osborne, F. H. (1973). Arsenite oxidation by a soil isolate of *Alcaligenes.* Ph.D. thesis. Rensselaer Polytechnic Institute, Troy, N.Y.

Osborne, F. H., and H. L. Ehrlich (1976). *J. Appl. Bacteriol.* 41:295-305.

Ottow, J. C. G. (1968). *Z. Allg. Mikrobiol.* 8:441-443.

Ottow, J. C. G. (1969a). *Z. Pflanzenernaehr. Bodenkd.* 124:238-253.

Ottow, J. C. G. (1969b). *Zentralbl. Bakteriol. Parasitenkd. Infektionskr. Hyg. Abt. II* 123:600-615.

Ottow, J. C. G. (1970a). *Z. Allg. Mikrobiol.* 10:55-62.

Ottow, J. C. G. (1970b). *Nature (Lond.)* 225:103.

Ottow, J. C. G. (1971). *Oecologia* 6:164-175.

Ottow, J. C. G., and H. Ottow (1970). *Zentralbl. Bakteriol. Parasitenkd. Infektionskr. Hyg. Abt. II* 124:314-318.

Ottow, J. C. G., and A. von Klopotek (1969). *Appl. Microbiol.* 18:41-43.

Padan, E. (1979). *Adv. Microb. Ecol.* 3:1-48.

Paine, S. G., F. V. Lingood, F. Schimmer, and T. C. Thrupp (1933). *R. Soc. Phil. Trans.* 222B:97-127.

Palacas, J. G., V. E. Swanson, and G. W. Moore (1966). Organic geochemistry of three north Pacific deep sea sediment samples. *U.S. Geol. Surv. Prof. Pap. 550C,* pp. C102-C107.

Palmer, A. R. (1974). *Am. Sci.* 62:216-224.

Panganiban, A., and R. S. Hanson (1976). *Abstr. Annu. Meet. Am. Soc. Microbiol.,* p. 121 (I59).

Panganiban, A. T., Jr., T. E. Patt, W. Hart, and R. S. Hanson (1979). *Appl. Environ. Microbiol.* 37:303-309.

Parès, Y. (1964a). *Ann. Inst. Pasteur (Paris)* 107:132-135.

Parès, Y. (1964b). *Ann. Inst. Pasteur (Paris)* 107:136-141.

Parès, Y. (1964c). *Ann. Inst. Pasteur (Paris)* 107:141-143.

Park, K. (1966). *Science* 154:1540-1542.

Park, P. K. (1968). *Science* 162:357-358.

Parker, C. D. (1947). *Nature (Lond.)* 159:439.

Paschinger, H., J. Paschinger, and H. Gaffron (1974). *Arch. Mikrobiol.* 96:341-351.

Patel, R. N., H. R. Bose, W. W. J. Mandy, and D. S. Hoare (1972). *J. Bacteriol.* 110:570-577.

Patel, R. N., C. T. Hon, and A. Felix (1978). *J. Bacteriol.* 136:352-358.

Patrick, W. H., S. Gotch, and B. G. Williams (1973). *Science* 179:564-565.

Patt, T. E., G. C. Cole, J. Bland, and R. S. Hanson (1974). *J. Bacteriol.* 120: 955-964.

Peeters, T., and M. I. H. Aleem (1970). *Arch. Mikrobiol.* 71:319-330.

Perfil'ev, B. V., and D. R. Gabe (1965). The use of the microbial-landscape method to investigate bacteria which concentrate manganese and iron in bottom deposits. In *Applied Capillary Microscopy: The Role of Microorganisms in the Formation of Iron-Manganese Deposits.* B. V. Perfil'ev, D. R. Gabe, A. M. Gal'perina, V. A. Rabinovich, A. A. Saponitskii, E. E. Sherman, and E. P. Troshanov, eds. Consultants Bureau, New York, pp. 9-54.

Perfil'ev, B. V., and D. R. Gabe (1969). *Capillary Methods of Studying Microorganisms.* English trans. by J. Shewan. University of Toronto Press, Toronto.

Perkins, E. C., and F. Novielli (1962). Bacterial leaching of manganese ores. *U.S. Bur. Mines Rep. Invest. 6102.* 11 pp.

Pettijohn, F. J. (1949). *Sedimentary Rocks.* Harper and Row, New York.

Pfennig, N. (1977). *Annu. Rev. Microbiol.* 31:275-290.

Pfennig, N., and H. Biebl (1976). *Arch. Microbiol.* 110:3-12.

Pflug, H. D. (1978). *Naturwissenschaften* 65:611-615.

Pflug, H. D., and H. Jaeschke-Boyer (1979). *Nature (Lond.)* 280:483-486.

Phillips, S. E., and M. L. Taylor (1976). *Appl. Environ. Microbiol.* 32:392-399.

Phillips, W. E., Jr., and J. J. Perry (1974). *J. Bacteriol.* 120:987-989.

Pierson, B. K., and R. W. Castenholz (1974). *Arch. Microbiol.* 100:5-24.

Pichinoty, F. (1963). *Ann. Inst. Pasteur (Paris)* 104:394-418.

Pinsent, J. (1954). *Biochem. J.* 57:10-16.

Piper, D. Z., and L. A. Codespoti (1975). *Science* 188:15-18.

Pirnik, M. P., R. M. Atlas, and R. Bartha (1974). *J. Bacteriol.* 119:868-878.

Poldervaart, A. (1955). Chemistry of the earth's crust. In *Crust of the Earth: A Symposium.* A. Poldervaart, ed. Spec. Pap. 62. Geological Society of America, Boulder, Colo., pp. 119-144.

Pol'kin, S. I., and Z. A. Tanzhuyanskaya (1968). *Izv. Vyssh. Uchebn. Zaved. Tsvetn. Metall.* 11:115-121.

Pol'kin, S. I., I. N. Yudina, V. V. Nanin, and D. Kh. Kim (1973). *Nauchno-Issled. Geologorazved. Inst. Tsvetn. Blagorodn. Metall.,* no. 107, pp. 34-41.

Pollack, J. R., and J. B. Neilands (1970). *Biochem. Biophys. Res. Commun.* 38: 989-992.

Popoff, L. (1875). *Arch. Ges. Physiol.* 10:142.

Postgate, J. R. (1951). *J. Gen. Microbiol.* 5:725-738.

Postgate, J. R. (1952). *Research* 5:189.

Postgate, J. R. (1963). *J. Bacteriol.* 85:1450-1451.

Praeve, P. (1957). *Arch Mikrobiol.* 27:33-62.

Pramer, D. (1964). *Science* 144:382-388.

Pringsheim, E. G. (1949). *Phil. Trans. R. Soc. (Lond.) Ser. B: Biol. Sci.* 233: 453-482.

Pringsheim, E. G. (1967). *Arch. Mikrobiol.* 59:247-254.

Puchelt, H., H. H. Schock, E. Schroll, and H. Hanert (1973). *Geol. Rundsch.* 62: 786-812.

Quastel, J. H., and P. G. Schoenfield (1953). *Soil Sci.* 75:279-285.

Quayle, J. R. (1972). *Adv. Microb. Physiol.* 7:119-203.

Quayle, J. R., and T. Ferenci (1978). *Microbiol. Rev.* 42:251-273.

Racz, G. J. and N. K. Savant (1972) *Soil Sci. Soc. Amer. Proc.* 36:678-682.

Rankama, K., and T. G. Sahama (1950). *Geochemistry.* University of Chicago Press, Chicago, pp. 657-676.

Rapp, G., Jr. (1972). Selenium: element and geochemistry. In *The Encyclopedia of Geochemistry and Environmental Sciences.* Encyclopedia of Earth Sciences Series, Vol. IVA. R. W. Fairbridge, ed. Van Nostrand Reinhold, New York, pp. 1079-1080.

Raven, P. H., and D. I. Axelrod (1972). *Science* 176:1379-1386.

Raymond, R. L., V. W. Jamison, and J. O. Hudson (1967). *Appl. Microbiol.* 15: 857-865.

Razzell, W. E., and P. C. Trussell (1963). *Appl. Microbiol.* 11:105-110.

Redden, J. A., and H. C. Porter (1962). *Min. Ind. J.* 9:1-4.

Reid, A. S. J., and M. H. Miller (1963). *Can. J. Soil Sci.* 43:250-259.

Reid, G. K. (1961). *Ecology of Inland Waters and Estuaries.* Van Nostrand Reinhold, New York.

Remsen, C., and D. G. Lundgren (1966). *J. Bacteriol.* 92:1765-1771.

Richmond, M. H., and M. John (1964). *Nature (Lond.)* 202:1360-1361.

Rickard, D. T. (1973). *Econ. Geol.* 68:605-617.

Rickard, P. A. D., and D. G. Vanselow (1978). *Can. J. Microbiol.* 24:998-1003.

Rittenberg, S. C. (1940). *J. Mar. Res.* 3:191-201.

Rittenberg, S. C. (1969). *Adv. Microb. Physiol.* 3:159-196.

Robbins, J. A., and E. Callender (1975). *Am. J. Sci.* 275:512-533.

Roberts, J. L. (1947). *Soil Sci.* 63:135-140.

Robinson, W. O. (1929). *Soil Sci.* 27:335-350.

Roemer, R., and W. Schwartz (1965). *Z. Allg. Mikrobiol.* 5:122-135.

Rogall, E. (1939). *Planta* 29:279-291.

Rogoff, M. H., I. Wander, and R. B. Anderson (1962). Microbiology of coal. *U.S. Bur. Mines Inform. Circ. 8075.*

Ronov, A. B., and A. A. Yaroshevsky (1972). Earth's crust geochemistry. In *The Encyclopedia of Geochemistry and Environmental Sciences.* Encyclopedia of Earth Sciences Series, Vol. IVA. R. W. Fairbridge, ed. Van Nostrand Reinhold, New York, pp. 243-254.

Rosenfeld, I., and O. A. Beath (1964). *Selenium, Geobotany, Biochemistry, Toxicity, and Nutrition.* Academic Press, New York.

Rossman, R., and E. Callender (1968). *Science* 162:1123-1124.

Rostikova, V. I. (1961). *Pochvovendenie* 4:82-90.

Roth, C. W., W. P. Hempfling, J. N. Connors, and W. V. Vishniac (1973). *J. Bacteriol.* 114:592-599.

Rotruck, J. T., A. L. Pope, H. E. Ganther, A. B. Swanson, D. G. Hafeman, and W. G. Hockstra (1973). *Science* 179:588-590.

Röttger, R. (1976). *Mikrokosmos* 65, no. 12.

Roy, A. B., and P. A. Trudinger (1970). *The Biochemistry of Inorganic Compounds of Sulphur.* Cambridge University Press, New York.

Rozanova, E. P. (1971). *Mikrobiologiya* 40:152-157.

Rozanova, E. P., and L. D. Shturm (1965). *Mikrobiologiya* 34:888-894.

Roze, E. (1896). *J. Bott. (Paris)* 10:325-330.

Rubenstein, I., O. P. Strausz, C. Spyckerelle, R. J. Crawford, and D. W. S. Westlake (1977). *Geochim. Cosmochim. Acta* 41:1341-1353.

Rudakov, K. J. (1927). *Zentralbl. Bakteriol. Parasitenkd. Infektionskr. Hyg. Abt. II* 79:229-245.

Sand, M. D., P. A. LaRock, and R. E. Hodson (1975). *Appl. Microbiol.* 29: 626-634.

Sandon, H. (1928). *Soil Sci.* 25:107-119.

Sapozhnikov, D. I. (1937). *Mikrobiologiya* 6:643-644.

Sartory, A., and J. Meyer (1947). *C.R. Acad. Sci. (Paris)* 225:541-542.

Sato, M. (1960). *Econ. Geol.* 55:1202-1231.

Schmalz, R. R. (1972). Calcium carbonate geochemistry. In *The Encyclopedia of Geochemistry and Environmental Sciences.* Encyclopedia of Earth Sciences Series, Vol. IVA. R. W. Fairbridge, ed. Van Nostrand Reinhold, New York, p. 110.

Schmidt, E. L., and R. O. Bankole (1965). *Appl. Microbiol.* 13:673-679.

Schnaitman, C. A., and D. G. Lundgren (1965). *Can. J. Microbiol.* 11:23-27.

Schnaitman, C. A., M. S. Korczinski, and D. G. Lundgren (1969). *J. Bacteriol.* 99:552-557.

Schoen, R., and G. G. Ehrlich (1968). Bacterial origin of sulfuric acid in sulfurous hot springs. *23rd Int. Geol. Congr.* 17:171-178.

Schoen, R., and R. O. Rye (1970). *Science* 170:1082-1084.

Schoettle, M., and G. M. Friedman (1971). *Geol. Soc. Am. Bull.* 82:101-110.

Schopf, J. W. (1978). *Am. Sci.* 238:110-139.

Schopf, J. W., and E. S. Barghoorn (1967). *Science* 156:508-512.

Schopf, J. W., and D. Z. Oehler (1976). *Science* 193:47-49.

Schopf, J. W., E. S. Barghoorn, M. D. Maser, and R. O. Gordon (1965). *Science* 149:1365-1367.

Schottel, J. A., D. Mandel, S. Clark, S. Silver, and R. W. Hodges (1974). *Nature (Lond.)* 251:335-337.

Schrauzer, G. N., J. H. Wever, T. M. Beckham, and R. K. Y. Ho (1971). *Tetrahedron Lett.* 3:275-277.

Schulz-Baldes, A., and R. A. Lewin (1975). *Science* 188:1119-1120.

Schwarz, J. R., and R. R. Colwell (1976). *Dev. Ind. Microbiol.* 17:299-310.

Schweisfurth, R. (1968). *Int. Verein. Limnol.* 14:179-186.

Schweisfurth, R. (1969). *Zentralbl. Bakteriol. Parasitenkd. Infektionskr. Hyg. Abt. I Orig.* 212:486-491.

Schweisfurth, R. (1971). *Z. Allg. Mikrobiol.* 11:415-430.

Schweisfurth, R. (1973). *Z. Allg. Mikrobiol.* 13:341-347.

Schweisfurth, R., and R. Mertes (1962). *Arch. Hyg. Bakteriol.* 146:401-417.

Segel, I. H. (1975). *Enzyme Kinetics: Behavior and Analysis of Rapid Equilibrium and Steady-State Enzyme Systems.* Wiley, New York.

Sellwood, B. W. (1971). *J. Sediment. Petrol.* 41:854-858.

Serkies, J., J. Oberc, and A. Idzekowski (1968). *Chem. Geol.* 2:217-232.

Shafia, F., and R. F. Wilkinson (1969). *J. Bacteriol.* 97:256-260.

Shafia, F., K. R. Brinson, M. W. Heinzman, and J. M. Brady (1972). *J. Bacteriol.* 111:56-65.

Shapiro, J. (1967). *Science* 155:1269-1271.

Sherman, G. D., J. S. McHargue, and W. S. Hodgkin (1942). *Soil Sci.* 54:253-257.

Shinano, H. (1972a). *Bull. Jap. Soc. Sci. Fish.* 38:717-725.

Shinano, H. (1972b). *Bull. Jap. Soc. Sci. Fish.* 38:727-732.

Shinano, H., and M. Sakai (1969). *Bull. Jap. Soc. Sci. Fish.* 35:1001-1005.

Shivvers, D. W., and T. D. Brock (1973). *J. Bacteriol.* 114:706-710.

Shrift, A. (1964). *Nature (Lond.)* 201:1304-1305.

Shturm, L. D. (1948). *Mikrobiologiya* 17:415-418.

Shum, A. C., and J. C. Murphy (1972). *J. Bacteriol.* 110:447-449.

Sieburth, J. McN. (1975). *Microbial Seascapes: A Pictorial Essay on Marine Microorganisms and Their Environments.* University Park Press, Baltimore, Md.

Sieburth, J. McN. (1976). *Annu. Rev. Ecol. Syst.* 7:259-285.

Sieburth, J. McN., J.-P. Willis, K. M. Johnson, C. M. Burney, D. Lavoie, K. R. Hinga, D. A. Caron, F. W. French, III, P. W. Johnson, and P. G. Davis (1976). *Science* 194:1415-1418.

Sienko, M. J., and R. A. Plane (1966). *Chemistry: Principles and Properties.* McGraw-Hill, New York.

Sillén, L. G. (1967). *Science* 156:1189-1197.

Silver, M., and A. E. Torma (1974). *Can. J. Microbiol.* 20:141-147.

Silverman, M. P., and H. L. Ehrlich (1964). *Adv. Appl. Microbiol.* 6:153-206.

Silverman, M. P., and D. G. Lundgren (1959a). *J. Bacteriol.* 77:642-647.

Silverman, M. P., and D. G. Lundgren (1959b). *J. Bacteriol.* 78:326-331.

Silverman, M. P., and E. F. Munoz (1970). *Science* 169:985-987.

Silverman, M. P., M. H. Rogoff, and I. Wender (1961). *Appl. Microbiol.* 9:491-496.

Simakova, T. L., Z. A. Kolesnik, N. V. Strigaleva, et al. (1968). *Mikrobiologiya* 37:233-238.

Singer, P. C., and W. Stumm (1970). *Science* 167:1121-1123.

Skerman, V. B. D., and S. M. Bromfield (1949). *Nature (Lond.)* 163:575.

Skerman, V. B. D., G. Dementyeva, and B. Carey (1957a). *J. Bacteriol.* 73:504-512 512.

Skerman, B. V. D., G. Dementyeva, and G. W. Skyring (1957b). *Nature (Lond.)* 179:742.

Skinner, F. A. (1968). The anaerobic bacteria of soil. In *Ecology of Bacteria.* T. R. G. Gray and D. Parkinson, eds. University of Toronto Press, Toronto, pp. 573-592.

Skirrow, G. (1975). The dissolved gases—carbon dioxide. In *Chemical Oceanography,* Vol 2, 2nd ed. J. P. Riley and G. Skirrow, eds. Academic Press, New York, pp. 144-152.

Skujins, J. J. (1967). Enzymes in soil. In *Soil Biochemistry,* Vol. 1. A. D. McLaren and G. H. Peterson, eds. Marcel Dekker, New York, pp. 371-414.

Smith, D. H. (1967). *Science* 156:1114-1116.

Smith, D. W., and S. C. Rittenberg (1974). *Arch. Mikrobiol.* 100:65-71.

Smith, K. L., Jr., and J. M. Teal (1973). *Science* 179:282-283.

Smith, M. R., and R. A. Mah (1978). *Appl. Environ. Microbiol.* 36:870-879.

Smith, R. L. (1968). *Mar. Biol. Annu. Rev.* 6:11-46.

Smith, W., D. H. Pope, and J. V. Landau (1975). *J. Bacteriol.* 124:582-584.

Soehngen, N. L. (1906). Het outstaan en verdwijnen van waterstof en methaan ouder den invloed van het organische leven. Thesis. Technical University, Delft (Vis, Jr., Delft, publisher).

Soehngen, N. L. (1914). *Zentralbl. Bakteriol. Parasitenkd. Infektionskr. Hyg. Abt. II* 40:545-554.

Sokolova, A. G. (1962). *Mikrobiologiya* 31:324-327.

Sokolova, G. A., and G. I. Karavaiko (1968). *Physiology and Geochemical Activity of Thiobacilli.* English transl. U.S. Department of Commerce. U.S. Clearinghouse, Fed. Sci. Tech. Inf., Springfield, Va.

Sokolova-Dubinina, G. A., and Z. P. Deryugina (1967a). *Mikrobiologiya* 36:535-542.

Sokolova-Dubinina, G. A., and Z. P. Deryugina (1967b). *Mikrobiologiya* 36:1066-1076.

Sokolova-Dubinina,G. A., and Z. P. Deryugina (1968). *Mikrogiologiya* 37:147-153.

Sorem, R. K., and A. R. Foster (1972). Internal structure of manganese nodules and implications in beneficiation. In *Ferromanganese Deposits on the Ocean Floor.* D. R. Horn, ed. Office of the International Decade of Ocean Exploration. National Science Foundation, Washington, D. C., pp. 167-181.

Sørensen, J., B. B. Jørgensen, and N. P. Revsbech (1979). *Microb. Ecol.* 5:105-115.

Sorokin, Yu. I. (1966a). *Nature (Lond.)* 210:551-552.

Sorokin, Yu. I. (1966b). *Mikrobiologiya* 35:761-766.

Sorokin, Yu. I. (1966c). *Mikrobiologiya* 35:967-977.

Sorokin, Yu. I. (1966d). *Dokl. Akad. Nauk SSSR* 168:199.

Sorokin, Yu. I. (1970). *Mikrobiologiya* 39:253-258.

Sorokin, Yu. I. (1971). *Mikrobiologiya* 40:563-566.

Spangler, W. J., J. L. Spigarelli, J. M. Rose, and H. M. Miller (1973a). *Science* 180:192-193.

Spangler, W. J., J. L. Spigarelli, J. M. Rose, R. S. Fillipin, and H. H. Miller (1973b). *Appl. Microbiol.* 25:488-493.

Spencer, W. F., R. Patrick, and H. W. Ford (1963). *Soil Sci. Soc. Am. Proc.* 27:134.

Sperber, J. I. (1958a). *Austr. J. Agric. Res.* 9:778-781.

Sperber, J. I. (1958b). *Nature (Lond.)* 181:934.

Stadtman, T. C. (1974). *Science* 183:915-922.

Standard Methods for the Examination of Water and Wastewater Including Bottom Sediments and Sludges. (1965). 12th ed. Prepared and published jointly by American Public Health Association, American Water Works Association, and Water Pollution Control Federation. American Public Health Association, New York.

Stanier, R. Y., E. A. Adelberg, and J. L. Ingraham (1976). *The Microbial World,* 4th ed. Prentice-Hall, Englewood Cliffs, N.J.

Stanton, R. L. (1972). Sulfides in sediments. In *The Encyclopedia of Geochemistry and Environmental Sciences.* Encyclopedia of Earth Sciences Series, Vol. IVA. R. W. Fairbridge, ed. Van Nostrand Reinhold, New York, pp. 1134-1141.

Starkey, R. L. (1934). *J. Bacteriol.* 28:387-400.

Starkey, R. L., and H. O. Halvorson (1927). *Soil Sci.* 24:381-402.

Steinmann, G. (1899-1901). *Ber. Naturforsch. Ges. Freiburg I. Br.* 11:40-45.

Stevenson, I. L. (1967). *Can. J. Microbiol.* 13:205-211.

Stewart, J. E., R. E. Kallio, D. P. Stevenson, A. C. Jones, and D. O. Schissler (1959). *J. Bacteriol.* 78:441-448.

Stoeckenius, W. (1976). *Sci. Am.* 234:38-46.

Stokes, J. L. (1954). *J. Bacteriol.* 67:278-291.

Stotzky, G. (1966a). *Can. J. Microbiol.* 12:831-848.

Stotzky, G. (1966b). *Can. J. Microbiol.* 12:1235-1246.

Stotzky, G., and L. T. Rem (1966). *Can. J. Microbiol.* 12:547-563.

Stotzky, G., and L. T. Rem (1967). *Can. J. Microbiol.* 13:1535-1550.

Straat, P. A., H. Woodrow, R. L. Dimmick, and M. H. Chatigny (1977). *Appl. Environ. Microbiol.* 34:292-296.

Strahler, A. N. (1977). *Principles of Physical Geology.* Harper & Row, New York.

Strohl, W. R., and J. M. Larkin (1978). *Appl. Environ. Microbiol.* 36:755-770.

Strom, T., T. Ferensi, and J. R. Quayle (1974). *Biochem. J.* 144:465-476.

Stutzer, A., and R. Hartleb (1899). *A. Agnew. Chem.* 12:402.

Stutzer, O. (1911). Die wichtigsten Lagerstaetten der Nicht-Erze. Part I. Berlin.

Sullivan, C. W. (1971). A silicic acid requirement for DNA polymerase, thymidylate kinase and DNA synthesis in the marine diatom *Cylindrotheca fusiformis.* Ph.D. thesis. University of California, Berkeley, Calif.

Sullivan, C. W., and B. E. Volcani (1973). *Biochim. Biophys. Acta* 308:212-229.

Summers, A. O., and E. Lewis (1973). *J. Bacteriol.* 113:1070-1072.

Summers, A. O., and S. Silver (1972). *J. Bacteriol.* 112:1228-1236.

Sutton, J. A., and J. D. Corrick (1963). *Min. Eng.* 15:37-40.

Sutton, J. A., and J. D. Corrick (1964). Bacteria in mining and metallurgy. Leaching selected ores and minerals; experiments with *Thiobacillus thiooxidans. U.S. Bur. Mines Rep. Invest. 5839.* 16 pp.

Suzuki, I. (1965). *Biochim. Biophys. Acta* 104:359-371.

Suzuki, I., and M. Silver (1966). *Biochim. Biophys. Acta* 122:22-33.

Swaby, R. J., and J. Sperber (1958). *Soils Fertil.* 22:Abstr. 286.

Tabita, R., and D. G. Lundgren (1971a). *J. Bacteriol.* 108:328-333.

Tabita, R., and D. G. Lundgren (1971b). *J. Bacteriol.* 108:334-342.

Tait, G. H. (1975). *Biochim. J.* 146:191-204.

Takai, Y., and T. Kamura (1966). *Folia Microbiol.* 11:304-315.

Tansey, M. R., and T. D. Brock (1972). *Proc. Natl. Acad. Sci. U.S.A.* 69:2426-2428.

Tappeiner, W. (1882). *Ber. Dtsch. Chem. Ges.* 15:999.

Taylor, R. M., and R. M. McKenzie (1966). *Austr. J. Soil Res.* 4:29-39.

Taylor, R. M., R. M. McKenzie, and K. Norrish (1964). *Austr. J. Soil Res.* 2:235-248.

Taylor, W. R. (1950). *Plants of Bikini and Other Northern Marshall Islands.* Michigan University Press, Ann Arbor, Mich.

Teh, J. S., and K. H. Lee (1973). *Appl. Microbiol.* 25:454-457.

Temple, K. L. (1964). *Econ. Geol.* 59:1473-1491.

Temple, K. L., and A. R. Colmer (1951). *J. Bacteriol.* 62:605-611.

Temple, K. L., and N. W. LeRoux (1964). *Econ. Geol.* 59:647-655.

Tezuka, T., and K. Tonomura (1976). *J. Biochem. (Tokyo)* 80:79-87.

Tezuka, T., and K. Tonomura (1978). *J. Bacteriol.* 135:138-143.

Thiel, G. A. (1925). *Econ. Geol.* 20:301-310.

Thresh, M. (1902). *J. Chem. Soc.* 82:567.

Tikhonova, G. V., L. L. Lisenkova, N. G. Doman, and V. P. Skulachev (1967). *Biokhimiya* 32:725-734.

Timonin, M. T. (1950a). Soil microflora and manganese deficiency. *Trans. 4th Int. Congr. Soil Sci.* 3:97-99.

Timonin, M. T. (1950b). *Sci. Agr.* 30:324-325.

Timonin, M. T., W. I. Illman, and T. Hartgerink (1972). *Can. J. Microbiol.* 18:793-799.

Tonomura, K., T. Makagami, F. Futai, and D. Maeda (1968). *J. Ferment. Technol.* 46:506-512.

Torma, A. E. (1971). *Rev. Can. Biol.* 30:209-216.

Torma, A. E. (1978). *Can. J. Microbiol.* 24:888-891.

Torma, A. E., and G. G. Gabra (1977). *Antonie van Leeuwenhoek J. Microbiol. Serol.* 43:1-6.

Torma, A. E., and F. Habashi (1972). *Can. J. Microbiol.* 18:1780-1781.

Tortoriello, R. C. (1971). Manganic oxide reduction by microorganisms in fresh water environments. Ph.D. thesis. Rensselaer Polytechnic Institute, Troy, N.Y.

Trautwein, K. (1921). *Zentralbl. Bakteriol. Parasitenkd. Infektionskr. Hyg. Abt. II* 53:513-518.

Tributsch, H. (1976). *Naturwissenschaften* 63:88.

Trimble, R. B. (1967). MnO$_2$-reduction by two strains of marine ferromanganese nodule bacteria. M. S. thesis. Rensselaer Polytechnic Institute, Troy, N.Y.

Trimble, R. B., and H. L. Ehrlich (1968). *Appl. Microbiol.* 16:695-702.

Trimble, R. B., and H. L. Ehrlich (1970). *Appl. Microbiol.* 19:966-972.

Trolldeiner, G. (1973). *Bull. Ecol. Res. Comm. (Stockholm)* 17:33-59.

Troshanov, E. P. (1968). *Mikrobiologiya* 37:934-940.

Troshanov, E. P. (1969). *Mikrobiologiya* 38:634-643.

Trost, W. R. (1958). The chemistry of manganese deposits. *Mines Branch, Res. Rep. R8.* Dept. of Mines and Technical Surveys, Ottawa, 125 pp.

Trudinger, P. A. (1970). *J. Bacteriol.* 104:158-170.

Tsubota, G. (1959). *Soil Plant Food* 5:10-15.

Tuovinen, O. H., and D. P. Kelly (1973). *Arch. Mikrobiol.* 88:285-298.

Tuovinen, O. H., S. I. Niemelae, and H. G. Gyllenberg (1971). *Antonie van Leeuwenhoek J. Microbiol. Serol.* 37:489-496.

Turner, A.W. (1949). *Nature (Lond.)* 164:76-77.

Turner, A. W. (1954). *Austr. J. Biol. Sci.* 7:452-478.

Turner, A. W., and J. W. Legge (1954). *Austr. J. Biol. Sci.* 7:479-495.

Turner, R. D. (1973). *Science* 180:1377-1379.

Tuttle, J. H., P. E. Holmes, and H. W. Jannasch (1974). *Arch. Mikrobiol.* 99: 1-14.

Tyler, P. A. (1970). *Antonie van Leeuwenhoek J. Microbiol. Serol.* 36:567-578.

Tyler, P. A., and K. C. Marshall (1967a). *Antonie van Leeuwenhoek J. Microbiol. Serol.* 33:171-183.

Tyler, P. A., and K. C. Marshall (1967b). *J. Am. Water Works Assoc.* 59:1043-1048.

Tyler, S. A., and E. S. Barghoorn (1954). *Science* 119:606-608.

Unz, R. F., and D. G. Lundgren (1961). *Soil Sci.* 92:302-313.

Van Delden, A. (1903). *Zentralbl. Bakteriol. Parasitenkd. Infektionskr. Hyg. Abt II* 11:81-94.

Van der Linden, A. C., and G. J. E. Thijsse (1965). *Adv. Enzymol.* 27:469-546.

Van Veen, W. L. (1972). *Antonie van Leeuwenhoek J. Microbiol. Serol.* 38: 623-626.

Van Veen, W. L. (1973). *Antonie van Leeuwenhoek J. Microbiol. Serol.* 39: 657-662.

Van Veen, W. L., E. G. Mulder, and M. H. Deinema (1978). *Microbiol. Rev.* 42:329-356.

Varentsov, I. M. (1972). *Usta. Miner. Petrogr. Szeged.* 20:363-381.

Varentsov, I. M., and N. V. Pronina (1973). *Miner. Deposita* 8:161-178.

Vavra, P., and L. Frederick (1952). *Soil Sci. Soc. Am. Proc.* 16:141-144.

Veeh, H. H., W. C. Burnett, and A. Soutar (1973). *Science* 181:844-845.

Vernadsky, V. I. (1908-1922). An attempt at descriptive mineralogy. *Izbrannye Trudy,* Vol. 2. Izdatel'stvo. Akad. Nauk SSSR, Moscor, 1955.

Vernon, L. P., J. H. Mangum, J. V. Beck, and F. M. Shafia (1960). *Arch. Biochem. Biophys.* 88:227-231.

Vine, F. J. (1966). *Science* 154:1405-1415.

Vishniac, W. (1952). *J. Bacteriol.* 64:363-373.

Vishniac, W., and M. Santer (1957). *Bacteriol. Rev.* 21:195-213.

Volesky, B., and J. E. Zajic (1970). *Dev. Ind. Microbiol.* 11:184-185.

Von Wolzogen-Kuehr, C. A. H. (1927). *J. Am. Water Works Assoc.* 18:1-31.

Vostal, J. (1972). Transport and transformation of mercury in nature and possible routes of exposure. In *Mercury in the Environment: An Epidemiological and Toxicological Appraisal.* L. Friberg and J. Vostal, eds. CRC Press, Cleveland, Ohio, pp. 15-27.

Wagner, C., and M. E. Levitch (1975). *J. Bacteriol.* 122:905-910.

Wagner, E., and W. Schwartz (1965). *Z. Allg. Mikrobiol.* 7:33-52.

Wainer, A. (1964). *Biochem. Biophys. Res. Commun.* 16:141-144.

Wainer, A. (1967). *Biochem. Biophys. Acta* 141:466-472.

Waksman, S. A. (1916). *Soil Sci.* 1:363-380.

Waksman, S. A. (1932). *Principles of Soil Microbiology,* 2nd ed. rev. Williams & Wilkins, Baltimore, Md.

Waksman, S. A. (1933). *Soil Sci.* 36:125-147.

Waksman, S. A., and M. Hotchkiss (1937). *J. Mar. Res.* 1:101-118.

Waksman, S. A., and R. L. Starkey (1931). *The Soil and the Microbe.* Wiley, New York.

Walker, J. D., and J. J. Cooney (1973). *J. Bacteriol.* 115:635-639.

Walsh, F., and R. Mitchell (1972a). *J. Gen. Microbiol.* 72:369-376.

Walsh, F., and R. Mitchell (1972b). *Environ. Sci. Technol.* 6:809-812.

Walter, M. R., R. Buick, and J. S. R. Dunlop (1980). *Nature (Lond.)* 284:443-445.

Ward, D. M., and T. D. Brock (1978). *Geomicrobiology J.* 1:1-9.

Webley, D. M., R. B. Duff, and W. A. Mitchell (1960). *Nature (Lond.)* 188: 766-767.

Webley, D. M., M. E. F. Henderson, and I. F. Taylor (1963). *J. Soil Sci.* 14: 102-112.

Wehmiller, J. (1972). Carbon cycle. In *The Encyclopedia of Geochemistry and Environmental Sciences*. Encyclopedia of Earth Sciences Series, Vol. IVA. R. W. Fairbridge, ed. Van Nostrand Reinhold, New York, pp. 124-129.

Weimer, P. J., and J. G. Zeikus (1978). *Arch. Microbiol.* 119:175-182.

Weimer, P. J., and J. G. Zeikus (1979). *J. Bacteriol.* 137:332-339.

Welch, P. H. (1952). *Limnology*, 2nd ed. McGraw Hill, New York.

Wellman, R. P., F. D. Cook, and H. R. Krouse (1968). *Science* 161:269-270.

Wenberg, G. M., F. H. Erbisch, and M. Bolin (1971). *Trans. Soc. Min. Eng. AIME* 250:207-212.

Werner, D. (1966). *Arch. Mikrobiol.* 55:278-308.

Werner, D. (1967). *Arch. Mikrobiol.* 57:51-60.

Whittenbury, R., K. C. Phillips, and J. F. Wilkinson (1970). *J. Gen. Microbiol.* 61:205-218.

Widdel, F., and N. Pfennig (1977). *Arch. Microbiol.* 112:119-122.

Williams, B. G., and W. H. Patrick, Jr. (1971). *Nature (Phys. Sci.) (Lond.)* 234: 16.

Williams, J. (1962). *Oceanography*. Little, Brown, Boston.

Williams, R. A. D., and D. S. Hoare (1972). *J. Gen. Microbiol.* 70:555-566.

Williamson, T. A. (1967). *Coal Mining Geology*. Oxford University Press, New York.

Wilson, R. D., P. H. Monaghan, A. Osanik, L. C. Price, and M. A. Rogers (1974). *Science* 184:857-865.

Winogradsky, S. (1887). *Bot. Ztg.* 45:489-600; 606-616.

Winogradsky, S. (1888). *Bot. Ztg.* 46:261-276.

Winogradsky, S. (1922). *Zentralbl. Bakteriol. Parasitendk. Infektionskr. Hyg. Abt. II* 57:1-21.

Wirsen, C. O., and H. W. Jannasch (1974). *Microb. Ecol.* 1:25-37.

Wirsen, C. O., and H. W. Jannasch (1975). *Mar. Biol.* 31:201-208.

Wirsen, C. O., and H. W. Jannasch (1978). *J. Bacteriol.* 136:765-774.

Woese, C. R., and G. F. Fox (1977). *Proc. Natl. Acad. Sci. U.S.A.* 74:5088-5090.

Wolfe, R. S. (1964). Iron and manganese. In *Principles and Applications in Aquatic Microbiology*. H. Heukelakian and N. C. Dondero, eds. Wiley, New York, pp. 82-97.

Wood, E. J. F. (1965). *Marine Microbial Ecology*. Van Nostrand Reinhold, New York.

Wood, J. M. (1974). *Science* 183:1049-1052.

Wood, J. M., F. S. Kennedy, and C. G. Rosen (1968). *Nature (Lond.)* 220: 173-174.

Woolfolk, C. A., and H. R. Whiteley (1962). *J. Bacteriol.* 84:647-658.

Yamada, M., and K. Tonomura (1972a). *J. Ferment. Technol.* 50:159-166.

Yamada, M., and K. Tonomura (1972b). *J. Ferment. Technol.* 50:589-900.

Yamada, M., and K. Tonomura (1972c). *J. Ferment. Technol.* 50:901-909.

Yang, S. H. (1974). Effect of manganese, nickel, copper, and cobalt ions on some bacteria from the deep sea. Ph.D. thesis. Rensselaer Polytechnic Institute, Troy, N.Y.

Yayanos, A. A., A. S. Dietz, and R. van Boxtel (1979). *Science* 105:808-810.

Yen, H.-C., and B. Marrs (1977). *Arch. Biochem. Biophys.* 181:411-418.

Youssef, M. I. (1965). *Econl. Geol.* 60:590-600.

Zajic, J. E. (1969). *Microbial Biogeochemistry*, Academic Press, New York.

Zalokar, M. (1953). *Arch. Biochem. Biophys.* 44:330-337.

Zappfe, C. (1931). *Econ. Geol.* 26: 799-832.

Zavarzin, G. A. (1961). *Mikrobiologiya* 30:393-395.

Zavarzin, G. A. (1962). *Mikrobiologiya* 31:586-588.

Zavarzin, G. A. (1972). *Mikrobiologiya* 41:369-370.

Zehnder, A. J. B., and T. D. Brock (1979). *J. Bacteriol.* 137:420-432.

Zehnder, A. J. B., B. A. Huser, T. D. Brock, and K. Wuhrmann (1980). *Arch. Microbiol.* 124:1-11.

Zeikus, J. G. (1977). *Bacteriol. Rev.* 41:514-541.

Zeikus, J. G., and R. S. Wolfe (1972). *J. Bacteriol.* 109:707-713.

Zeikus, J. G., P. J. Weimer, D. R. Nelson, and L. Daniels (1975). *Arch. Mikrobiol.* 104:129-134.

Zeikus, J. G., G. Fuchs, W. Kenealy, and R. K. Thauer (1977). *J. Bacteriol.* 132:604-613.

Zimmerley, S. R., D. G. Wilson, and J. D. Prater (1958). Cyclic leaching process employing iron oxidizing bacteria. U.S. patent 2,829,964 (Cl. 75-104).

ZoBell, C. E. (1942). *Sci. Mon.* 55:320-330.

ZoBell, C. E. (1946). *Marine Microbiology*. Chronica Botanica Co., Waltham, Mass.

ZoBell, C. E. (1952). *J. Sediment. Petrol.* 22:42-49.

ZoBell, C. E. (1963). The origin of oil. *Int. Sci. Technol.*, August, pp. 42-48.

ZoBell, C. E., and R. Y. Morita (1957). *J. Bacteriol.* 73:563-568.

Glossary

Acidophilic bacteria Bacteria that need an acid environment in which to grow

Adenosine triphosphatase (ATPase) An enzyme involved in ATP synthesis or degradation

ADP Adenosine $5'$-diphosphate

Adenosine $5'$-triphosphate An energy storage compound important in metabolism

Adventitious organisms Organisms introduced naturally from an adjacent habitat; they may or may not be able to grow or survive in the new habitat

Aerobe An oxygen-requiring organism

Aerobic heterotroph An organism that uses organic substances as carbon and energy sources in an oxygen-containing atmosphere

Agar (also agar agar) A polysaccharide derived from the walls of certain red algae and used for gelling bacterial culture media

Agar-shake culture A bacterial culture method in which the bacterial inoculum is completely mixed in an agar medium in a test tube

Allochthonous Introduced from another place

Aluminosilicate A mineral containing a combination of aluminum oxide and silicate

Amictic lake A lake that never turns over

Ammonifying Capable of enzymatic release of ammonia from organic nitrogen compounds such as proteins and amino acids

Amorphous Noncrystalline

AMP Adenylic acid

Amphibole A ferromagnesian mineral with two infinite chains of silica tetrahedra linked to each other; the double chains are cross-linked by Ca, Mg, and Fe

Amphoteric Having both acidic and basic properties

Anabolism That part of metabolism which deals with synthesis and polymerization of biomolecules

Anaerobe An oxygen-intolerant organism

Anaerobic heterotrophy A form of nutrition using organic energy and carbon sources in the absence of atmospheric oxygen

Anaerobic respiration A process of respiration in which nitrate, sulfate, sulfur, carbon dioxide, ferric oxide, and so on, substitute for oxygen as terminal electron acceptor

Anhydrite A calcium sulfate mineral, $CaSO_4$

APS Adenosine $5'$-phosphosulfate

Aragonite A calcium carbonate mineral with orthorhombic structure

Archaebacteria A group of bacteria, including methanogens, *Sulfolobus, Halobacterium,* and *Thermoplasma,* which have a unique cell envelope and plasma membrane structure and a unique type of ribosomal RNA that distinguishes them from all other bacteria (eubacteria).

Aridisol A mature desert soil

Arsenopyrite An iron-arsenic sulfide mineral (FeAsS)

Arsine AsH_3

Ascomycetes Fungi that deposit their sexual spores in sacs (asci) (e.g., *Neurospora*)

Asparagine The amide of aspartic acid, a dicarboxylic amino acid

Assimilation Uptake and incorporation of nutrients by cells

Asthenosphere Upper portion of the earth's mantle, which is thought to have a plastic consistency and upon which the crustal plates float

Atmosphere The gaseous envelope around the earth

ATP Adenosine $5'$-triphosphate

Augite A pyroxene type of mineral

Authigenic Formed de novo from dissolved species, in the case of minerals

Autochthonous Generated in place; indigenous

Autotroph An organism capable of growth exclusively at the expense of inorganic nutrients

Bacterioneuston The bacterial population located in a thin film at the air-water interface in a natural body of water

Bacterioplankton Unattached bacterial forms in an aqueous environment

Bacterium   A prokaryotic single- or multicelled organism. Single cells may appear as rods, spheres, or spirals

Baltica   A continent encompassing Russia west of the Urals, Scandinavia, Poland, and northern Germany

Banded Iron Formation   A sedimentary deposit featuring alternating cherty iron oxide (Fe_2O_3 and Fe_3O_4)-rich layers and iron oxide-poor layers; thought to have originated 3.3-2 billion years ago

Barium psilomelane   A complex manganese(IV) oxide

Barophile   An organism capable of growth at elevated hydrostatic pressure

Barren solution   Pregnant solution after its valuable metals have been removed

Basaltic rock   Rock of volcanic origin showing very fine crystallization due to rapid cooling. Basalt is rich in pyroxene and feldspars

Basidiomycetes   Fungi that form sexual spores on basidia and feature septate mycelium (e.g., mushrooms)

Benthic   Located at the bottom of a body of water

Betaine $HOOCCH_2N^+(CH_3)_3$

Binary fission   Cell division in which one cell divides into two cells of approximately equal size

Biosphere   The portion of the earth's surface inhabited by living organisms

Birnessite   A manganese(IV) oxide mineral, MnO_2

Bisulfite HSO_3^-

Calcareous ooze   A sediment having calcareous structures from foraminifera, coccolithophores, or other $CaCO_3$-depositing organisms as major constituents

Calcite   A calcium carbonate mineral with rhombohedral structure

Capillary culture method   The use of a glass capillary with optically flat sides inserted into soil or sediment for culturing microbes from these sources; developing microbes in the capillaries may be observed directly under a microscope

Catabolism   That part of metabolism which involves energy derivation from and breakdown of nutrients

Catalase   An enzyme capable of catalyzing the reaction $H_2O_2 = H_2O + \frac{1}{2}O_2$; it may also catalyze the reduction of H_2O_2 with an organic hydrogen donor

Celestite   A strontium sulfate mineral, $SrSO_4$

Cellulytic   Capable of enzymatic hydrolysis of cellulose

Centric geometry   Cylindrical, in reference to diatoms

Chalcopyrite A copper-iron sulfide mineral, $CuFeS_2$

Chasmolithic Living inside preformed pores or cavities in rock

Chemocline A chemical gradient zone in a water column that separates a more dilute and less dense phase from a more concentrated and denser phase

Chemolithotroph An autotroph deriving energy from the oxidation of inorganic matter

Chemostat A culture system permitting microbial growth under steady-state conditions

Chlorophyll A light-harvesting and energy-transducing type of pigment of photosynthetic organisms; it is linked to specific proteins

Chloroplasts Photosynthetic organelles in eukaryotic cells

Choline $HOCH_2CH_2N^+(CH_3)_3$

Coccolithophore A chrysophyte alga whose surface is covered with $CaCO_3$ platelets (coccoliths)

Colony counting A method of microbial enumeration by counting colonies (visible aggregates of microbes) on and/or in an agar medium in a Petri dish

Conjugation A method of transfer of genetic information between cells that requires cell to cell contact

Connate water Saline water trapped in rock strata in the geologic past, usually having undergone chemical alteration through reaction with the enclosing rock

Constitutive enzyme An enzyme that is always present in an active form in a cell, whether needed or not

Contaminants Organisms introduced during experimental manipulation of a habitat

Continental drift Migration of continents on the earth's surface as a result of plate motions

Continental margin The edge of the continent

Continental rise Gently sloping sea floor at the base of the continental slope

Continental shelf Gently sloping sea floor between the shore and the continental slope

Continental slope Steeply sloping sea floor at the outer edge of the continental shelf

Convergence The confluence of two water masses

Co-oxidation Simultaneous microbial oxidation of two compounds which may be quite unrelated and only one of which supports growth

Coriolis force An apparent force that seems to deflect a moving object to the right in the northern hemisphere and to the left in the southern hemisphere of the earth

Crustal plates Portions of the earth's crust which have irregular shapes and sizes and which contact and interact with each other while floating on the asthenosphere

Cyanobacteria Blue-green algae

Cysteine $HSCH_2CH(NH_2)COOH$

Cytochrome system An electron transport system used in biological oxidations (respiration) which includes iron porphyrin proteins called cytochromes

Dehydrogenase An enzyme that catalyzes removal or addition of hydrogen atoms

Denitrifying Capable of reduction of nitrate to dinitrogen, nitrous oxide, and nitric oxide

Deoxyribonucleic acid A polymer that has genetic information encoded into it

Deuteromycetes Fungi that do not form sexual spores

Diagenesis A process of transformation or alteration or rocks or minerals

Diatom A chrysophyte alga grouped with the Bacillarophyceae, which is encased in a siliceous wall

Diatomaceous ooze A sediment having diatom frustules as a major constituent

Dimethyl arsinate $CH_3AsO(OH)$, the acid form; also known as cacodylic acid
$$\overset{|}{CH_3}$$

Dimethylarsine $(CH_3)_2AsH$

Dimethylmercury $(CH_3)_2Hg$

Dimictic lake A lake that turns over twice a year

Dithiothreitol $HSCH_2CH(OH)CH_2SH$

Divergence The separation of two water masses

DNA Deoxyribonucleic acid; a biopolymer that has the genetic information of a cell encoded in its structure

Dolomite A $CaMg(CO_3)_2$ mineral

Dunite An ultrabasic rock rich in olivine

Dystrophic Referring to waters with an oversupply of organic matter that is only incompletely decomposed because of an insufficiency of oxygen, phosphorus, and/or nitrogen

Earth's core The innermost portion of the earth, consisting mostly of Fe and Ni

Earth's mantle The portion of the earth overlying the core, consisting mainly of O, Mg, and Si and lesser amounts of Fe, Al, Ca, and Na

Enargite A copper-arsenic sulfide mineral (Cu_3AsS_4)

Endolithic Living inside rock (limestone) as a result of boring into it

Endosymbionts Cells that live inside other cells for mutual benefit

Enrichment A culture method that selects for a desired organism(s) by providing special nutrients and/or physical conditions that favor its (their) development

Entisol An immature desert soil

Epigenetic Referring to emplacement of a mineral in cracks and fissures of pre-existing rock

Epilimnion The portion of a lake above the thermocline

Epiphytes Organisms attached to the surface of other living organisms or inanimate objects

Eubacteria Typical bacterial prokaryotes

Eukaryotic cell A cell with a true nucleus, mitochondria, and chloroplasts (if photosynthetic)

Euphotic zone That part of a water column which is penetrated by sunlight in sufficient quantity to permit photosynthesis

Euryhaline Capable of growth over a wide range of salinities

Eutrophic Referring to a nutrient-rich state of natural water

Facultative chemolithotroph An organism that can grow heterotrophically or chemolithotrophically, depending on growth conditions

Facultative organism An organism capable of living with or without oxygen

Fecal pellet Compacted fecal matter packaged in a membrane by the organism that excretes it

Feldspar A type of mineral consisting of anhydrous aluminosilicates of Na, K, Ca, and Ba

Fermentation A process of intramolecular oxidation/reduction operating without an externally supplied terminal electron acceptor

Fluorescence microscopy A microscopy method making use of natural or artificial fluorescence of objects upon irradiation with UV light

Foraminifer(a) Amoeboid protozoan(s) that mostly form a calcareous test (shell) about them; some form tests by cementing sand grains or other inorganic detrital structures to their cell surface (arenaceous foraminifera)

Fungus(i) A mycelial or occasionally single-celled eukaryotic organism possessing a cell wall but no chloroplasts; yeast, molds, mildews, and mushrooms are examples.

Galena A lead sulfide mineral, PbS

Garnet A silicate mineral of Ca, Mg, Fe, or Mn; it is hard and vitreous

Generation time   The average time required for cell division

Geomicrobiology   The study of the role microbes have played and are playing in a number of fundamental geologic processes

Glutathione
$$H_2NCHCH_2CH_2CONHCH$$
with CH_2SH on the upper branch, $COOH$ and $CONHCH_2COOH$ below

Goethite   An iron oxide mineral, $Fe_2O_3 \cdot H_2O$

Gondwana   A continent encompassing Africa, South America, Australia, Antarctica, and India

Gram-negative   A differential staining reaction of bacteria in which a counter-stain, usually safranin, is retained by the cell

Gram-positive   A differential staining reaction of bacteria in which crystal violet is retained by the cell

Granite   Rock of volcanic origin showing coarse crystallization due to slow cooling of magma; granite is rich in quartz and feldspars

Granodiorite   A type of rock intermediate between granite and diorite, showing coarse crystallization

Gravitational water   A film of water surrounding pellicular water, which moves by gravity, responds to hydrostatic pressure, and may freeze

GSH   Reduced glutathione

GSSH   Oxidized glutathione

Guyots   Flat-topped seamounts

Gypsum   A calcium sulfate mineral, $CaSO_4 \cdot 2H_2O$

Halophile   A microbe that grows preferentially at a high salt concentration

Hematite   An iron oxide mineral, Fe_2O_3

Heterotroph   An organism requiring one or more organic nutrients for carbon and for energy for growth

Heulandite   A type of zeolite mineral

Histosol   Organic soil

Holozoic   Feeding on living cells; predatory

Hornblende   A type of amphibole mineral

Humic acid   A humus fraction that is acid- and alcohol-insoluble

Humus   In soil, a mixture of substances derived from partial decomposition of plant, animal, and microbial remains and from microbial syntheses; in marine sediment of the open ocean, a mixture of substances derived from phytoplankton remains

Hydrogenase   An enzyme catalyzing the reaction $H_2 \rightleftharpoons 2H^+ + 2e$

Hydrosphere The aqueous portion of the earth's surface, including the oceans, seas, lakes, and rivers

Hydrothermal Referring to hot, metal-laden solutions generated in the lithosphere in regions of cooling magma

Hygroscopic water A thin film of water covering a soil particle, which never freezes or moves as a liquid

Hypersthene A type of pyroxene mineral

Hypha(e) A branch of a mycelium It is filamentous

Hypolimnion The portion of a lake located below the thermocline

Hypophosphite HPO_2^{2-}

Igneous rock Referring to rock of volcanic or magmatic origin

Illite A group of micalike clay minerals, having a three-layered structure like montmorillonite in which Al may substitute for Si, and containing significant amounts of Fe and Mg

Indigenous organisms Organisms native to a habitat

Inducible enzyme An enzyme that is formed by the cell only when needed

Juvenile mineral A mineral formed from source material from within the earth that had never before reached the earth's surface

Kaolinite A type of clay $[Al_4Si_4O_{10}(OH)_8]$ having alternating aluminum oxide and tetrahedral silica sheets

Karstic Referring to a landscape with sinkholes or cavities due to local dissolution of limestone

Kazakhstania A continent encompassing present-day Kazakhstan

Labradorite A type of feldspar mineral related to plagioclase

Laterization A soil transformation in which iron and aluminum oxides, silicates, and carbonates are precipitated, cementing soil particles together and thus destroying the porosity of the soil

Laurasia A continent encompassing North America, Europe, and most of Asia

Laurentia A continent encompassing most of North America, Greenland, Scotland, and the Chukotski Peninsula of the eastern U.S.S.R.

Lentic waters Static waters

Lichen An organism consisting of an intimate association of a fungus and a green alga or cyanobacterium

Lignin A polymer of units of substituted phenylpropane derivatives; an abundant constituent of wood

Limestone A rock type rich in $CaCO_3$

Limonite An iron oxide mineral, $Fe_2O_3 \cdot nH_2O$

Lithification A process of rock formation by compaction and/or cementation

Lithosphere The crust of the earth

Lotic waters Flowing waters

Macrofauna Large animals

Magma Molten rock

Mannitol A polyhydric alcohol which may be formed by the reduction of fructose or mannose

Mercaptoethanol $HOCH_2CH_2SH$

Mesophile A bacterium capable of growth in a temperature range from 10 to 45°C (optimal range between 25 and 40°C)

Mesotrophic Referring to a nutritional state of natural water between oligotrophic and eutrophic

Metabolism The cellular biochemical activities collectively

Metabolite A product of metabolism

Metamorphic rock Rock produced by alteration of igneous or sedimentary rock through the action of heat and pressure

Molybdenite A molybdenum disulfide mineral, MoS_2

Monomethyl arsinate $CH_3AsO(OH)$, the acid form
$$OH$$

Monomictic lake A lake that turns over once a year

Montmorillonite A type of clay $[Al_2Si_4O_{10}(OH)_2 \cdot xH_2O]$ formed of successive aluminum oxide sheets, each sandwiched between two sheets of silica tetrahedra

Mycelium A network of hyphae produced by most fungi and some bacteria

Nepheline A sodium aluminum silicate

Nitrifying Capable of oxidizing ammonia to nitrate; nitrification is a two-step process involving two different kinds of bacteria

Nitrogen-fixing Referring to microbial conversion of dinitrogen to amino nitrogen via ammonia

Nucleic acid Polymers of purines, pyrimidines, pentose, or deoxypentose and phosphoric acid found in chromosomes, plasmids, ribosomes, plastids, and cytoplasm of cells

Nucleotides Polymer units of nucleic acids

Ocean trench Deep cleft in the ocean floor. A site of subduction of an oceanic crustal plate below a continental crustal plate

Ochre An iron oxide ore

Oligotrophic Referring to a nutrient-poor state of natural water

Olivine A mineral consisting of orthosilicates of magnesium and iron

Organic soil A soil formed from accumulation and slow and incomplete decomposition of organic matter in a sedimentary environment

Orogeny Mountain building

Orpiment An arsenic mineral (As_2S_3)

Orthoclase A feldspar mineral

Orthophosphate Monomeric phosphate (H_3PO_4)

Orthosilicate Monomeric silicate (H_4SiO_4)

Oxidative phosphorylation A process of ATP synthesis coupled to electron transport

Oxisol A soil type in tropical and subtropical humid climates

Pangaea A supercontinent taking in all major continents of today; existed from about 250 to 200 million years ago

PAPS 3'-Phosphoadenosine phosphosulfate

Pectinolytic Capable of enzymatic hydrolysis of pectin

Pedoscope A system of glass capillaries with optically flat sides for insertion into soil and subsequent microscopic inspection for microbial development in the capillary lumen

Pellicular water A film of water surrounding hygroscopic water, which moves by intermolecular attraction and which may freeze

Peloscope A system of capillaries with optical flat sides for insertion into sediment and subsequent microscopic inspection for microbial development in the capillary lumen

Pennate geometry Symmetrical about a long and a short axis, in reference to diatoms

Peptone A mixture of peptides from a digest of beef muscle by pepsin; used in bacterial culture media

Peridotite An igneous granitoid rock, rich in olivines but lacking in feldspars

Peroxidase An enzyme that catalyzes the reduction of H_2O_2 by an oxidizable organic molecule

Phagotrophic Consuming whole cells by engulfment (phagocytosis)

Phosphatase An enzyme catalyzing hydrolysis of phosphate esters

Phosphine PH_3

Phosphite HPO_3^{2-}

Phosphorite A calcium phosphate mineral; apatite

Photolithotroph An autotroph deriving energy from sunlight

Photophosphorylation A light-dependent process of ATP synthesis associated with photosynthesis

Photosynthesis A metabolic process using sunlight energy for the assimilation of carbon in the form of CO_2, HCO_3^-, or CO_3^{2-}

Phycomycetes Aquatic and terrestrial fungi whose vegetative mycelium shows no septation (e.g., *Rhizopus*)

Phytoplankton Photosynthetic plankton

Plankton Free-floating biota in an aqueous habitat

Plasmid An extrachromosomal bit of genetic substance (DNA)

Plutonic water Deep, anoxic underground water, likely to contain significant amounts of sulfate and/or chloride

Podzolic soil A type of spodosol

Pregnant solution A metal-laden effluent from an ore-leaching operation

Prokaryotic cell A cell lacking a true nucleus, mitochondria, and chloroplasts

Proteolytic Capable of enzymatic hydrolysis of proteins

Psychrophile A bacterium capable of growth in a temperature range from 0 to 20°C (optimum at 15° or below)

Psychrotolerant Capable of surviving but not growing at a temperature in the psychrophilic range

Psychrotroph An organism capable of growth in a temperature range from 0 to 30°C (optimum about 25°C)

Purines A group of organic bases having the purine-ring structure in common

Pyridine nucleotide Nicotinamide adenine dinucleotide or nicotinamide adenine dinucleotide phosphate; hydrogen-carrying coenzyme

Pyrimidines A group of organic bases having the pyrimidine ring structure in common

Pyrite (iron) An iron disulfide mineral, FeS_2

Pyroxene A ferromagnesium mineral with silica tetrahedra linked in single chains and cross-linked mainly by Ca, Mg, and Fe

Quartzite A metamorphic rock derived from sandstone

Radiolarian ooze A sediment having radiolarian tests as a major constituent

Red-Bed deposit A sedimentary deposit rich in ferric oxide; first appeared when the atmosphere of the earth became oxidizing

Respiration Biological oxidations utilizing an electron transport system that may operate with either oxygen or another inorganic compound as terminal electron acceptor

Reverse electron transport The transfer of electrons by an electron transport system against the redox gradient

Rhodanese An enzyme capable of catalyzing the reaction $CN^- + S_2O_3^{2} = SCN^- + SO_3^{2-}$ and the reductive cleavage of $S_2O_3^{2-}$

Rhodochrosite A mineral form of $MnCO_3$

Rhyolite An igneous rock rich in plagioclase feldspar

Ribosome Submicroscopic, intracellular particle consisting of ribonucleic acids and proteins which is part of the protein-synthesizing system of cells

Ribulose biphosphate carboxylase An enzyme that can catalyze carboxylation or oxygenation of ribulose diphosphate in autotrophs

RNA Ribonucleic acid: Some acts as a template in protein synthesis; some is found as part of the ribosome structure; some acts as an amino acid activator in protein synthesis

Rock Massive, solid inorganic matter, usually consisting of two or more intergrown minerals

Rusticyanin A copper protein enzyme involved in catalyzing Fe^{2+} oxidation in *Thiobacillus ferrooxidans*

Saccharolytic Capable of the enzymatic hydrolysis or fermentation of sugars

Salinity A measure of the salt content of seawater based on its chlorinity

Salt dome The cap rock composed of anhydrite, gypsum, and calcite at the top of a salt plug—a geologic formation

Sandstone A rock formed from compacted and cemented sand

Saponite A montmorillonite type of clay in which Mg replaces Al

Saprozoic Feeding on dead organic matter

Satellite organism In the case of mixed bacterial cultures, an organism not identical to the dominant organism in the culture; will give rise to distinctive colonies on appropriate solid media

Sclerotium(a) A vegetative, resting food storage body in certain higher fungi, composed of a compact mass of hardened mycelium

Sediment Finely divided mineral and organic matter that has settled to the bottom in a body of water

Sedimentary rock Rock formed from compaction and/or cementation of sediment

Seismic activity Earth tremors

Shale A laminate sedimentary rock formed from mud or clay

Sheath An organic tubular structure around some bacterial organisms

Siderite A mineral form of $FeCO_3$

Siderophore An iron-chelating substance produced by certain microbes

Silica Silicon dioxide; quartz is an example

Silicate A salt of silicic acid

Slime molds A group of organisms that have a life cycle including a motile swarmer stage and a multinucleate aggregational phase leading to formation of a sessile fruiting body

Sodium azide NaN_3, an inhibitor of cytochrome oxidase

Soil horizon A soil stratum

Soil profile A vertical section through a soil

Solfataras Fumarolic hot springs that yield sulfuretted waters

Spent culture medium Culture medium after microbial growth has taken place in it

Spodosol Forest soil types in temperate climates

Stenohaline Capable of growth in only a narrow range of salinities

Stromatolite A laminated structure formed by filamentous organisms that grew in mats which have either entrapped inorganic detrital material or formed $CaCO_3$ deposits in which the organisms became embedded; the organisms are most commonly cyanobacteria; in ancient stromatolites, the organic remains have frequently disappeared

Subduction A process in which the edge of an oceanic crustal plate slips under a continental crustal plate; manifested in the form of deep ocean trenches

Substrate-level phosphorylation A process of ATP synthesis involving high-energy phosphate bond formation on the substrate being oxidized

Sulfate-reducing bacteria Bacteria that convert sulfate to sulfide

Sulfhydryl compound An organic compound with an $-SH$ functional group

Superoxide dismutase An enzyme that catalyzes the transformation of superoxide (O_2^-) to hydrogen peroxide (H_2O_2)

Synergism The interaction of two or more organisms, resulting in a reaction that none of the organisms could carry out by itself

Syngenetic Referring to deposition of an ore mineral contemporaneously with the enclosing sediment or rock

Talc A hydrous magnesium silicate mineral

Tectonic activity Crustal transformation

Teichoic acid A glycerol- or ribitol-based polymeric constituent of the cell walls of gram-positive bacteria

Tetrathionate $S_4O_6^{2-}$

Thermocline A zone in a water column with a steep temperature gradient

Thermophilic bacteria Bacteria that grow at temperatures above 45°C; some grow near the boiling point of water

Thiobacilli Gram-negative rod-shaped bacteria, mostly chemolithotrophic, which can use H_2S, S^o and $S_2O_3{}^{2-}$ as energy sources

Thiosulfate $S_2O_3{}^{2-}$

Todorokite A complex manganese(IV) oxide mineral

Transduction A method of transfer of genetic information between bacteria, involving a bacterial virus as the transmitting agent

Transpiration Loss of water by evaporation through the stomata (pores) of leaves

Travertine A porous limestone that may be formed by rapid $CaCO_3$ precipitation by cyanobacteria

Tricarboxylic acid cycle A cyclic sequence of biochemical reactions in which acetate is completely oxidized in one turn of the cycle

Trithionate $S_3O_6{}^{2-}$

Tundra soil A soil type occurring in high northern latitudes

Turbidity current A strong current of a sediment suspension moving rapidly downslope; may exert scouring action as it moves over rock surfaces

Ultramafic rock An igneous rock, usually rich in olivine and pyroxene

Upwelling An upward movement of a mass of deep, cold ocean water, which may bring nutrients (nitrate, phosphate) into surface water

Vermiculite A micaceous mineral

Wad A complex manganese(IV) oxide

Water potential A measure of water availability in soil

Weathering A process of breakdown of rock

Wollastonite A calcium silicate mineral ($CaSiO_3$)

Zeolite A hydrated silicate of aluminum containing alkali metals

Zooplankton Nonphotosynthetic plankton

Zygospore A sexual spore formed by certain algae and phycomycetous fungi

Index of Organisms

In most cases the names of microorganisms cited in this book are those used by the investigators in their particular studies. Some of these names may have been changed more recently because of changes in classification and/or because of changes in the rules of nomenclature. In cases where names have been changed, the current, more acceptable but often less familiar names are given in parentheses in this index.

Bacteria

Acetobacter acetophilum, 171
Acholeplasma laidlawii, 184
Achromatium oxaliferum, 101
Achromobacter (Alcaligenes?), 149, 160, 297
Achromobacter (Alcaligenes) aqua-marinus, 95
Achromobacter arsenoxydans-tres (Alcaligenes faecalis), 149
Acinetobacter, 160, 185
Actinomyces albus Gasper, 108
Actinomyces roseolus (Streptomyces roseolus?), 108
Actinomyces verrucosis, 108
Aerobacter (Enterobacter) sp., 187, 218
Aerobacter (Enterobacter) aerogenes, 145, 166, 187, 254
Alcaligenes, 153, 160
Alcaligenes faecalis, 149, 156
Archangium, 185

Arthrobacter, 30, 138, 141, 205, 206, 207, 212, 213, 214-217, 220, 223, 231, 242, 321
Arthrobacter siderocapsulatus, 185, 206, 211

Bacillus, 160, 185, 212, 218
Bacillus (29), 69, 191, 223, 224-226, 247
Bacillus arsenoxydans, 148
Bacillus caldolyticus, 133, 144
Bacillus centrosporus, 189, 223
Bacillus cereus, 189, 223
Bacillus circulans, 118, 187, 189, 223, 236
Bacillus filaris, 223
Bacillus firmus, 290
Bacillus malabarensis, 138
Bacillus manganicus, 230
Bacillus megaterium, 119, 138, 141, 187, 263, 315
Bacillus mesentericus, 119, 189, 223, 236

Cyanobacteria (Blue-Green Algae)

Subject Index